CORROSION OF ALUMINIUM

Elsevier Internet Homepage- http://www.elsevier.com

Consult the Elsevier homepage for full catalogue information on all books, journals and electronic products and services including further information about the publications listed below.

Elsevier titles of related interest

Books

KASSNER
Fundamentals of Creep in Metals and Alloys
2004. ISBN: 0080436374

HUMPHREYS AND HATHERLEY
Recrystallization and Related Annealing Phenomena, 2nd Edition
2004. ISBN: 008-044164-5

GALE
Smithells Metals Reference Book, 8th Edition
2003. ISBN 0-7506-7509-8

Elsevier author discount

Elsevier authors (of books and journal papers) are entitled to a 30% discount off the above books and most others. See ordering instructions below.

Journals

Sample copies of all Elsevier journals can be viewed online for FREE at www.sciencedirect.com, by visiting the journal homepage.

Journals of Alloys & Compounds
Corrosion Science
International Journal of Fatigue
Electrochimica Acta

To contact the Publisher:

Elsevier welcomes enquiries concerning publishing proposals: books, journal special issues, conference proceedings, etc. All formats and media can be considered. Should you have a publishing proposal you wish to discuss, please contact, without obligation, the publisher responsible for Elsevier's material science programme:

David Sleeman
Publishing Editor
Elsevier Ltd
The Boulevard, Langford Lane   Phone: +44 1865 843265
Kidlington, Oxford   Fax: +44 1865 843920
OX5 1GB, UK   E.mail: d.sleeman@elsevier.com

General enquiries, including placing orders, should be directed to Elsevier's Regional Sales Offices-please access the Elsevier homepage for full contact details www.elsevier.com

CORROSION OF ALUMINIUM

Christian Vargel

Consulting Engineer,
Member of the Commission of Experts within the
International Chamber of Commerce, Paris, France
http://www.corrosion-aluminium.com

Foreword by

Michel Jacques

President, Alcan Engineered Products

Translated by

Dr. Martin P. Schmidt

Patent Attorney, Lyon, France

2004

ELSEVIER

Amsterdam – Boston – Heidelberg – London – New York – Oxford – Paris
San Diego – San Francisco – Singapore – Sydney – Tokyo

ELSEVIER B.V.
Sara Burgerhartstraat 25
P.O.Box 211, 1000 AE
Amsterdam, The Netherlands

ELSEVIER Inc.
525 B Street, Suite 1900
San Diego, CA 92101-4495
USA

ELSEVIER Ltd
The Boulevard, Langford Lane
Kidlington, Oxford OX5 1GB
UK

ELSEVIER Ltd
84 Theobalds Road
London WC1X 8RR
UK

First published by Dunod, Paris 1999 (French Edition)
First edition 2004 (English Edition)

Library of Congress Cataloging in Publication Data
A catalog record is available from the Library of Congress.

British Library Cataloguing in Publication Data
A catalogue record is available from the British Library.

ISBN: 0 08 044495 4

♾ The paper used in this publication meets the requirements of ANSI/NISO Z39.48-1992 (Permanence of Paper).
Printed in The Netherlands.

For my grandchildren:

ALEXANDRE
ROMAIN
ALEXANDRA

Foreword

With an annual consumption of 35 million metric tons, aluminium is one of the world's most commonly used metals. Its lightness coupled with its strength, conductivity, barrier properties and its excellent corrosion resistance have all been and continue to be its most important advantages and the main reasons for the continued growth in the usage of aluminium.

Aluminium's lightweight performance delivers great benefit in transport applications such as aerospace, cars, ships, trains and buses. The metal's intrinsic characteristics help give automotive and other transport users improved driving performance as well as increasing fuel economy and reducing emissions.

Another significant advantage of aluminium is its corrosion resistance. This characteristic is valuable for products used in architecture, construction, civil engineering, transport, heat exchangers and many other applications.

From its early industrial days aluminium was used in many areas of transportation, however it was not until scientists developed commercially available aluminium magnesium alloys in the 1930's that aluminium was used in shipbuilding and other harsh operating conditions. The alloys developed by these metallurgists as well as having excellent corrosion resistance are also weldable, which led to them being firmly established as a key construction material by ship and high speed ferry builders.

New alloy development has been critical throughout the history of aluminium, establishing many new market applications. Improving corrosion performance has been a key focus of this metallurgical development. In the 1970's alloys were developed with very good thermal conductivity, combined with excellent resistance to engine coolants. This combination means that it was possible to manufacture heat exchangers that were cheaper and of course lighter than the traditional copper versions. More recently new alloy developments have included automotive, aerospace, desalination of sea water, renewable energy and maritime applications, as well as many civil engineering and industrial applications.

Christian Vargel is renowned as one of the Aluminium industry's leading experts in aluminium corrosion. During his long and successful career within the Pechiney group, his expertise in corrosion was valuable in the product development of key markets such as automotive, marine and other transport applications. He has also given many presentations on the corrosion resistance of aluminium, and has contributed to many of Pechiney's technical documents and brochures, such as "Aluminium and the Sea" and "Aluminium in Industrial Vehicles". His first book *Le comportement de l'aluminium et de ses alliages* (*The behaviour of aluminium and its alloys*) was published by Dunod in 1979.

Christian Vargel's second book *Corrosion de l'aluminium* was first published in French in 1999 and I am delighted that it is now being published in English, thus enabling this excellent work to be more accessible to a wider audience. The author's practical approach has resulted in what I believe will become a standard reference for aluminium users seeking to understand corrosion characteristics. The work provides an excellent manual, advising of the best ways of working with aluminium including the selection of alloys, design principles and operating conditions. I believe the book will become a valuable text for many engineers and technicians who work and use aluminium.

MICHEL JACQUES
President, Alcan Engineered Products

April 2004

Foreword to the Original French Edition

With an annual consumption of 25 million metric tons, aluminium is the second most commonly used metal in the world after steel. Its lightness is very often the most important advantage for the commercial development of aluminium, which explains why it is extensively used for ground transport, aerospace and shipbuilding. This is also the reason why the automotive industry is currently very interested in aluminium: lightness is becoming a priority.

The second advantage of aluminium is its corrosion resistance. This explains its important position in construction, civil engineering, transport, heat exchangers, etc.

Already in 1890, naval architects had considered aluminium for reducing weight in vessels. But in order for aluminium to be usable for shipbuilding, metallurgists and corrosion specialists in the 1930s first had to develop aluminium magnesium alloys. These alloys have excellent corrosion resistance in the marine environment, and they are weldable. Since 1960, all high-speed ferries have been built in these alloys.

A similar trend was observed with heat exchangers: aluminium was recognised as an obvious solution, especially for automotive heat exchangers since 1970. In fact, several alloys have very good thermal conductivity, excellent resistance to engine coolants, making it possible to manufacture heat exchangers that are cheaper and of course lighter than traditional heat exchangers in copper alloys.

Projects for developing renewable energy sources (solar, etc.) have often been based on the use of aluminium heat exchangers, for several reasons: a much lower cost than titanium, good thermal conductivity, and excellent corrosion resistance.

Christian Vargel, throughout his long career within Pechiney, has been a practitioner of aluminium corrosion and a recognised expert in this field. His first book *Le comportement de l'aluminium et de ses alliages* (*The behavior of aluminium and its alloys*) was published by Dunod in 1979.

Since then, his experience has grown steadily. He has followed marine applications and automotive heat exchangers and has participated in many damage assessments involving corrosion in service. He has also given many talks on the corrosion resistance of aluminium, and has contributed to many of Pechiney Rhenalu's technical documents and brochures, such as "Aluminium and the Sea" and "Aluminium in Industrial Vehicles". We therefore encouraged him in his project to write a second book: his recognised experience in the field of aluminium corrosion deserved to be more widely known and disseminated. This book will certainly contribute to meeting this goal.

Corrosion is a difficult topic. I am deeply convinced that the practitioner's approach, based on expertise and experience, is best for assessing the corrosion resistance of aluminium – an assessment which is obviously one of the main conditions for the development of many uses of aluminium in transport, construction, power transmission, etc.

Christian Vargel's book presents the reader with a global approach to corrosion, comprising the selection of alloys, design principles and service conditions. I am convinced that it will contribute to the development of aluminium in those fields where resistance to corrosion is an essential property.

BERNARD LEGRAND
Former Deputy Chief Executive Officer,
Pechiney, September 1998

Introductory Remark

It is customary and convenient to refer to aluminium although in most cases what is meant is aluminium alloys. It should be remembered that unalloyed aluminium amounts to not more than 10% of the annual worldwide consumption, all uses and products combined.

For the sake of simplicity, I use here the term "aluminium" instead of "aluminium and aluminium alloys". However, it should not be concluded that the corrosion resistance of all aluminium is similar in all types of environment! There are similar behaviours, but also substantial differences between the alloys of the series 2000 and 7000 and the alloys belonging to the other series. For this reason, I repeat this distinction whenever appropriate.

When one or more alloys have been used for corrosion testing in a given environment, this is mentioned, because they are a part of the testing protocol chosen by the scientist and have been included in the validation of this protocol. For the same reason, it is necessary to indicate the alloys which are commonly used for a given application. These references contribute to collecting data on the use of aluminium alloys and form a basis for the choice to be made by designers and users.

The designation of aluminium alloys changed substantially during the 20th century. Only since the 1970s has a truly international designation been developed, when the Aluminum Association's designation system for wrought alloys, based on four digits, came into wide use. There is still no corresponding worldwide designation system for casting alloys.

In former times, trade marks such as Silumin, Duralinox, as well as national designations, more or less standardised, were in use. For the sake of clarity and whenever possible, I have transposed the outdated designations into their corresponding modern designations, even if the compositions are slightly different.

In order to simplify the reading of Parts E and F on the corrosion resistance of aluminium in contact with chemicals, I have used the ADR numbers[1] when the product was included in the RTMDR[2] on July 1, 1997. The reader should refer to this regulation for all details on the storage and transportation of products included in the RTMDR.

The designation of organic chemicals in Part F refers as far as possible to the IUPAC designation rules[3] while retaining commonly used names such as ethyl alcohol for ethanol. For products included in the RTMDR, I have used this designation and I have mentioned their IUPAC designation.

---

[1] European Agreement concerning the International Carriage of Dangerous Goods by Road.
[2] Regulations concerning the International Carriage of Dangerous Goods by Rail.
[3] International Union of Pure and Applied Chemistry

Certain products are excluded from the scope of the present book: aluminium powders and granules, whose properties and uses have no relationship with the metallurgy and uses of aluminium in the form of castings or rolled, extruded or forged semi-products. Sintered aluminium powder (SAP) products, which have been studied since the 1970s, and aluminised steels are also excluded.

Preface to the Original French Edition

After a career of forty years with Pechiney, an important part of which was dedicated to the corrosion of aluminium, I had the desire to come back to my first book, *Le comportement de l'aluminium*, published in 1979 by Dunod.

Since the publication of that book, my experience has grown because of my involvement in developing applications in fields where resistance to corrosion is essential: desalination of sea water, renewable energy (solar energy, thermal energy of oceans, etc.), condensation gas boilers, automotive heat exchangers, shipbuilding, maritime applications, civil engineering, as well as in the assessment of corrosion under service conditions.

The study of the corrosion of aluminium is a long story, almost a century old. It has been enriched by the considerable work done since 1925 by the corrosion experts of the major European and American aluminium producers: A. Guilhaudis at Pechiney, D. Altenpohl at Alusuisse, C. Panseri at ISML, P. Brenner at VAW, V.E. Carter and H.S. Campbell at BNFRMA, E.H. Dix, R.B. Mears, R.H. Brown at Alcoa, P.M. Aziz, H. Godard at Alcan, W.H. Ailor at Reynolds, and others.

The documentation tools currently available have taken advantage of data bases such as Aluminium Industry Abstracts, set up during the 1970s. More than 8000 references on the corrosion of aluminium are available in the Chemical Abstracts since its first volume was published in 1907.

As in my first book, I have chosen a practical and realistic approach to the corrosion of aluminium: a practitioner's approach. The corrosion resistance of an alloy depends not only on its chemical composition and metallurgical temper, but also on the joining method and the conditions of service.

Corrosion is a complex phenomenon that depends on a number of parameters, related both to the environment and the metal. Sometimes, corrosion damage is difficult to explain, either because the origin of corrosion has not been identified or because the theoretical foundations are inadequate for giving a convincing answer.

Corrosion indicators should therefore specify the limits in the performance that can be achieved with a given material, both in terms of lifetime and acceptable stress conditions, as well as the extent to which electrochemical theory and experience are sufficient to support such a prediction.

Even so, some aspects remain unclear, partly because real environments are often more complex than laboratory conditions. They compel corrosion experts to be cautious, but also sometimes to be bold, for the sake of progress. This requires taking into account experiences that may sometimes seem outdated.

My professional experience with Pechiney and the means Pechiney has made available to me have allowed me to undertake the writing of the present book.

I am very grateful to Bernard Legrand, Deputy Chief Executive Officer of Pechiney, and Yves Farge, Senior Vice President for Research and Developement at Pechiney, for their constant encouragement, to Dieter Gold, Martine Courtois and Isabelle Pellissier from the document center at Pechiney CRV for their invaluable help with documentation, to Ronan Dif, Philippe Gimenez, Jacques Lefebvre, Robert Macé, Daniel Robert, Evelyne Hank and Arlette Seywert, who have made valuable comments on the present book.

I would also like to express my deep gratitude to André Guilhaudis who was, from 1945 to 1980, Pechiney's corrosion expert. He welcomed me to Pechiney's Research Centre in Chambéry in 1957, and shared his passion for aluminium and his experience with corrosion.

C. VARGEL
Engineer ENSEEG

July 15, 1998

Acknowledgements

I am very grateful to Dr. Christian Leroy, Technical and Educational Officer at the European Aluminium Association, Brussels, for his contribution to the promotion of my book, to Dr. Martin P. Schmidt, my former colleague, who did the translation, and to Andrée for her assistance during the preparatory stages of my work.

The original French edition of this book was published by Dunod (Paris) in 1999 "Corrosion de l'aluminium".

Contents

Part E: The Action of Inorganic Products   353

Part A

Aluminium and Its Alloys

It was the chemist Louis Guyton de Morveau (1736–1816), a co-worker of Antoine Laurent Lavoisier (1743–1794), who coined the word "alumine" for one of the sulphates contained in alum. "Alumine" is derived from the Latin word *alumen*, which is said to have been used for potassium alum $KAl(SO_4)_2 \cdot 12H_2O$ during Roman period. Aluminium compounds were used in large quantities in antique pottery, as dyestuff and as an astringent in medicine [1].

It is not the word "alumine" which has come to designate aluminium ore, but the word "bauxite". This is because in 1821 Pierre Berthier (1782–1867), a mining engineer, discovered that the red soil of the village Les-Baux-de-Provence contained 40–50% alumina, the rest being comprised essentially of iron oxide, Fe_3O_4 (the source of its red colour), and silica, SiO_2.

Even though it is widely distributed in the earth's crust, aluminium did not become an industrial metal before the end of the 19th century. Alumina is one of the most stable of all oxides, with an enthalpy ΔG of $-1582\ kJ \cdot mol^{-1}$ (the enthalpy of iron oxide is $-1015\ kJ \cdot mol^{-1}$). It is very difficult to reduce this oxide [2].

The discovery of metallic aluminium is attributed to Sir Humphrey Davy (1778–1829). He referred to it using the term "aluminium" in 1809. By electrolysis of molten aluminium salts, he obtained an alloy of aluminium with iron, because he had used an iron cathode [3].

The chemist Hans Christian Œrsted (1777–1851) and later Friedrich Wöhler (1800–1882) chose to reduce aluminium chloride with potassium. The chloride had been prepared by chlorination of bauxite in the presence of carbon.

It was Wöhler who, in 1827, succeeded in producing a sufficiently pure metal to determine some of its properties, most notably its low density.

In 1854, Henri Sainte-Claire Deville (1817–1881) improved Wöhler's process. He replaced potassium with sodium, for two reasons: the reduction of 1 mol of Al uses 3 mol of Na (sodium), totalling 60 g, instead of 3 mol of K (potassium) amounting to 117 g. At that time, sodium was less expensive than potassium. He also replaced aluminium chloride, which is rather volatile, with a sodium aluminium chloride.

The first plant was created in Paris, in 1856, in the La Glacière area, but soon it was shut down: "The small plant of La Glacière, located in an inner suburb of Paris, amidst houses and market gardens, releasing into the atmosphere smoke laden with soda and chlorine, was forced to cease its aluminium production after numerous complaints [4]".

In the spring of 1857, Sainte-Claire Deville moved the plant to Nanterre, far away from residential areas. In 1859, production reached 500 kg. That same year, when Louis Le

Chatelier had patented a reduction process of alumina with sodium carbonate, a plant was built in Salindres, close to Alès in the Gard country, not far from the bauxite supply and the salt fields of La Camargue. The production in Salindres has varied between 505 kg, when the plant was started in 1860, and 2959 kg in 1880, when this process was discontinued.

The first kilograms of aluminium manufactured in 1856 were sold at a price slightly higher than silver, around 300 Germinal Francs, corresponding to about F15 000 in 1998. During the 1880s, the metal from the Salindres plant was sold at a price of F60–70 per kg.

Aluminium, which Sainte-Claire Deville liked to compare to silver, was mainly used for silverware and jewellery. Charles Cristofle (1805–1863), the celebrated Parisian silversmith, produced cast artwork made of aluminium alloyed with 2% copper. In 1858, the son of Napoleon III was offered a rattle made of aluminium.

In 1871, Zénobe Gramme (1826–1901) invented the first revolving machine called a "dynamo". The use of powerful sources of direct current made it possible to envision production methods based on electrolysis. Sainte-Claire Deville had tried unsuccessfully to electrolyse molten aluminium chloride.

The manufacturing process of aluminium by electrolysis of molten alumina was developed in France by Paul Louis Toussaint Héroult (1863–1914), who filed a patent on April 23, 1886, and in the United States by Charles Martin Hall (1863–1914), who filed his patent on July 9, 1886. Both had succeeded in dissolving alumina (melting point 2030 °C) in cryolithe $AlF_3 \cdot 3NaF$, which melts at 977 °C; the industrial melt contained about 2 or 3% alumina.

In 1887, Bayer filed a patent for a method to extract alumina from bauxite based on the attack of bauxite by hot caustic soda. Héroult went to Switzerland, to Neuhausen, in order to set up his process; the year after, he came back to France and in 1888 created the Froges plant near Grenoble. In 1889, 1100 kg were produced in Froges, and sold at F50 per kg

Table A1. World production of primary aluminium

Year	Annual world production (kt)
1900	6
1910	44
1920	125
1930	270
1940	780
1950	1500
1960	4540
1970	10 300
1980	16 080
1990	19 830

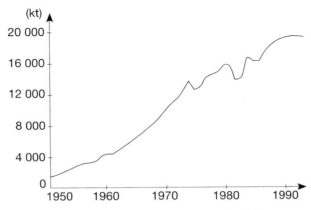

Figure A1. World production of primary aluminium.

(corresponding to F2500 in 1998). And in 1893, the production reached 86 300 kg, while the selling prices dropped to F5 per kg (F250 in 1998) [5].

Several plants were set up in France, Switzerland and the United States. Six thousand metric tons were produced in 1900, and the sales price of aluminium stabilised around F3 per kg (F15 in 1998). This was the start of the industrial adventure of aluminium. The world production of the so-called primary aluminium amounted to 6000 tonnes in 1900 (see Table A1) and has steadily increased (see Figure A1), especially since 1950.

REFERENCES

[1] Rey A., (dir.), *Dictionnaire historique de la langue française*, Le Robert, Paris, 1993.
[2] Talbot J., *Les éléments chimiques et les hommes*, SIRPE, 1995.
[3] Pascal P., *Nouveau traité de chimie minérale*, Masson et Cie, Paris, 1961.
[4] Sainte-Claire Deville H., *De l'aluminium, Ses properiétés, sa fabrication, et ses applications*, Mallet-Bachelier, Paris, 1859, p. 8.
[5] Bocquentin J., *La fabrication de l'aluminium par électrolyse, Histoire technique de la production de l'aluminium*, PUG, Grenoble, 1992.

Chapter A.1

The Advantages of Aluminium

Chapter A.1
The Advantages of Aluminium

With an annual world consumption of 25 million tons, aluminium is the leader in the metallurgy of non-ferrous metals. The production of aluminium has been increasing steadily since 1950 (see Figure A1).

The development of applications for aluminium and its alloys, as well as the sustained rise in consumption can be attributed to several of its properties which are decisive criteria in users' choice of metals, especially in the fields of transport, building, electrical engineering and packaging.

These advantageous properties are

–   lightness,
–   thermal conductivity,
–   electrical conductivity,
–   suitability for surface treatments,
–   corrosion resistance,
–   diversity of aluminium alloys,
–   diversity of semi-products,
–   functional advantages of extruded and cast semi-products,
–   ease with which aluminium can be formed,
–   ease of recycling.

1.1.   LIGHTNESS

The discoverers of aluminium were particularly impressed by the low density of this metal: "Aluminium is much lighter than any other common metal, and the kind of sensation which it gives you to carry an ingot of this metal is always amazing, even if you already know about this peculiar aspect [1]."

Lightness is the property of aluminium that first springs to mind, so much so that for a long time the term "light alloy" was used for what is now called "aluminium alloys". Aluminium is the lightest of all common metals (Table A.1.1). Its density is 2700 kg·m^{-3}, which is almost three times less than that of steel. The density of aluminium alloys ranges from 2600 to 2800 kg·m^{-3} (Tables A.3.5 and A.3.9).

Experience has shown that an aluminium alloy structure can be up to 50% lighter than its equivalent made from mild steel or stainless steel. This takes into account the modulus

Table A.1.1. Comparison of typical properties

Property	Alloy 1050A H14	Alloy 3003 H14	Alloy 2017A T4 T451	Alloy 5086 H116 H321	Alloy 6005A T5 T6	Alloy 7075 T6 T651	Steel E24 A37	Steel E36 A52	Stainless steel Z7CN 18-09 (304)	Stainless steel Z7CN 18-12 (316)	Rolled copper, temper M20	Brass CuZn4 OPb3	Polyamide PA6
Melting point or range (°C)	654–658	640–655	510–645	585–640	605–655	475–635	1400–1350	1400–1530	1400–1450	1375–1400	1083	875–890	220
Volumic mass, ρ (kg·m^{-3})	2700	2730	2790	2670	2710	2810	7820	7820	7900	7900	8940	8500	1130
Coefficient of linear expansion, α_1, 20 at 20–100 °C (10^{-6} K^{-1})	23.5	23.2	22.9	23.8	23.5	23.5	13.5	13.5	17.5	19	17.0	21.0	100–150
Thermal conductivity, λ at 20 °C (W·m^{-1}·K^{-1})	229	155	134	126	193	130	54	40	15	152	391	121	0.24
Electric conductivity at 20 °C (MS·m^{-1})	34.5	24.4	19.7	18.0	28.7	19.1	5	5	1.37	1.35	57	14.3	$<10^{-17}$
Proof stress, $R_{p0.2}$ (MPa)	80	140	275	250	260	505	240	360	>230	>280	69	196–245	
Tensile stress, R_m (MPa)	115	155	425	330	285	570	410	550	≤700	570–690	235	432–500	80
Elongation, A (%)	6	8	21	17	12	10	24	20	≥50	≥40	45	≥15	50
Elastic modulus (MPa)	69 000	69 000	72 500	71 000	69 500	72 000	210 000	210 000	200 000	200 000	115 000	95 000	
Brinell hardness (HB)	35	46	110	80	90	160	110	155			45	105–150	

Mean values, given for guidance.

of elasticity (one-third of the modulus for steel) and the fatigue limits of welded or bolted structures made of aluminium alloys. It is not appropriate to simply transpose the rules for steel design to aluminium. Rather, the specific properties of aluminium need to be taken into account.

Several areas of technology take advantage of aluminium's lightness:

– *Transport*. Since 1930, all aeroplanes have been made of aluminium alloys. They account for approximately 75% of the empty weight of modern aeroplanes. In commercial vehicles, the use of aluminium leads to a weight saving of 30–50%, depending on the type of structure. Aluminium plays an important role in urban railways (underground, tramcars), and more recently in high-speed trains: the new double-decker high-speed train (TGV) of the French National Railway Company (SNCF) is made of aluminium, especially extruded profiles.

In order to achieve weight savings in high-speed ferries, shipyards have turned to aluminium alloys of the 5000 and 6000 series. Experience has shown that weight savings in hull structures can also be as high as 40–50% compared to an equivalent hull structure made of steel.

The need to limit carbon dioxide emissions and other polluting gases requires reducing the fuel consumption of cars. The target is a weight saving of 100–300 kg, depending on the model. Aluminium is the most promising material for taking up this challenge in the automotive industry.
– *Mechanical engineering*. The replacement of a steel component by one of the same size made from a 2000, 7000, 5000 or 6000 series alloy leads to a weight saving on the order of the density ratio, i.e. 60%. Aluminium is widely used for moving parts, for example in robots, in order to minimise inertia.
– *Power transmission*. Given the respective electrical resistivities of aluminium and copper, for the same electrical resistance of a conductor or an equivalent drop in voltage, the 1370 conductor is half the weight of the copper conductor.

Lightness is not only an advantage for the application itself, but it also affects factory operations and working conditions. Handling semi-products and components made of aluminium alloys is easier, potentially cutting the capital cost of handling equipment.

1.2. THERMAL CONDUCTIVITY

Unalloyed aluminium is an excellent heat conductor, with roughly 60% of the thermal conductivity of copper, the optimum performer among common metals. The thermal conductivity of aluminium alloys depends on their composition and metallurgical temper (Tables A.3.5 and A.3.9).

As early as the end of the 19th century, this property led to replacing tin-plated copper with aluminium alloys in the manufacture of kitchen utensils, both for domestic and professional use.

Whenever there is a problem related to heat exchange, the use of aluminium is always taken into consideration, under the condition, of course, that the medium is appropriate when liquid–liquid or liquid–gaseous exchange is envisioned. There are many applications of aluminium heat-exchangers: cars, commercial vehicles, refrigerators, air conditioning, desalination of seawater, solar energy, coolers in electronic devices, etc.

1.3. ELECTRICAL CONDUCTIVITY

The electrical conductivity of aluminium is around two-thirds that of copper, which it is replacing in many electrical applications. Overhead power transmission lines made of aluminium or aluminium alloy of the Almelec type, on the market in France since 1927 [2] are used throughout the world. Aluminium bars and tubes are also widely used in connecting stations for high- and medium-voltage outdoor networks.

Aluminium is used for protecting underground and undersea telephone cables and for the construction of sulphur hexafluoride (SF_6) insulated, sealed converters, where it provides protection from electrical or magnetic fields (hardening).

1.4. RESISTANCE TO CORROSION

Sainte-Claire Deville observed that aluminium had good resistance to atmospheric corrosion, which included the particular atmosphere of gas lamps (used for street lighting in the Second Empire), an atmosphere laden with hydrogen sulphide (H_2S). He also recognised the very good resistance of aluminium in contact with water.

Many decades of experience with its use in buildings, public works, shipbuilding, etc. have confirmed the observations of the 19th century chemists. Aluminium and the alloys of the 1000, 3000, 5000, 6000 and 8000 series have excellent resistance to atmospheric corrosion in the marine, urban and industrial environments (see Part C).

This very good resistance to corrosion, as much as lightness, explains the development of numerous aluminium applications and offers users a number of major advantages:

–   Equipment and components can have a very long service life. It is not uncommon to find roofing (e.g. a church in Rome, see Section C.5.1), wall cladding panels, marina installations, ships, etc. with decades of service behind them. This also applies to the field of transport and many other applications.

–   Maintenance is minimal, even when no extra protection (painting, anodising) is provided. When aluminium is painted, replacement of the paint is less frequent and less urgent because the underlying metal generally has good resistance to corrosion. However, the use of aluminium does not make maintenance unnecessary, especially in buildings (see Chapters C.5 and G.6).

–   Appearance is preserved longer, because of the very good resistance to corrosion. This has become a strong sales argument, especially for applications where the user wishes to maintain a good surface appearance at the lowest possible cost. Examples are commercial vehicles, outdoor municipal facilities, traffic signs (indicator boards, gantries), etc.

The corrosion products of aluminium are white. They do not stain uncoated or coated metal surfaces, contrary to rust on steel. This is appreciated in certain areas of the chemical industry (textile fibres, etc.), because in case of a corrosive attack of the reactor, the corrosion products of aluminium do not alter the appearance of the products.

Finally, it should be mentioned that anodising to a depth of a few micrometers helps to preserve optical properties, reflectance, and the decorative features of such products as luxury packaging for cosmetics and decorative panels for buildings.

1.5.   SUITABILITY FOR SURFACE TREATMENTS

Aluminium surface treatments can serve several purposes, including (see Chapter B.5)

–   protecting certain alloys if their natural corrosion resistance is deemed insufficient, often the case with copper-containing alloys of the 2000 and 7000 series,
–   preserving the surface aspect, in order to avoid pitting corrosion or blackening,
–   modifying certain surface properties such as superficial hardness, and
–   decorating the metal.

The industrial availability of several types of surface treatments such as anodising and lacquering (as a continuous coil-coating process) has contributed to the increasing use of aluminium in the building industry; it is thus possible to guarantee a remarkable durability of the surface aspect, while offering a wide range of colours.

1.6.   THE DIVERSITY OF ALUMINIUM ALLOYS

With eight series or families, aluminium alloys are very numerous and offer a wide range of compositions, properties and uses. The continuing progress in the metallurgy of aluminium has produced high-performance alloys that are well suited to all types of applications, using conventional or special fabrication techniques among others.

While alloys in the same series share common properties, one series can differ greatly from another, and certain properties can vary widely. Thus alloys in the 5000 series are weldable and generally have good corrosion resistance, while alloys in the 2000 series have better mechanical properties, but cannot be welded using conventional techniques, and their resistance to atmospheric corrosion is poor. However tempting the prospect, this means that it is not always possible to switch from one series to another in the search for better mechanical properties. For example, one would not replace 6061 T6 by 2017A T4 without a thorough analysis of the prevailing service conditions. Otherwise, a choice based on a single criterion, here mechanical properties, may well penalise the user on other properties such as corrosion resistance. There is no shortage of examples where a poor choice of the alloy has resulted in severe corrosion problems.

1.7. THE DIVERSITY OF SEMI-PRODUCTS

The transformation techniques of aluminium—rolling, extrusion and casting—present designers and manufacturers with a very wide range of semi-products:

-   *castings*: sand castings for small series, mould castings for large series;
-   *flat rolled products*: plate and sheet, tread plate, coil-coated sheet, etc.;
-   *extrusions*: hollow or full profiles, in standard shapes or customised shapes. In fact, it is possible to manufacture extrusion dies at a reasonable cost, which makes it possible to extrude customised profiles for specific purposes;
-   die-forged or hand-forged.

Several alloys are compatible for welding purposes. Rolled or extruded semi-products of the 3000, 5000 and 6000 series can be joined with castings in alloys 42100 (A-S7G03), 42200 (A-S7G06), 43000 (A-S10G) and 71000 (A-Z5G) by means of TIG and MIG welding.

This diversity of semi-products makes it possible to

-   select the right location for stresses on components of bolted or welded structures;
-   simplify finishing processes by using coil-coated or preanodised sheet;
-   save on the time needed for assembly. This can compensate for the added raw material cost of structures made from aluminium alloy compared with equivalent steel structures.

1.8. THE FUNCTIONALITY OF CASTINGS AND EXTRUSIONS

Casting makes it possible to manufacture pieces with complex shapes and several functions, reducing complex machining to simple machining or surface milling.

Extrusion allows manufacturing profiles with a very wide range of dimensions and shapes, profiles that are well suited to the needs of designers who need to select the right location for stresses on structures. Extrusion dies are normally easy to manufacture at a moderate cost.

1.9. THE EASE OF USE

Provided that certain rules specific to aluminium alloys are observed, aluminium alloys can be processed using the same conventional techniques of shaping, bending, fabrication, deep drawing and machining as used for other common metals such as mild or stainless steel.

Aluminium alloys can usually be processed without the need for specific equipment or machine tools. It is advisable, however, to set up a workshop dedicated to aluminium alloy processing; this workshop should be separated from the workshop processing steel and, especially, copper alloys.

Like all other common metals, aluminium alloys lend themselves to joining techniques such as

- welding,
- bolting,
- riveting,
- clinching,
- adhesive bonding, and
- brazing.

1.10. RECYCLING

Aluminium recycling is very attractive, both in the context of energy savings and for economic reasons. Aluminium remelting requires only 5% of the energy that is needed to extract the metal from its ore.

Decades of experience with scrap collecting have shown that aluminium scrap always has a higher market value than steel scrap. Today in France, scrap collecting has reached about 70% of the available resources. The recycling rate of end-of-life aluminium is roughly

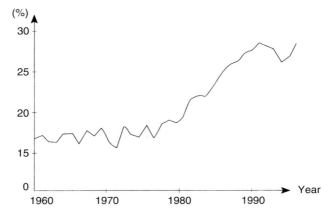

Figure A.1.1. Share of recycled aluminium in total aluminium consumption.

–   85% in the building industry and public amenities,
–   80% in the transport sector,
–   70% in mechanical and electrical engineering, and
–   65% in household appliances.

   Also significant is the steady rise in the consumption of recycled metal over the last 20 years; it now stands at about 30% of the production of primary aluminium (Figure A.1.1).

REFERENCES

[1] Sainte-Claire Deville, ibidem, p. 15.
[2] Suhr J., Un alliage d'aluminium–magnésium–silicium: l'Almelec, *Revue de l'Aluminium*, vol. 18, avril 1927.

Chapter A.2
Physical Properties of Aluminium

Chapter A.2
Physical Properties of Aluminium

The principal physical properties of unalloyed aluminium are listed in Table A.2.1.

Table A.2.1. Properties of unalloyed aluminium

Property	Unit	Value	Note
Atomic number		13	
Density, ρ	$kg \cdot m^{-3}$	$2698^{(1)}$	
Melting point	°C	660.45	$< 1013 \times 10^{-3}$ bar
Boiling point	°C	2056	$< 1013 \times 10^{-3}$ bar
Vapour pressure	Pa	3.7×10^{-3}	at 927 °C
Mass internal energy, u	$J \cdot kg^{-1}$	3.98×10^5	
Mass thermal capacity, C_p	$J \cdot kg^{-1} \cdot K^{-1}$	897	at 25 °C
Thermal conductivity, λ	$W \cdot m^{-1} \cdot K^{-1}$	237	at 27 °C
Linear expansion coefficient, α_1	$10^{-6} \, K^{-1}$	23.1	at 25 °C
Electrical resistivity, ρ	$10^{-9} \, \Omega \cdot m$	26.548	at 25 °C
Magnetic susceptibility, K		0.6×10^{-3}	at 25 °C
Longitudinal elasticity modulus, E	MPa	69 000	
Poisson's ratio, ν		0.33	

$^{(1)}$This is the generally accepted value for the density of metal between 99.65 and 99.99% pure. At 700 °C, the density of molten metal of 99.996% purity is 2357 kg·m^{-3}.

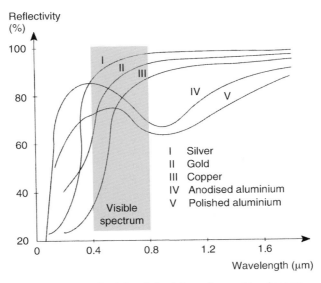

Figure A.2.1. Reflectivity of aluminium, silver, gold and copper.

19

Corrosion of Aluminium

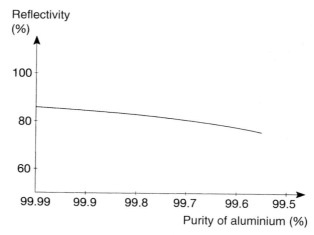

Figure A.2.2. Effect of purity on the reflectivity of anodised aluminium (5-μm-thick anodisation layer).

The reflectivity of bare and anodised aluminium depends both on the surface aspect and on the wavelength (Figure A.2.1). It increases with purity: with bright metal anodised to an oxide thickness of 5 μm, reflectivity increases from 75% on metal with a purity of 99.6 to 85% reflectivity for a 99.99% pure metal (Figure A.2.2).

Chapter A.3

The Metallurgy of Aluminium

Chapter A.3
The Metallurgy of Aluminium

Unalloyed metals generally have only few applications: while they exhibit very particular properties, these properties are often limited to a very narrow field of application.

The art of the metallurgist is to create alloys from a given base metal, be it copper, iron or aluminium, etc. by adding controlled amounts of other metals (or metalloids) in order to improve or modify certain properties such as mechanical properties, formability, weldability, etc. This is how, more than 5000 years ago, the most ancient metallurgists discovered that by adding tin to copper, they produced an alloy that was easy to mould and that offered an outstanding resistance to marine corrosion.

Sainte-Claire Deville prepared an aluminium alloy containing 10% silicon. The silversmith Charles Cristofle used an alloy containing 2% copper, thus harder and easier to chisel, for the objects that he manufactured in aluminium during the Second Empire. Immediately after the development of electrolysis of molten salts by Heroult in France and Hall in the United States in 1866, metallurgists tried to improve the properties of aluminium such as mechanical resistance, resistance at high temperatures, machinability, corrosion resistance, etc.

Research and development work carried out for over 100 years on the composition of alloys, their transformation processes, the heat treatment processes, and on optimising all these factors has resulted in a wide range of alloys, from which users can select those best suited to their specific requirements.

The diversity of alloys and the wide range of certain properties such as their tensile strength ranging from 100 to 700 MPa, explains the growth in applications, as numerous and as varied, from aeronautics to packaging (Figure A.3.1).[1]

3.1. ALLOY SERIES

All aluminium products belong to one of eight alloy series. They are available as

– castings,
– wrought semi-products that are flat rolled, extruded or forged.

---

[1] The reader will find the basic principles of the metallurgy of aluminium in this chapter and should refer to specific technical books or catalogues such as Pechiney Rhenalu's catalogue of semi-products, which has been a source of inspiration for Chapters 3 and 4 of Part A of this book.

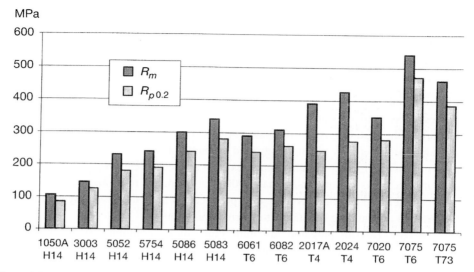

Figure A.3.1. Amplitude of the mechanical properties of aluminium alloys according to standard EN 485-2: guaranteed minimum values on sheets 1.5–3 mm thick.

Alloys belonging to the same series exhibit a set of common properties such as castability, mechanical properties, extrudability, corrosion resistance, etc. These properties can vary considerably from one series to another. For a given use, it is therefore not always possible, or desirable, to switch to another series.

The same applies to metallurgical tempers. A strain-hardened temper (temper H1X) will not have the same deformation capacity as a soft temper (temper O) in strain-hardenable alloys, and as an aged temper (temper T4) in age-hardenable alloys.

Industrial alloys comprise

- alloying elements,
- additives, and
- impurities.

3.1.1. Alloying elements

Alloying elements are added to wrought alloys in quantities ranging from 1 to 7% (in mass per cent), and in higher quantities, up to 20% silicon, to casting alloys. These elements are copper, magnesium, manganese, silicon, and zinc.

Some of these elements may be added simultaneously: silicon and magnesium for casting alloys of the series 40000, magnesium and silicon for wrought alloys of the 6000 series, and zinc and copper for those of the 7000 series. Alloying elements determine the common basic properties of alloys belonging to the same series.

The metallurgy of industrial aluminium alloys is, therefore, based on six systems:

- aluminium–copper,
- aluminium–manganese,
- aluminium–silicon (with or without magnesium),
- aluminium–magnesium,
- aluminium–magnesium–silicon, and
- aluminium–zinc (with or without copper).

The influence of the main alloying elements is explained in Table A.3.1.

■ **Designation of aluminium alloys**

The designation of aluminium alloys depends on the alloying element.

For wrought alloys, the numerical designation of the Aluminum Association (AA), based on a four-digit number, has become more and more popular in Europe since 1970. The first figure always specifies the alloy series (Table A.3.2).

This designation was adopted in the European Standard EN 573, which is based on two different designation systems:

- the first system is the numerical designation using the AA system: the four-digit number (and any suffix letter A or X) is preceded by EN AW (European Norm Wrought Product), for example, EN AW-3003;
- the second is based on chemical symbols: these are put in square brackets, e.g. [AlMn1Cu]. This system, reminiscent to the ISO designation, is being phased out.

Formally therefore, according to the standard EN 573, the complete designation of a wrought alloy is

EN AW-3003 [AlMn1Cu]

For the sake of simplification, wrought aluminium alloys are designated in this volume only by the four-digit number, followed, if necessary, by a letter, almost always "A", which designates a national variant, such as 1050A.

Only the first digit has practical importance. It designates the series to which the alloy belongs. For example, all the alloys of the series 5XXX are alloys with magnesium.

Recently, the European standard, EN 1780, has introduced a numerical designation system for casting alloys, which is based on the same principles as that for wrought alloys, but uses five-digit numbers without a space (Table A.3.2). This new designation is to replace the alphanumerical designation indicated in Table A.3.3, which is used only in France.

Table A.3.1. Influence of alloying elements

Element	Influence	Main alloys	
		Casting alloys	Wrought alloys
Copper	Age-hardenable alloys	21000 (A-U5GT)	2017A
	Improves mechanical properties		2024
	Decreases resistance to corrosion		2014
	Decreases electrical conductivity		2618
	Improves machinability		2219
	Improves creep resistance		2030
	Makes welding very difficult		
	Forming is possible in soft tempers or		
	immediately after quenching		
	Good suitability for surface treatments		
Manganese	Strain-hardenable alloys		3003
	Improves mechanical properties		3004
	Facilities deep drawing		
	Good resistance to corrosion		
Magnesium	Strain-hardenable alloys		
	Increase in mechanical properties depends on	51000 (A-G3T)	5052
	the magnesium content (Figure A.4.2)	51300 (A-G6)	5454
	Improves resistance to corrosion		5754
	Weldable alloys		5083
	Good suitability for surface treatments		5086
Silicon	Age-hardenable alloys, if containing		
	0.3–0.6% magnesium	41000 (A-S2GT)	
	Improves castability of aluminium;	45100 (A-S5U3)	
	castability is maximum at 13%	42000 (A-S7G)	
	silicon (eutectic composition)	42100 (A-S7G03)	
	Decreases machinability, since silicon	42200 (A-S7G06)	
	is a hard and abrasive element		
	Decreases linear expansion coefficient		
	Welding possible for certain casting alloys		
	without copper		
	Good resistance to corrosion for alloys		
	without copper		
Silicon +	Age-hardenable alloys		6060
Magnesium	Improves mechanical properties		6005A
	Improves extrudability		6061
	Very good resistance to corrosion		6082
	Very good suitability for		6106
	surface treatments		6262
Zinc	Improves mechanical properties; alloys of the	71000 (A-Z5G)	7020
	7000 series have the highest mechanical		7075
	properties of all aluminium alloys		7049A
	Weldable alloys, if no copper is added		
	Poor resistance to corrosion		

Table A.3.2. Series of aluminium alloys

Alloying element	Series of casting alloys	Series of wrought alloys
None	10000	1000
Copper	20000	2000
Manganese		3000
Silicon	40000	4000
Magnesium	50000	5000
Magnesium and silicon		6000
Zinc (and copper)	70000	7000

According to the old French standard, NF A 02-400, each alloying element and additive was designated by a letter.

After the letter "A", the alloying elements are placed in order of decreasing concentration. Their symbol is followed by the nominal concentration of that element. When this concentration is comprised between 0 and 1%, a 0 precedes the corresponding figure.

For example, A-S7G03 used to designate an alloy containing 7% silicon and 0.3% magnesium, and A-G3T designated an alloy containing 3% magnesium and a small added amount of titanium.

3.1.2. *Additives*

Additives are added in smaller quantities, usually less than 1%. Their task is to improve certain properties such as a small grain size after casting, quenchability, and weldability.

Table A.3.3. Alphanumeric designation of casting alloys

Element	Designation
Beryllium	Be
Chromium	C
Cobalt	K
Copper	U
Magnesium	G
Manganese	M
Nickel	N
Silicon	S
Titanium	T
Zinc	Z

The traditional additions are chromium, manganese (these two elements improve weldability), nickel (improves resistance at high temperatures), titanium (refining the as-cast structure), beryllium, zirconium, lead (free machining alloys), etc. An alloy can contain more than one additive, and their concentrations may exceed 1% in certain cases. All the alloying elements can also be additives in another series of alloys.

3.1.3. *Impurities*

Iron and silicon are the two main impurities of unalloyed aluminium in the 1000 series; their total concentration determines the purity of the metal. The iron/silicon ratio is close to two, for most grades, unless it is deliberately modified, as with the 8000 series. The concentration of impurities can vary, depending on the alloy, from a few ppm (parts per million) in refined aluminium (1199) up to 1000–2000 ppm in most wrought alloys. The impurity level of casting alloys can be higher for alloys based on secondary aluminium.

3.2. CASTING ALLOYS

There is a distinction between

–   primary aluminium alloys, elaborated with primary aluminium resulting from electrolysis, and
–   secondary aluminium alloys (also called refined alloys). They are elaborated by fusion of collected scrap. They are used, in particular, for the manufacture of engine components for motor vehicles.

3.2.1. *Principal casting alloys*

The foremost use of casting alloys is in the automotive industry (60% of the tonnage world-wide). These are mainly engine components elaborated from alloys with silicon, most of which also contain at least 3% copper: 45100 (A-S5U3), 46200 (A-S9U3), etc.

   The main applications outside the automotive industry are in mechanical construction, electrical engineering, transport, household electrical appliances and ironmongery.

■ Unalloyed aluminium, series 10000

Unalloyed cast aluminium is almost exclusively used for electrical applications.

■ Alloys with silicon, series 40000

These alloys are easy to cast because their eutectic temperature is 575 °C. Most aluminium casting alloys are alloys with silicon (40000 series). They are divided into two categories:

–   copper-containing alloys, which are used only in the automotive industry. They are beyond the scope of this book. Due to their high copper content, their resistance to corrosion is rather low;
–   alloys without copper:

    •   44100 (A-S13), formerly called Alpax,
    •   44100 (A-S7G03),
    •   42200 (A-S7G06),
    •   43300 (A-S10G).

These alloys have many applications as structural components in mechanical engineering, in electrical industry and automotive industry. Alloys that contain magnesium are age-hardenable alloys. Their resistance to atmospheric corrosion is excellent.

■ **Alloys with copper, series 20000**

The age-hardenable alloy 21000 (A-U5GT) is the most common one. Its resistance to corrosion is poor, and it should not be exposed to outdoor conditions or any other aggressive environment.

■ **Alloys with magnesium, series 50000**

Two alloys are commonly used: alloy 51000 (A-G3T) and alloy 51300 (A-G6). Because of their high magnesium content, they are rather difficult to cast.
   Their resistance to atmospheric corrosion is excellent, especially in marine environments. They are, therefore, used for the manufacture of components for upperworks.

■ **Alloys with zinc, series 70000**

Only alloy 71000 (A-Z5G) needs to be mentioned here. It is difficult to cast. Its resistance to corrosion is fair.

3.2.2.   *Alloy compositions and properties*
The standardised compositions and the properties of the most commonly used casting alloys are given in Tables A.3.4 and A.3.5, respectively.

Table A.3.4. Standardised compositions of castings according to EN 1706 (in wt%)

Alloy														Others	
NF EN 1706 [1]	NF A 57-702	Si	Fe	Cu	Mn	Mg	Ni	Zn	Ti	Cr	Pb	Sn		Each	Total
21 100	A-U5GT	0.20	0.35	4.2–5.0	0.10	0.15–0.35	0.05	0.10	0.15–0.30		0.05	0.05		0.03	0.10
41 000	A-S2GT	1.6–2.4	0.60	0.10	0.30–0.50	0.45–0.65	0.05	0.10	0.05–0.20		0.05	0.05		0.05	0.15
42 100	A-S7G03	6.5–7.5	0.19	0.05	0.10	0.25–0.45		0.07	0.08–0.25					0.03	0.10
42 200	A-S7G06	6.5–7.5	0.19	0.05	0.10	0.45–0.70		0.07	0.08–0.25					0.03	0.10
43 300	A-S10G	9.0–10.0	0.19	0.05	0.10	0.25–0.45		0.07	0.15					0.03	0.10
44 100	A-S13	10.5–13.5	0.65	0.15	0.55	0.10	0.10	0.15	0.20		0.10	0.05		0.05	0.15
51 100	A-G3	0.55	0.55	0.05	0.45	2.5–3.5		0.10	0.20					0.05	0.10
51 300	A-G6	0.55	0.55	0.10	0.45	4.5–6.5		0.10	0.20					0.05	0.15
71 000	A-Z5G	0.30	0.80	0.15–0.35	0.40	0.40–0.70	0.05	4.50–6.0	0.10–0.25	0.15–0.60	0.05	0.05		0.05	0.15

Note: a single value is to be understood as a maximum value.

[1] The equivalency between the compositions in these two standards is more or less precise: for certain elements, rather substantial differences may occur.

Table A.3.5. Physical properties of the most important aluminium casting alloys

Alloy		Melting range (°C)	Density, ρ (kg·m^{-3})	Linear expansion coefficient, α_1 (μm·m^{-1}·K^{-1}) 20/100 °C	Young's modulus, E (MPa)	Thermal conductivity at 20 °C, λ (W·m^{-1}·K^{-1})	Electrical resistivity at 20 °C, ρ (10^{-3} $\mu\Omega$·m)
10 000	A5	658	2700	24	67 500	210	28
21 100	A-U5GT	535–650	2800	23	72 000	140	60
41 000	A-S2GT	555–640	2700	22	70 000	160	40
42 100	A-S7G03	555–615	2680	21.5	74 000	160	40
42 200	A-S7G06	555–615	2680	21.5	76 000	160	40
43 300	A-S10G	555–600	2650	20.5	76 000	160	45
44 100	A-S13	575–580	2650	20	76 000	165	45
51 100	A-G3T	590–640	2670	24	69 000	145	60
51 300	A-G6	450–625	2640	24	69 000	125	70
71 000	A-Z5G	600–650	2800	23	72 000	140	60

3.2.3. Methods of elaboration and thermal treatments

Due to the widespread use of aluminium casting alloys, the casting techniques have been highly developed [1]. Casting techniques can be grouped into four categories, each of which can have several variants:

–   sand casting, i.e. casting by gravity in a sand mould formed from a model made in wood, resin or metal;
–   investment casting, i.e. casting by gravity in a ceramic mould formed around a pattern in wax or expanded polystyrene that is lost during the process;
–   permanent mould casting by gravity or under reduced pressure in a permanent metal mould; and
–   pressure die-casting in a permanent metal mould.

The choice of the casting technique has no influence on the corrosion resistance of casting alloys.

Alloys with silicon-containing magnesium such as alloys 42XXX and 43XXX, as well as alloy 21100 (A-U5GT) can undergo age-hardening, the principle of which is outlined in Table A.3.6. The designations of the tempers which result from these thermal treatments are also given in Table A.3.6.

In European standardisation, casting alloys are designated by five figures, followed by the metallurgical temper designation, such as 42100 T6 (formerly designated as A-S7G03 Y23).

Table A.3.6. Standardised temper designations of casting alloys

Thermal treatment	NF EN 1706	AFNOR NF A 57 702 [1]	AFNOR NF A 57 703 [2]
As cast	F	Y20, Y30	Y40
Annealed (above 350 °C)	O		
Controlled cooling after unmoulding and natural ageing	T1		
Solution heat treatment, quenching and natural ageing	T4	Y24, Y34	
Controlled cooling after unmoulding and ageing or over-ageing	T5		
Solution heat treatment, quenching and peak ageing	T6 [3]	Y23, Y33	
Solution heat treatment, quenching and under-ageing	T64	Y23, Y33	
Solution heat treatment, quenching and over-ageing (stabilisation)	T7	Y26, Y36	

[1] Y2X designates sand casting, Y3X die casting.

[2] Y4X designates pressure die-casting.

[3] In order to preserve minimum elongation, T6 ageing is never carried out as peak ageing but slightly below peak strength. Temper T64 designates such an under-ageing treatment, which favours elongation.

3.3. WROUGHT ALLOYS

The concentrations of alloying elements and additives are given in Table A.3.7. The standardised compositions of the most common alloys are given in Table A.3.8, and their physical properties in Table A.3.9.

From a metallurgical point of view, the series of wrought aluminium alloys are split into two groups that are very distinct, both in processing route and in certain properties. These groups are

- strain-hardenable alloys
- age-hardenable alloys.

3.3.1. *Strain-hardenable alloys*

Strictly speaking, all metals and alloys can be strain hardened. However, in the field of aluminium metallurgy, this designation is used only for alloys of those series that cannot be age-hardened.

These alloys belong to the series 1000, 3000, 5000 and 8000. Their processing route is a sequence of hot forming steps, possibly followed by cold forming steps with intermediate or final annealing.

Table A.3.7. Series of wrought alloys

Mode of hardening	Series	Alloying element	Concentration range (%)	Additives	Mechanical strength R_m (MPa)
Strain hardening	1000	None		Cu	50–160
	3000	Manganese	0.5–1.5	Mg, Cu	100–240
	5000	Magnesium	0.5–5	Mn, Cr	100–340
	8000	Iron and silicon	Si: 0.30–1 Fe: 0.6–2		130–190
Age hardening	6000	Magnesium and silicon	Mg: 0.5–1.5 Si: 0.5–1.5	Cu, Cr	200–320
	2000	Copper	2–6	Si, Mg	300–480
	7000	Zinc and magnesium	Zn: 5–7 Mg: 1–2	Cu	Without copper 320–350 With copper 430–600
	4000	Silicon	0.8–1.7		150–400

■ **Strain-hardenable alloys**

Strain hardening involves a modification of the structure due to plastic deformation. It occurs not only during the manufacturing of semi-products in the course of rolling, stretching, drawing, etc., but also during subsequent manufacturing steps such as forming, bending or fabricating operations.

Strain hardening increases the mechanical resistance and hardness, but decreases ductility (Figure A.3.2).

The level of mechanical properties that can be attained depends on the alloying element. As an example, the alloys of the 5000 series that have a high magnesium content have a potential level of mechanical properties that is superior to that of alloys of the other series: 1000, 3000 and 8000. However, the gradual increase in mechanical strength always reaches a point beyond which any further working becomes difficult if not impossible. In this case, if further deformation is desired, it is necessary to carry out a thermal annealing treatment.

■ **Softening by thermal annealing**

The hardening caused by working can be eliminated or mitigated by annealing. Depending on the selected time–temperature combination, softening can be (Figure A.3.3):

– *partial*: this is called a softening or recovery annealing;
– *complete*: this is called recrystallisation annealing, during which a new grain structure develops (Figure A.3.4).

Table A.3.8. Composition of wrought alloys. Extracted from standard EN 573 (1000 series)

Alloy	Si	Fe	Cu	Mn	Mg	Cr	Ni	Zn	V	Ti	Others [1]		Remarks
											Each	Total [2]	
1200	Si + Fe 1.00		0.05	0.05				0.10		0.05	0.05	0.15	Aluminium minimum: 99.00 [3]
1100	Si + Fe 0.95		0.05 0.20	0.05				0.10			0.05	0.15	Aluminium minimum: 99.00 [3]
1050A	0.25	0.40	0.05	0.05	0.05			0.07		0.05	0.03		Aluminium minimum: 99.50 [3]
1198	0.010	0.006	0.006	0.006				0.010		0.006	0.003		Ga 0.006 Aluminium minimum: 99.98 [3]
1199	0.006	0.006	0.006	0.002	0.006			0.006	0.005	0.002	0.002		Aluminium minimum: 99.99 [3]

[1] Includes the listed elements for which no specific limit is given.

[2] The total of these other metallic elements, each with a percentage of 0.010 or more, is given to two decimal places before addition.

[3] For unalloyed aluminium which is obtained by a refining process, the percentage of aluminium is equal to the difference between 100% and the total of all the other metallic elements present in an amount of 0.001% each, given to three decimal places before addition, which is rounded to the second decimal place before subtraction.

Table A.3.8. 2000 series

Alloy	Si	Fe	Cu	Mn	Mg	Cr	Ni	Zn	V	Ti	Others [1]		Remarks
											Each	Total [2]	
2618A	0.15–0.25	0.9–1.4	1.8–2.7	0.25	1.2–1.8		0.8–1.4	0.15		0.20	0.05	0.15	Zr + Ti: 0.25. The remainder is aluminium
2017A	0.20–0.8	0.7	3.5–4.5	0.40–1.0	0.40–1.0	0.10		0.25			0.05	0.15	Zr + Ti: 0.25. The remainder is aluminium
2024	0.50	0.50	3.8–4.9	0.30–0.9	1.2–1.8	0.10		0.25		0.15	0.05	0.15	The remainder is aluminium [3]
2124	0.20	0.30	3.8–4.9	0.30–0.9	1.2–1.8	0.10		0.25		0.15	0.05	0.15	The remainder is aluminium
2014	0.50–1.2	0.70	3.9–5.0	0.40–1.2	0.20–0.80	0.10		0.25		0.15	0.05	0.15	The remainder is aluminium
2214	0.50–1.2	0.30	3.9–5.0	0.40–1.2	0.20–0.8	0.10		0.25		0.15	0.05	0.15	The remainder is aluminium
2219	0.20	0.30	5.8–6.8	0.20–0.40	0.02			0.10	0.05–0.15	0.02–0.10	0.05	0.15	Zr: 0.10–0.25. The remainder is aluminium
2030	0.8	0.7	3.3–4.5	0.20–1.0	0.50–1.3	0.10		0.50		0.20	0.10	0.30	Bi: 0.20; Pb: 0.8–1.5. The remainder is aluminium
2011	0.40	0.7	5.0–6.0					0.30			0.05	0.15	Bi: 0.20–0.6; Pb: 0.20–0.6. The remainder is aluminium

[1] Includes the listed elements for which no specific limit is given.

[2] The total of these other metallic elements, each with a percentage of 0.010 or more, is given to two decimal places before addition.

[3] A limit of 0.20 max. for Ti + Zr can be used for extruded and forged products by mutual agreement between the supplier or the producer and the client.

Table A.3.8. 3000 series

Alloy	Si	Fe	Cu	Mn	Mg	Cr	Ni	Zn	V	Ti	Others [1] Each	Total [2]	Remarks
3015	0.6	0.7	0.30	0.30–0.8	0.20–0.8	0.20		0.40		0.10	0.05	0.15	The remainder is aluminium
3103	0.50	0.7	0.10	0.9–1.5	0.30	0.10		0.20			0.05	0.15	Zr + Ti: 0.10 The remainder is aluminium
3003	0.6	0.7	0.05–0.20	1.0–1.5				0.10			0.05	0.15	The remainder is aluminium
3004	0.30	0.70	0.25	1.0–1.5	0.8–1.3			0.25			0.05	0.15	The remainder is aluminium
3005	0.6	0.7	0.30	1.0–1.5	0.20–0.6	0.10		0.25		0.10	0.05	0.15	The remainder is aluminium

[1] Includes the listed elements for which no specific limit is given.
[2] The total of these other metallic elements, each with a percentage of 0.010 or more, is given to two decimal places before addition.

Table A.3.8. 4000 series

| Alloy | Si | Fe | Cu | Mn | Mg | Cr | Ni | Zn | V | Ti | Others [1] | | Remarks |
											Each	Total [2]	
4043A [3]	4.5												The remainder is aluminium [4]
	6.0	0.6	0.30	0.15	0.20			0.10		0.15	0.05	0.15	

[1] Includes the listed elements for which no specific limit is given.
[2] The total of these other metallic elements, each with a percentage of 0.010 or more, is given to two decimal places before addition.
[3] Cladding alloy for brazing and filler wire for welding.
[4] Be: 0.008 for welding electrodes and filler wires only.

Table A.3.8. 5000 series

Alloy	Si	Fe	Cu	Mn	Mg	Cr	Ni	Zn	V	Ti	Others[1]		Remarks
											Each	Total[2]	
5005	0.30	0.7	0.20	0.20	0.50–1.1	0.10		0.25			0.05	0.15	The remainder is aluminium
5657	0.08	0.10	0.10	0.03	0.6–1.0			0.05	0.05		0.02	0.05	Ga: 0.03
5049	0.40	0.50	0.10	0.50–1.0	1.6–2.5	0.30		0.20			0.05	0.15	The remainder is aluminium
5052	0.25	0.40	0.10	0.10	2.2–2.8	0.15–0.35	0.10	0.10			0.05	0.15	The remainder is aluminium
5454	0.25	0.40	0.10	0.50–1.0	2.4–3.0	0.05–0.20		0.25		0.20	0.05	0.15	The remainder is aluminium
5754	0.40	0.40	0.10	0.50	2.6–3.6	0.30		0.20		0.15	0.05	0.15	Mn + Cr: 0.1–0.6 The remainder is aluminium
5154A	0.50	0.50	0.10	0.50	3.1–3.9	0.25		0.20		0.20	0.05	0.15	Mn + Cr: 0.10–0.50 The remainder is aluminium
5086	0.40	0.50	0.10	0.20–0.7	3.5–4.5	0.05–0.25		0.25		0.15	0.05	0.15	The remainder is aluminium
5182	0.20	0.35	0.15	0.20–0.50	4.0–5.0	0.10		0.25		0.10	0.05	0.15	The remainder is aluminium
5083	0.40	0.40	0.10	0.40–1.0	4.0–4.9	0.05–0.25		0.25		0.15	0.05	0.15	The remainder is aluminium
5019[3]	0.40	0.50	0.10	0.10–1.0	4.5–5.6	0.20		0.20		0.20	0.05	0.15	Mn + Cr: 0.10–0.6 The remainder is aluminium
5356[4]	0.25	0.40	0.10	0.05–0.20	4.5–5.5	0.05–0.20		0.10		0.06–0.20	0.05	0.15	The remainder is aluminium [5]

[1] Includes the listed elements for which no specific limit is given.
[2] The total of these other metallic elements, each with a percentage of 0.010 or more, is given to two decimal places before addition.
[3] Formerly called 5056A.
[4] Welding wire.
[5] Be: 0.008 for welding electrodes and filler wires only.

Table A.3.8. 6000 series

Alloy	Si	Fe	Cu	Mn	Mg	Cr	Ni	V	Zn	Ti	Others [1]		Remarks
											Each	Total [2]	
6060	0.30–0.6	0.30	0.10	0.10	0.35–0.6	0.05			0.15	0.10	0.05	0.15	The remainder is aluminium
6005A	0.50–0.9	0.35	0.30	0.50	0.40–0.7	0.30			0.20	0.10	0.05	0.15	Mn + Cr: 0.12–0.50; The remainder is aluminium
6106	0.30–0.6	0.35	0.25	0.05–0.20	0.40–0.8	0.20			0.10		0.05	0.10	The remainder is aluminium
6063	0.20–0.6	0.35	0.10	0.10	0.45–0.9	0.10			0.10	0.10	0.05	0.15	The remainder is aluminium
6082	0.7–1.3	0.50	0.10	0.40–1.0	0.6–1.2	0.25			0.20	0.10	0.05	0.15	The remainder is aluminium
6056	0.7–1.3	0.50	0.50–1.1	0.40–1.0	0.6–1.2	0.25			0.10–0.7		0.05	0.15	Zr + Ti: 0.20 max; The remainder is aluminium
6061	0.40–0.8	0.7	0.15–0.40	0.15	0.8–1.2	0.04–0.35			0.25	0.15	0.05	0.15	The remainder is aluminium
6012	0.6–1.4	0.50	0.10	0.40–1.0	0.6–1.2	0.30			0.30	0.20	0.05	0.15	Bi: 0.7; Pb: 0.40–0.20; The remainder is aluminium
6262	0.40–0.8	0.7	0.15–0.40	0.15	0.8–1.2	0.04–0.14			0.25	0.15	0.05	0.15	Bi:0.40; Pb: 0.40–0.7; The remainder is aluminium

[1] Includes the listed elements for which no specific limit is given.

[2] The total of these other metallic elements, each with a percentage of 0.010 or more, is given to two decimal places before addition.

Table A.3.8. 7000 series

Alloy	Si	Fe	Cu	Mn	Mg	Cr	Ni	Zn	V	Ti	Others [1]		Remarks
											Each	Total [2]	
7039	0.30	0.40	0.10	0.10–0.40	2.3–3.3	0.15–0.25		3.5–4.5		0.10	0.05	0.15	The remainder is aluminium
7020	0.35	0.40	0.20	0.05–0.50	1.0–1.4	0.10–0.35		4.0–5.0			0.05	0.15	The remainder is aluminium [3]
7075	0.40	0.50	1.2–2.0	0.30	2.1–2.9	0.18–0.28		5.1–6.1		0.20	0.05	0.15	The remainder is aluminium [4]
7475	0.10	0.12	1.2–1.9	0.06	1.9–2.6	0.18–0.25		5.2–6.2			0.05	0.15	Zr: 0.08–0.15 The remainder is aluminium
7050	0.12	0.15	2.0–2.6	0.10	1.9–2.6	0.04		5.7–6.7		0.06	0.05	0.15	Zr: 0.08–0.15 The remainder is aluminium
7010	0.12	0.15	1.5–2.0	0.10	2.1–2.6		0.05	5.7–6.7		0.06	0.05	0.15	Zr: 0.10–0.16 The remainder is aluminium
7049A	0.40	0.50	1.2–1.9	0.50	2.1–3.1	0.05–0.25		7.2–8.4		0.06	0.05	0.15	Zr + Ti: 0.25 The remainder is aluminium
7072 [5]	Si + Fe 0.7		0.10	0.10	0.10			0.8–1.3			0.05	0.15	The remainder is aluminium

[1] Includes the listed elements for which no specific limit is given.

[2] The total of these other metallic elements, each with a percentage of 0.010 or more, is given to two decimal places before addition.

[3] Zr: 0.08–0.20; Zr + Ti: 0.08–0.25.

[4] A limit of 0.20 max. for Ti + Zr can be used for extruded and forged products by mutual agreement between the supplier or the producer and the client.

[5] Cladding alloy.

Table A.3.8. 8000 series

Alloy	Si	Fe	Cu	Mn	Mg	Cr	Ni	Zn	V	Ti	Others [1] Each	Total [2]	Remarks
8011A	0.40 0.8	0.50 1.0	0.10	0.10	0.10	0.10		0.10		0.05	0.05	0.15	The remainder is aluminium
8006	0.40	1.2 2.0	0.30	0.30 1.0	0.10			0.10			0.05	0.15	The remainder is aluminium

[1] Includes the listed elements for which no specific limit is given.

[2] The total of these other metallic elements, each with a percentage of 0.010 or more, is given to two decimal places before addition.

Corrosion of Aluminium

Table A.3.9. Physical properties of the most common wrought alloys

Alloy	Melting range (°C)	Density, ρ (kg·m⁻³)	Linear expansion, α_l (μm·m⁻¹·K⁻¹) 20–100°C	Mass thermal capacity, c (J·kg·C⁻¹) 0–100°C	Modulus of elasticity (Young's modulus), E (MPa)	Temper	Thermal conductivity, λ (W·m⁻¹·K⁻¹) at 20°C	Electrical resistivity, ρ (10⁻³ μΩ·m) at 20°C
1200	645–657	2720	23.4	898	69 000	all	225	33.9
1050A	645–658	2700	23.5	899	69 000	O	229	29.2
1199	660	2700	23.6	900	62 000	O	243	26.7
2618A	549–638	2760	22.3	875	74 000	T6	146	47
2017A	510–640	2790	23.6	920	74 000	T4	134	51
2024	502–638	2770	21.1	875	73 000	O	193	34
						T3, T4	121	56
						T6	151	45
2014	507–638	2800	22.5	920	74 000	O	192	34
						T4	134	51
						T6	155	43
2214	502–538	2770	21.1	882	73 000	T4	135	51
2219	545–645	2840	22.5	864	73 000	O	172	38
						T3	113	62
						T6, T8	121	59
2030	510–640	2820	23.0	864	73 000	T3, T4	134	51
2011	541–638	2820	23.1	864	71 000	T3, T4	152	44
3105	638–657	2710	23.6	897	69 000	all	172	38
3103	640–655	2730	23.1	892	69 000	O	190	34
						H14	160	41
3003	643–654	2730	23.2	893	69 000	O	193	34
						H12	163	41
						H14	159	42
						H18	155	43
3004	629–634	2720	23.2	893	69 000	all	166	39
3005	632–653	2730	23.2	897	69 000	all	166	39

4043A	575–630	2680	22.0			O	163	40
5005	632–652	2700	23.7	900	69 000	all	201	33
5657	638–657	2690	23.7	900	69 000	all	205	32
5049	620–650	2710	23.7					
5052	605–650	2680	23.8	900	70 000	all	138	50
5454	602–646	2660	23.7	900	70 000	all	134	50
5754	590–645	2670	23.8	900	70 000	all	132	53
5086	585–640	2660	23.8	900	71 000	all	126	56
5182	577–638	2650	24.1	904	69 500	O	123	56
5083	574–638	2660	24.2	900	71 000	O	117	59
5019	568–638	2640	24.1	904		O	117	59
5356	574–630	2640	24.2	904		O	117	59
6060	615–655	2700	23.4	945	69 500	T5	200	33
6101	621–654	2690	23.5	895	69 000	T6	218	30
6005A	607–654	2700	23.6	940	69 500	T1	180	37
6106	610–655	2700	23.5		69 500	T5	188	36
6063	615–655	2690	23.4	900	69 500	T5	180	35
						O	218	29
6082	570–645	2710	23.5	960	69 500	T6, T8	201	32
						T6	174	42
6061	582–652	2700	23.6	896	69 500	O	180	37
						T4	153	43
						T6	167	40
6262	582–652	2710	23.4		69 500	T6	172	38
7020	604–645	2780	23.1	875	71 500	O	166	40
						T6	137	49
7075	477–635	2800	23.4	960	72 000	T6	130	52
						T73	155	43
7050	490–635	2830	23.5	860	71 500	O	180	36
						T76	154	43
7049A	477–627	2820	23.4	960	72 000	T6	154	43

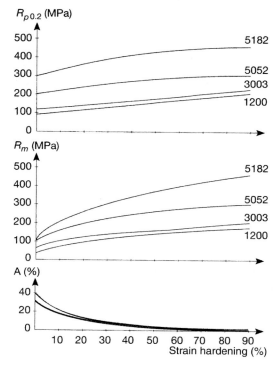

Figure A.3.2. Influence of strain hardening on mechanical properties.

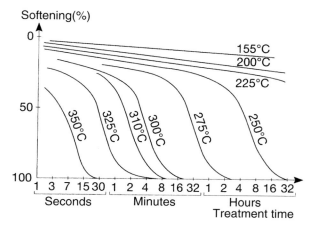

Figure A.3.3. Isothermal annealing curves of alloy 5754.

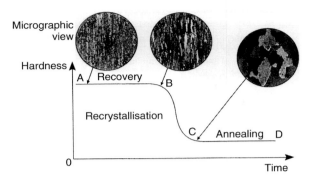

Micrographic view

Hardness

A / Recovery   B

Recrystallisation

C / Annealing   D

0

Time

Figure A.3.4. Change in hardness and structure during annealing.

The time and temperature parameters are specific to each alloy and depend on the amount of strain hardening that the metal has undergone prior to annealing.

As with other metals and alloys, there is a critical strain-hardening zone (Figure A.3.5). If annealing is applied to a strain-hardening rate within that critical zone, uncontrolled grain growth may take place. This makes subsequent forming operations such as deep drawing and bending more difficult. After deformation, the surface of the metal may then take on a peculiar appearance known as the orange peel effect.

The level of mechanical properties of a semi-product, and in particular the compromise between mechanical resistance (R_m) and ductility ($A\%$), are controlled by the working conditions and any subsequent annealing operations (intermediate or final). This is the so-called transformation route, which is chosen by the manufacturer.

It should be noted that for a given level of mechanical resistance, the ductility is higher in partially annealed (recovered) metal (H2X) than in strain-hardened metal (H1X).

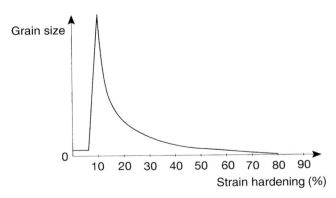

Grain size

0

10   20   30   40   50   60   70   80   90

Strain hardening (%)

Figure A.3.5. Change in grain size as a function of strain hardening.

Table A.3.10. Standard conditions of annealing for strain-hardenable alloys

Series or alloy	Annealing of strain-hardened metal			Recovery annealing	
	Temperature (°C)	Time [1] (h)	Minimum strain hardening before annealing (%)	Temperature (°C)	Time [1] (h)
1000	330–400	0.5–2	>20	240–280	1–4
3105, 3004, 3005	330–380	0.5–2	>20	25–300	1–4
3003	400–430	0.5–2	>20	260–300	1–4
5000	330–380	0.5–2	>20	240–280	1–4

[1]All parts of the batch should be checked to be sure that they actually reach the minimum temperature of the heat treatment.

Partially annealed tempers will thus be preferred when a maximum formability is the prime consideration such as for deep drawing.

When semi-products are formed, it is possible to perform soft annealing to achieve an adequate working capacity under the conditions given in Table A.3.10.

H3X tempers aim at stabilising the microstructure of an alloy. When applied to 5000 series alloys with a high magnesium content, they improve resistance to intergranular corrosion over time.

■ Concept of metallurgical tempers

According to standard EN 515, there are three basic tempers:

–   F: as manufactured, with no guarantee of properties;
–   O: annealed, thus the capacity of deformation is maximum;
–   H: strain hardened.

Most of the common H tempers are designated by two digits:

–   The first digit, 1, 2 or 3, denotes the specific combination of basic operations, i.e.
  - H1X: strain hardened only; this is the strain-hardened temper,
  - H2X: strain hardened and partially annealed; this is the partially annealed temper,
  - H3X: strain hardened and stabilised by a low-temperature heat treatment or by heating carried out during working; this is the stabilised temper, which applies mainly to alloys of the 5000 series.
–   The second digit, which is usually 2, 4, 6 or 8, indicates the final degree of strain hardening, as characterised by a minimum value for the ultimate tensile strength (UTS). This digit is defined as explained below.

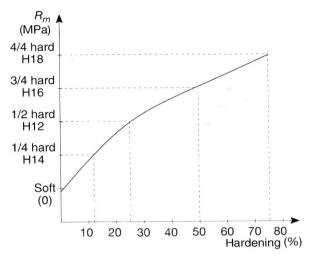

Figure A.3.6. Metallurgic states and work hardening.

The digit 9 designates tempers in which the UTS exceeds by at least 10 MPa, the tensile strength of HX8 tempers.

To define the four main strain-hardened tempers, the range of R_m between the value of maximum strain hardening and the value for the annealed temper is divided into four zones (Figure A.3.6):

–   hard temper, indicated by the number 8. H18 denotes 75% strain hardening by cold rolling. H28 (like all H2X tempers) indicates the level of mechanical properties equivalent to that of H18.
–   half hard temper, indicated by the number 4. This temper indicates an UTS halfway between the annealed and the hard temper, and corresponds to strain hardening on the order of 25%.
–   three-quarter hard, indicated by the number 6, equivalent to strain hardening on the order of 50%. The UTS is halfway between half hard and hard.
–   quarter hard, indicated by 2, equivalent to strain hardening of around 12%. UTS is halfway between the half-hard and the annealed tempers.

Several tempers have a three-digit designation:

–   H111 differs from O in that rolled products in this temper have been levelled after annealing to improve their dimensional characteristics;
–   H112 is used for products whose level of mechanical properties is obtained by hot working or limited cold working;

–   H116 denotes 5000 series semi-products with more than 4% magnesium (5083, 5086, etc.). In this temper, products display no sign of exfoliation corrosion following the ASTM G66 test (see Table B.4.1);
–   HXX4 is used for tread plates embossed from the corresponding HXX temper.

H4X is used for products that are strain hardened and that may undergo partial annealing during the curing operation following lacquering or painting.

All these tempers have their own minimum mechanical properties that are prescribed in European standards, a list of which is provided in Table A.3.11.

3.3.2.   Age-hardenable alloys

These belong to the 2000, 6000 and 7000 series.

■ The principle of age hardening

The maximum mechanical properties of these alloys are obtained by a process comprising three steps (Figure A.3.7):

–   heating to a high temperature which is defined for each alloy, in order to put into solution the alloying element contained in the aluminium;
–   rapid cooling, called quenching, usually achieved by immersion in water at room temperature. The effect of this is to keep the alloying elements and additives in a supersaturated solid solution within the aluminium matrix;
–   an ageing step:
    • at room temperature, i.e. around 20 °C. This is called natural ageing, or
    • at elevated temperature, between 100 and 200 °C. This is called artificial ageing and leads to the formation of hardening precipitates from the supersaturated solid solution. The nature, size and volume fractions of these hardening precipitates determine the level of mechanical properties.

□ Solution heat treatment

Given the high temperature of solution heat treatment—it is very close to the solidus temperature—it is of paramount importance to stick to the temperatures given in Table A.3.12 for some common alloys. If the temperature is too low, the resulting mechanical properties will be low, whereas too high a temperature will overheat the product (formation of hot spots), leading to irreversible damage. This is due to partial fusion of complex eutectics and elements precipitated at the grain boundaries.

Table A.3.11. The most important European Standards on aluminium

No	Title of CEN standard	Part	Title of the part	Equivalent standards	
				France (AFNOR)	Unites States (ASTM)
1706	Castings. Chemical composition and mechanical properties				
573	Chemical composition and form of wrought products	1	Numerical designation system	NF A 02-104	B 275
		2	Chemical symbol based designation system		
		3	Chemical composition		
		4	Forms of products	NF A 02-006	B 296 ANSI H35.1
515	Wrought products. Temper designations				B 209
485	Sheet, strip and plate	1	Technical conditions for delivery and inspection		
		2	Mechanical properties	NF A 50-451	
		3	Tolerances on shape and dimensions for hot-rolled products	NF A 50-751	
		4	Tolerances on shape and dimensions for cold-rolled products	NF A 50-761	
755	Extruded rod/bar, tube and profiles	1	Technical conditions for inspection and delivery	NF A 01-101	B 221
				NF A 50-411	
		2	Mechanical properties	NF A 50-411	B 483
					B 221
					B 241
					B 429
		3	Round bar, tolerances on dimensions and form	NF A 50-702	ANSI H35.2
		4	Square bars, tolerances on dimensions and form	NF A 50-703	ANSI H35.2
		5	Rectangular bars, tolerances on dimensions and form	NF A 50-705	ANSI H35.2
		6	Hexagonal bars, tolerances on dimensions and form	NF A 50-704	ANSI H35.2
		7	Seamless tubes, tolerances on dimensions and form	NF A 50-737	B 483
					B 241
					B 429
					B 491
		8	Porthole tubes, tolerances on dimensions and form		B 483

(*continued*)

Table A.3.11. (continued)

No	Title of CEN standard	Part	Title of the part	France (AFNOR)	United States (ASTM)
754	Cold-drawn rod/bar and tube	9	Profiles, tolerances on dimensions and form		B 429 B 491
		1	Technical conditions for inspection and delivery	NF A 01-101 NF A 50-411	B 211
		2	Mechanical properties	NF A 50-411	B 211 B 234
		3	Round bar, tolerances on dimensions and form	NF A 50-702	ANSI H35.2
		4	Square bars, tolerances on dimensions and form	NF A 50-703	ANSI H35.2
		5	Rectangular bars, tolerances on dimensions and form	NF A 50-705	ANSI H35.2
		6	Hexagonal bars, tolerances on dimensions and form	NF A 50-704	ANSI H35.2
		7	Seamless tubes, tolerances on dimensions and form	NF A 50-711	B 241 B 491
		8	Porthole tubes, tolerances on dimensions and form		B 491
1301	Drawn wire	1	Technical conditions for inspection and delivery	NF A 01-101 NF A 50-411	
		2	Mechanical properties	NF A 50-411	
		3	Tolerances on dimensions	NF A 50-735 NF A 50-736	
1386	Tread plate				
1396	Coil coated sheet and strip for general applications		Specifications	NF A 50-452	B 632
601	Castings. Chemical composition of castings for use in contact with food		Specifications	NF A 57-105	
602	Wrought products. Chemical composition of semi products used for the fabrication of articles in contact with food			NF A 57-105	

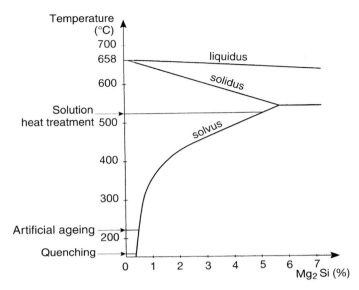

Figure A.3.7. Sequences of age hardening.

The duration of a solution heat treatment, from the moment the core of the semi-product reaches the set-point temperature, must be adequate. It will depend on the thickness (or diameter) of the semi-product and on the type of furnace (Table A.3.13).

□ **Quenching**

Quenching must immediately follow the discharge of the products from the solution furnace. The quenching rate must be sufficiently high; it depends on the quenching technique:

–   immersion in water,
–   spray quenching, and
–   forced-air quenching.

In order to obtain maximum properties after quenching, the quenching rate must exceed a certain value called the critical quenching rate which depends on the alloy: it ranges from $0.5\ °\text{C·s}^{-1}$ for an 7020 type alloy to more than $100\ °\text{C·s}^{-1}$ for 7075.

With increasing product thickness, quenching becomes slower and hence less efficient in the centre of the product. This can lead to a gradient of mechanical properties between

Table A.3.12. Conditions for age-hardening of alloys

Alloy	Temper	Solution heat treatment temperature [1] (°C)	Quenching medium	Waiting time between artificial ageing and natural ageing	Artificial ageing Temperature (°C)	Artificial ageing Duration (h)	Natural ageing duration mini. (d)
2618A	T8	530 ± 5	Water ≤ 80 °C	24 h max or 10 d mini	190 ± 3	19–21	
2017A	T4	500 ± 5	Water ≤ 40 °C				4
2024	T3	495 ± 5	Water ≤ 40 °C				4
2014	T6	505 ± 5	Water ≤ 40 °C		160 ± 3 or 175 ± 5	18–22 10	
2219	T81	535 ± 5	Water ≤ 40 °C		175 ± 3	18	
2030	T4	475 ± 5	Water ≤ 40 °C				4
2011	T4	510 ± 5	Water ≤ 40 °C				4
	T8	510 ± 5	Water ≤ 40 °C		160 ± 5		14
6005A	T4	530 ± 5	Water ≤ 40 °C				8
	T6	530 ± 5	Water ≤ 40 °C		175 ± 5 [2] or 185 ± 5	8 6	
6082	T4	535 ± 5	Water ≤ 40°C [3]				8
	T6	535 ± 5	Water ≤ 40°C [3]	2 h maxi or 5 d mini	165 ± 5 [2] or 175 ± 5	16 8	
6061	T4	535 ± 5	Water ≤ 40 °C				8
	T6	535 ± 5	Water ≤ 40 °C		175 ± 5 [2] or 185 ± 5	8 6	
7020	T4	450 ± 10	Water or forced air [3]				4
	T6	450 ± 10	Water or forced air [3]	4 d mini	100 ± 5 and 140 ± 5	4–6 24–26	4
7075	T6	465 ± 5	Water ≤ 40 °C		135 ± 3	12–16	
	T73 Sheets	465 ± 5	Water ≤ 40 °C		108 ± 5 and 161 ± 3	8 24–30	
	T73 Bars	465 ± 5	Water ≤ 40 °C		108 ± 5 and 177 ± 5	6–8 8–12	
7049A	T6	465 ± 5	Water ≤ 40 °C		135 ± 5	12	
	T73	465 ± 5	Water ≤ 40 °C		120 ± 5 and 165 ± 5	24 12–24	

[1] The holding time is given in table A.3.13.

[2] This treatment leads to optimal mechanical properties, associated with the highest values for *A%*.

[3] Because of the low critical quenching rate of these alloys, thin semi-products can be quenched by forced-air quenching.

the surface and the centre, and also to a gradient in corrosion resistance, since this parameter depends on the quenching rate.

Quenching can cause internal stresses in the products, leading to geometrical distortion or deformation during subsequent machining operations. In order to relax these internal stresses, products are stress-relieved immediately after quenching by controlled plastic

Table A.3.13. Solution heat treatment times

Thickness or diameter (mm)	Minimum holding time at temperature (min)	
	Air furnace	Salt bath furnace
≤0.5	20	10
0.5–0.8	25	15
0.8–1.6	30	20
1.6–2.3	35	25
2.3–3.0	40	30
3.0–6.25	50	35
6.25–12.0	60	45
12.0–25.0	90	60
25.0–37.5	120	90
37.5–50	150	108
50–62.5	180	120
62.5–75	210	150
75–90	240	165
90–100	270	180

deformation on the order of 2%, either by stretching or by compression. These are the T451 and T651 tempers (stress-relieved by stretching) and the T452 and T652 tempers (stress-relieved by compression).

For certain alloys, the hot rolling or extrusion temperatures are on the same level as the solution heat treatment temperatures. If the critical quenching rate is sufficiently low, it may be possible to quench immediately after hot working. This is current practice for several extrusion alloys of the series 6000 (6060, 6005A, 6106) that are press-quenched by forced air-cooling or water spray. These tempers are referred as T1 (quenched and naturally aged) and T5 (quenched and artificially aged).

☐ Natural ageing

The process of natural ageing that takes place at room temperature is accompanied by a change in mechanical properties as a function of time (Figure A.3.8), and reduces the product's capacity to be worked.

For deep forming, it is therefore advisable to work in the as-quenched condition, say within 1 h after quenching for alloy 2024. The conventional method of delaying natural ageing for a number of days consists in maintaining the product at a low temperature, below − 10 °C.

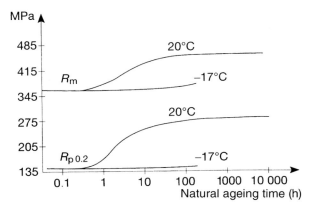

Figure A.3.8. Effect of temperature on the natural ageing of 2024.

☐ Artificial ageing

During artificial ageing, abundant precipitation of hardening elements, which are intermetallic compounds, occurs:

–   Al₂Cu in 2017A,
–   Al₂CuMg in 2024,
–   Mg₂Si in 6000 series alloys,
–   MgZn₂ in 7000 series alloys without copper, and
–   Mg(ZnAlCu₂) in 7000 series alloys with copper.

This precipitation leads to an improvement in mechanical properties. The degree of precipitation hardening is a function of temperature and holding time.

Artificial ageing is usually carried out in air furnaces. As with solution heat treatment, it is essential to stick to the times and temperatures recommended to obtain the desired level of properties.

The conditions of artificial ageing, i.e. temperature, time, waiting time between quenching and artificial ageing, are specific to each alloy. They are given in Table A.3.12 for some selected alloys.

During isothermal ageing, the change in tensile strength over time can be plotted as a bell-shape curve (Figure A.3.9), with a maximum value for the duration t. This maximum tensile strength defines the T6 temper (duration t), whereas the under-aged temper suggests a treatment time shorter than t. These are the T51, T61, T63, T64 and T65 tempers, according to standard EN 515. Under-aged tempers maintain a sufficient capacity of plastic

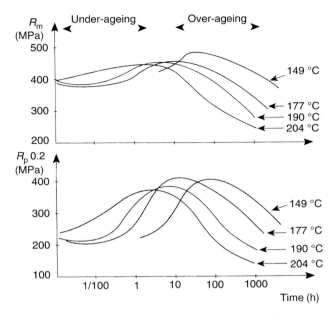

Figure A.3.9. Typical ageing curves: alloy 2024.

deformation, as well as a good formability. However, they are generally less corrosion resistant than over-aged tempers.

The over-aged temper corresponds to a treatment longer than t. These are the T73, T74, T76 and T79 tempers, according to standard EN 515.

In order to improve the corrosion resistance of certain alloys of the 7000 series, such as 7075 or 7049A, a two-step ageing treatment, a so-called duplex ageing, is recommended. These are the tempers:

– T73, in which the artificial ageing conditions improve the stress corrosion resistance and the level of mechanical properties;
– T76, where "over-ageing is limited so as to obtain maximum tensile strength compatible with good resistance to exfoliation corrosion" (standard EN 515).

☐ Intermediate (soft) annealing

When semi-products of the series 2000, 6000 and 7000 are formed and shaped, it may be necessary to restore plasticity by one or more intermediate annealing treatments. Three types of annealing treatments need to be distinguished:

Table A.3.14. Annealing treatments of age-hardenable alloys

Alloy	Coalescence annealing [1] Temperature (°C)	Duration (h)	Restoration annealing Temperature (°C)	Duration (h)	Recrystallisation annealing [1] Temperature (°C)	Duration (h)	Strain hardening [2] (%)
2618A	400–430	1–3	270–300	2–8	350–400	0.5–2	>20
2017A	400–430	1–3	270–300	2–8	350–400	0.5–2	>20
2024	400–430	1–3	270–300	2–8	350–400	0.5–2	>20
2014	400–430	1–3	270–300	2–8	350–400	0.5–2	>20
2030	400–430	1–3			350–400	0.5–2	
2011	400–430	1–3			350–400	0.5–2	
6060	400–430	1–3	240–280	1–4	330–380	0.5–2	
6005A	400–430	1–3	250–280	1–4	330–380	0.5–2	
6106	380–420	1–3	240–280	1–4	330–380	0.5–2	
6082	400–430	1–3	250–280	1–4	330–380	0.5–2	
6061	400–430	1–3	250–280	1–4	330–380	0.5–2	
7020	250–280	4–6	250–280	1–4	340–420	0.5–3	>30
7075	360–430	1–3	270–300	2–8	320–380	0.5–2	>30
7049A	360–430	1–3	270–300	2–8	320–380	0.5–2	

[1]Cooling after coalescence and recrystallisation annealing must be slow (25–30 °C·h^{-1}) from annealing temperatures down to 250 °C for all the alloys listed in this table except 7020, which must be cooled at 20°C·h^{-1} from annealing temperature down to 200 °C.
[2]This is the minimum strain-hardening percentage before annealing to avoid grain growth.

– coalescence annealing, which is performed directly after the solution heat treatment,
– partial (recovery) annealing or recrystallisation annealing, which are performed during forming operations. A minimum degree of strain hardening is required to prevent anomalous grain growth during recrystallisation annealing.

After these annealing treatments, the cooling rate must be controlled in order to avoid a quenching effect. The annealing temperature and time will vary according to the alloy (Table A.3.14).

3.3.3. *Designation of metallurgical tempers*

The tempers of age hardenable alloys are all designated by the letter T, followed by one to five digits, the precise definition of which can be found in standard EN 515. The most common tempers are listed in Table A.3.15. As with H tempers for strain hardenable alloys, the minimum mechanical properties can be standardised.

Table A.3.15. Definition of metallurgical tempers according to the standard EN 515

Temper [1]	Solution heat treatment in furnace	Controlled stretching [2]	Strain hardening	Natural ageing	"Normal" artificial ageing	Over-aging	Extrusions [3]
T1 [4]				×			
T3	×		×	×			
T351	×	×		×			
T3510	×	×		×			×
T3511	×	×		×			×
T4	×			×			
T451	×			×			
T4510	×			×			×
T4511	×			×			×
T5 [4]					×		×
T6	×				×		
T651	×	×			×		
T6510	×	×			×		×
T6511	×	×			×		×
T73	×					×	
T7351	×	×				×	
T73510	×	×				×	×
T73511	×	×				×	×
T76	×					×	
T76510	×	×				×	×
T76511	×	×				×	×
T8	×		×		×		
T8510	×	×			×		×
T8511	×	×			×		×

[1]Designation according to EN 515 "Aluminium and aluminium alloys—Wrought products—Designation of tempers".

[2] In the TX51 temper, the standard specifies that stretching must achieve "a permanent set of 0.5–3% for sheet, 1.5–3% for plate and 1–3% for rolled or cold or cold finished bar". In the TX510 temper, the standard specifies that stretching must achieve "a permanent set of 1–3% for bar, shapes and extruded tubes and 0.5–3% for drawn tube"; the products are not straightened. TX511 is the same as TX510 "except that slight straightening after stretching is permitted to satisfy the tolerances in the standards."

[3]Temper affecting bars, shapes and extruded and drawn tubes.

[4]These tempers relate mainly to extruded semi-products, and involve solution heat treatment during hot working and separate cooling after hot working.

The most important European Standards on aluminium, which were published at the end of 1997, can be found in Table A.3.11.

REFERENCE

[1] *Technologie de fonderie en moules métalliques. T.1, Considérations théoriques et fonderie par gravité. T.2, Fonderie sous pression. Conception et calcul des systèmes d'alimentation en fonderie sous pression*, Éditions techniques des industries de fonderie, CTIF, Sèvres, 1984.

Chapter A.4

The Most Common Wrought Aluminium Alloys

Chapter A.4
The Most Common Wrought Aluminium Alloys

This chapter gives a brief description of the most widely used aluminium alloys. No indications on very specific alloys will be given, especially on those used for the construction of aircraft or heat exchangers in cars, for example. Specialist books [1] or technical documents from suppliers [2] should be consulted for these alloys.

4.1. THE 1000 SERIES

This series includes two categories of grades: the refined alloys and the others such as 1050A and 1200.

The refined alloys (1199, 1198) have a degree of purity between 99.90 and 99.999%. Depending on their purity, they are used in the manufacture of electrolytic condensers (so-called etched metal), lighting devices (bright-trim quality), and for decorative applications in the building sector and luxury packaging (cosmetics, perfumes). The metal is usually anodised.

The 1050A alloy is more than 99.50% pure and is one of the most widely used grades. It is a good compromise between mechanical resistance, capacity for plastic deformation and decorative appearance. It has a wide range of applications: packaging, buildings, sheet metal working, fins and tubes for heat exchangers, electrical conductors, etc.

The 1200 alloy is between 99 and 99.5% pure, and replaces 1050A whenever its plastic formability is adequate (packaging, circles for kitchen utensils).

4.2. THE 3000 SERIES

Industrial alloys of the 3000 series (Figure A.4.1) contain between 1 and 1.5% manganese. This alloying element significantly enhances the mechanical properties of aluminium and adds between 40 and 50 MPa to the minimum guaranteed tensile strength values, while retaining good formability.

The 3003 alloy is the most representative alloy in this series. Adding up to 0.20% copper provides a further increase in mechanical resistance, and adding up to 0.7% copper makes it possible to obtain a fine-grained structure.

As with all the alloys in this series, it has the highest capacity plastic deformation in the annealed temper O.

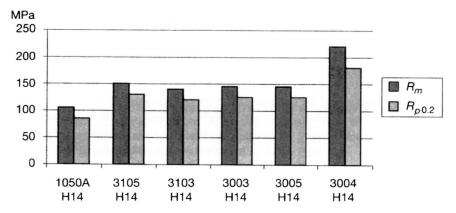

Figure A.4.1. Mechanical properties of 3000 series alloys.

The main applications of 3003 are in the building sector (cladding panels, roofing sheet), fabrication, sheet metal work, heat exchanger tubing, circles for kitchen utensils, etc.

3103 is a variant of 3003, with no copper added.

The 3004 alloy, with roughly 1% magnesium added, offers slightly better mechanical properties, while retaining the overall properties of 3003. It is chiefly used for cans (food cans), circles for kitchen utensils, buildings (coil-coated sheets), etc.

Alloys 3005 and 3105 are two alloys whose mechanical properties and formability fall between those of 3003 and 3004. They are used in the fields of building, fabrication, sheet metal work, thermal insulation, capsuling, etc.

4.3. THE 5000 SERIES

The mechanical properties of the alloys in this series increase with increasing magnesium content (Figure A.4.2). Industrial wrought alloys rarely contain more than 5% magnesium, because above this level, the stability of the alloy decreases, particularly under the influence of temperature.

Prolonged holding at a high temperature leads to the precipitation of the intermetallic compound Al_3Mg_2 at the grain boundaries. The possible effects of this are described in Section B.6.4. If required by the application, a stabilisation heat treatment can be carried out on alloys containing 3% magnesium or more (H321 and H116 tempers).

Most alloys of the 5000 series contain other additions such as manganese, chromium and titanium, which provide a further increase in tensile strength and/or certain properties such as corrosion resistance, weldability, and others.

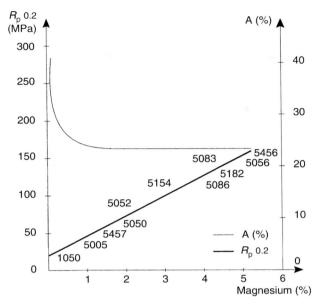

Figure A.4.2. Effect of magnesium content of mechanical properties of the 5000 series alloys in the annealed (○) temper.

These alloys

– are very suitable for welding, except those alloys containing between 1.8 and 2.2% magnesium. The tensile strength of a welded joint is approximately equal to that measured on the parent plate in the annealed temper;
– have good properties at low temperatures; and
– have good corrosion resistance, whether welded or not.

Surface treatments such as brightening or anodising can give them a very attractive surface appearance, especially when the alloy is derived from base metal that is low in iron and silicon; this is the case of alloy 5657 (base metal 1080).

The most common alloys of the 5000 series are listed below (Figure A.4.3):

Alloy 5005 contains about 0.6% Mg and replaces 1050A or 1200 whenever a slight improvement in mechanical properties is required. Anodised (OAB quality, anodic oxidation quality for building) or coil-coated 5005 is very widely used for building applications (external cladding panels, etc.).

Alloy 5657 is a variant of 5005 using a purer base metal (1085). This means the so-called bright-trim quality for cosmetics packaging, lighting and decorative purposes can be obtained.

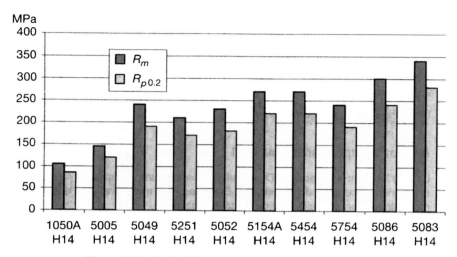

Figure A.4.3. Mechanical properties of alloys of the 5000 series.

Alloy 5052, with 2.5% magnesium and added chromium, is a good compromise between mechanical resistance, formability, fatigue resistance and corrosion resistance. It is widely used in the H28 temper for food cans and in a large number of applications in fabricating, commercial vehicle bodies, road signs, etc.

Alloy 5049 is a variant of 5052 containing manganese but no chromium. Coil in 5049 is widely used for thermal insulation and for sheet metal forming.

Alloys 5454, 5754 and 5154A, combining a magnesium level between 2.5 and 4% with minor additions of manganese or chromium, are widely used in the building sector, civil engineering, transport and mechanical industries. Wire in 5154A is frequently used as rivet stock, and in thinner gauges for mosquito screens.

Alloys 5086 and 5083, containing between 3.5 and 5% magnesium with added manganese and chromium, offer the highest mechanical properties of all semi-products of the 5000 series, including those at cryogenic temperatures. They are suitable for welding and have outstanding corrosion resistance, especially in maritime environments. They have enjoyed growing popularity in naval construction and industrial fabrication.

Alloy 5182, with 4–5% magnesium but with less iron and silicon, provides a good compromise between mechanical resistance and formability in the annealed temper. It is used for reinforcements in cars. Lacquered 5182 in H28 temper retains a good level of mechanical resistance and residual formability, which means it can be used for the manufacture of can ends for beverage cans.

Alloy 5019, with 5% magnesium, is devoted to specific products such as rivet wire, zip fasteners, clips and staples for use in contact with foodstuffs.

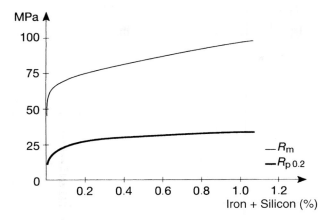

Figure A.4.4. Effect of iron and silicon content in the annealed (○) temper.

4.4. THE 8000 SERIES

The simultaneous addition of iron (which yields a fine-grained structure) and silicon improves the mechanical properties of aluminium (Figure A.4.4). With their fine grain structure and good isotropy, these alloys have good formability under difficult conditions, even as foil (between 50 and 200 μm thick). This explains their increasing use as fins for heat exchangers, spiral tubes, dishes, thin foil, etc. The two most commonly used alloys are 8006 (with added manganese) and 8011.

4.5. THE 2000 SERIES

The main addition is copper, and, in lesser quantities, magnesium and silicon. The selection criteria for these alloys are generally

–   the mechanical strength in the T6 temper,
–   the low crack propagation rate in the T4 temper,
–   the heat resistance, and
–   the aptitude for free machining.

Alloys of this series are used either in T4 temper, or in T6 or T8 temper.

4.5.1. The quenched and naturally aged tempers T4 and T451
Alloy 2017A (formerly called Duralumin) provides average tensile strength but good machinability. It is widely used in mechanical applications (as extrusions or thick plates).

Alloy 2024, with its higher magnesium content, is an improved variant of 2017A, with superior mechanical properties, good fracture toughness and good resistance to crack propagation. It is used mainly in aircraft construction as sheet in thin and medium gauges (temper T351) and as extrusion.

4.5.2. The artificially aged tempers T6 (T651) and T8 (T851)

Alloy 2014, with its higher silicon content (0.5–1.2%) has particularly high mechanical properties in the T6 temper. This alloy is mainly used for aircraft construction and in the mechanics industry.

Alloy 2214 is a variant of 2014 containing less iron, which improves its fracture toughness and resistance to crack propagation in thick plates for aircraft construction. It is also used in the mechanics industry and for the manufacture of video recorder drums.

In thick products, alloy 2024 in the T8 temper provides a good level of mechanical strength and satisfactory resistance to intergranular corrosion in products with a thickness of less than about 10 mm. Strain hardening between quenching and artificial ageing leads to an increase in mechanical strength (temper T8). The main applications of this alloy are in aeronautics and armaments.

Further additions of iron, nickel, manganese and vanadium increase the mechanical properties at temperatures between 100 and 300 °C. These are alloys 2618A and 2219, which in the T6 temper provide good stability and creep resistance up to 100–150 °C.

Alloy 2618A is used in aircraft construction (including supersonic Concorde aircraft) and in the mechanics industry.

Alloy 2219 has the highest copper content (6%) of all industrial alloys and additions of manganese, vanadium, zirconium, and titanium. Strain hardening after quenching increases the mechanical properties of this alloy, which is available in the tempers T3, T6 and T8. It has several interesting properties:

– creep resistance and tensile yield strength in the temperature range between 200 and 300 °C,
– mechanical properties at cryogenic temperatures,
– arc weldability (TIG and MIG), and
– resistance to stress corrosion in the T6 temper.

It is used for the manufacture of welded tanks for space launcher rockets.

Finally, adding lead and/or bismuth presents a good aptitude for free machining, due to the fragmentation of turnings; this is the case of alloys 2011 and 2030.

It should be recalled that with the exception of 2219, these alloys cannot be welded by conventional TIG and MIG techniques. Furthermore, their resistance to corrosion is poor, and thus they need to be protected if they are to be used in humid environments, and all the more in aggressive environments.

4.6. THE 6000 SERIES

The two alloying elements of the 6000 series are magnesium and silicon (Figure A.4.5). These alloys show

−   good aptitude for hot transformation by rolling, extrusion and forging,
−   good resistance to corrosion, especially atmospheric corrosion,
−   a high level of mechanical properties which can be further improved by adding silicon (above the stoichiometric amount in the hardening precipitate Mg_2Si) or copper,
−   a good aptitude to arc welding and brazing,
−   good cold formability (bending of profiles, deep drawing of sheets) in the O temper and, to a lesser extent, the T4 temper, and
−   attractive surface appearance after brightening or anodising.

All these properties account for the extensive use of this alloy series, especially in the field of metallic fittings.[1]

These alloys are available

−   either as extrusions only: 6005A, 6106, 6056, 6060, 6262,
−   or as rolled or extruded semi-products: 6061 and 6082.

Alloy 6060 is the extrusion alloy par excellence. Very complex shapes can be obtained, and press quenching (T5 temper) can be achieved. Several variants of this alloy with variable magnesium and silicon content are available, which aim at improving certain properties or optimising a set of properties such as extrudability, surface appearance, suitability for anodising, and mechanical properties.

Alloy 6005A is easily extrudable and can be press quenched. In the T5 temper, its mechanical strength is in the order of 290 MPa. This alloy has remarkable fracture toughness, and can therefore be used as a structural member, for instance in commercial road and railway vehicles, and in mechanical applications.

Alloy 6106 is an extrusion alloy that has been specially developed by Pechiney for many applications related to lightweight constructions. It is very easy to extrude, has a good aptitude to press quenching and good mechanical strength, in the order of 265 MPa, which is midway between 6060 and 6005A.

Alloy 6056 in the T6 temper has the highest mechanical properties of this series of alloys: R_m ranges from 450 to 470 MPa. It is used as reinforcement for car doors.

Alloy 6262, a free machining alloy, contains additions of lead and bismuth which facilitate fragmentation of turnings.

---

[1]   The world consumption of extrusions in 6000 series alloys is in the order of 6 million tons per year.

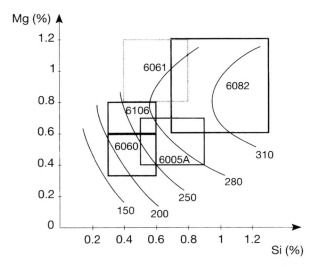

Figure A.4.5. Alloys of the 6000 series. Iso-curves of for ultimate tensile strength R_m.

Alloy 6082 has high mechanical properties of 320–340 MPa in the T6 temper. Like 6005A, it is widely used in commercial vehicles, railcars, shipbuilding, in the mechanics industry and as forging stock.

The mechanical resistance of alloy 6016 in the T4 temper is at a medium level: 220 MPa. It has a good aptitude to deep drawing and is widely used for car bodies.

Alloy 6061 has medium mechanical strength (310 MPa in T6 temper). It is frequently used for extruded products (bars, profiles, tubes), cold-drawn, as well as rolled and forged products. It is used for structures requiring both average mechanical strength and good resistance to corrosion, in diverse fields of technology: transport (railcars, commercial vehicles), tubing (pipelines), welded boilerwork, mechanics applications, and tubular furniture.

4.7.   THE 7000 SERIES

This series comprises two subgroups: alloys with and without copper.

4.7.1.   *7000 alloys without added copper*
Alloy 7020 is the most widely used alloy for rolled or extruded semi-products in different fields such as transport, the mechanics industry, and armaments.

In the quenched and artificially aged condition (T5 or T6), i.e. its normal use condition, its mechanical strength is in the order of 360–400 MPa. Its corrosion resistance is satisfactory, unless it is welded.

Its heat resistance is rather poor as soon as the temperature exceeds 120–130 °C. Exposure to temperatures above 200 °C can sensitise the alloy to exfoliation corrosion.

This alloy has satisfactory arc weldability (with filler metal 5356), and after welding, mechanical properties equivalent to those of the parent metal in the T4 temper can be achieved. However, it shows a strong sensitivity to exfoliation corrosion in the heat-affected zone, on either side of the welding bead (See Section B.6.4.2 and Figure B.2.11). This handicap severely limits its use in welded structures to very specific applications that are subject to frequent inspection.

4.7.2.   7000 alloys with added copper

Adding copper to the aluminium–zinc–magnesium system produces the aluminium alloys with the highest mechanical resistance in the T6 temper.

Products capable of resisting corrosion in aggressive atmospheres, which are particularly resistant to stress corrosion in the short-transverse direction can be obtained by duplex ageing (tempers T73 and T76); however, this leads to a decrease in mechanical strength of about 20%.

Alloy 7075 is most often used as rolled, extruded, die-forged and hand-forged products in the fields of aerospace, mechanics, and sport and leisure equipment.

The need for products with a thickness above 80 mm has triggered the development of 7050 and 7010, two variants, which differ from 7075 in that

–   chromium has been replaced by zirconium, which facilitates quenching, and
–   the copper content has been increased, which, after a T73 or T76-type duplex ageing process, provides a good compromise between mechanical properties and corrosion resistance.

Alloy 7049A can be considered as one of the commercial aluminium alloys with the best mechanical properties in the T6 temper. Its average mechanical properties are

– $R_{p0.2}$: 570 MPa
– R_m: 650 MPa
–   A: 10%

Extrusions and forgings in 7049A are used in the aircraft and armaments industries.

REFERENCES

[1] Develay, R., Propriétés de l'aluminium et des alliages d'aluminium corroyés. *Techniques de l'ingénieur*, fasc. M. 438 (1992), M 439 (1992), M 440 (1992), M 441 (1992), M443 (1989), M 445 (1985), M 446 (1989), M 447 (1989), M 448 (1989).
[2] *Demi-produits en aluminium*, Pechiney Rhenalu, 1997.

Chapter A.5
Selection Criteria for Aluminium Alloys

Chapter A.5
Selection Criteria for Aluminium Alloys

The metallurgy of aluminium is highly developed and offers a vast choice of alloys among the eight series of alloys capable of satisfying highly varied requirements and applications. However, this advantage may sometimes create confusion among users, and may lead to an erroneous choice.

This often happens when an application is transposed from steel to aluminium. Too often, only the mechanical properties are taken into account as the sole criterion of choice of the alloy, leaving out aspects such as corrosion resistance, which may well be an essential factor in certain applications.

In cases like this, one would almost certainly opt for 2017A or another alloy of the 2000 series. However, if used in an aggressive environment, serious corrosion problems may occur.

It is, therefore, essential to take into account as many selection criteria as possible. In the above example, two approaches are possible. First, one can stick to 2017A, because the use conditions require a high level of mechanical properties; in this case it will be necessary to provide for adequate protection against corrosion. Alternatively, a thorough analysis may lead to the conclusion that a lower level of mechanical properties is acceptable, and one may opt for another alloy in the 5000 or 6000 series of alloys, which are more corrosion-resistant.

5.1. SELECTION CRITERIA

The selection criteria are very closely related to the application, as can be seen from the following examples:

- in the construction industry, the criteria of durability of the surface appearance and corrosion resistance are preponderant;
- for sheet-metal structures and fabricating, formability and ease of assembly are more important;
- in mechanical engineering, the mechanical properties as well as suitability for machining are important;
- for heat exchangers, the metal's thermal conductivity determines the efficiency of the heat transfer, and resistance to corrosion is an important aspect.

The requirements must, therefore, be defined very carefully. For example, it is difficult to talk about corrosion resistance without specifying the environment, or to discuss fatigue strength without identifying stresses and load cycles: 10^6, 10^7.

5.2. SELECTING AN ALLOY

Whenever possible, the choice of an alloy should be based on the experience that has been gained, often over decades, with aluminium applications. First, this experience provides a valuable guidance for selection, once the decision to use aluminium has been taken. Second, experience forms the basis for the development of new applications, by comparison with what is being done, or what has been done.

The choice of an alloy is a match between its properties, its engineering capabilities, and the conditions of service. The most common selection criteria are

–   mechanical properties,
–   formability,
–   suitability for machining,
–   weldability,
–   suitability for anodising, and
–   corrosion resistance.

Certain alloys (or groups of alloys within the same series of alloys) specialise in certain applications, either by tradition, or because they have been developed for a specific use. This is the case with alloys 2011, 2030 and 6262 for free machining, alloy 5005 for construction, and alloy 3104 for canstock. However, these applications are not necessarily exclusive.

5.2.1. *Selecting an alloy series*

One of the first selection criteria must be in regard to the alloy series, because performance and workability are quite different for strain-hardenable and age-hardenable alloys (Table A.5.1).

This is why it does not seem sensible to compare the properties of alloys in all eight series in a single table. Both metallurgically and in terms of applications, there is no much similarity between 1050A or 8011 and 7049A. For this reason, the alloy series 4000 and 8000, which are devoted to very specific applications, are not included in Table A.5.1.

Table A.5.1. Selection criteria

Criteria	Strain-hardenable alloys			Age-hardenable alloys		
	1000	3000	5000	6000	2000	7000
UTS (MPa)	50–150	100–260	100–340	150–310	300–450	320–600
TIG or MIG welding	Possible	Possible	Possible [1]	Possible	No [2]	No [2]
Decorative anodising	Possible	No	Possible	Possible	No	No
Corrosion resistance [3]	Good	Good	Very good	Very good	Poor	Poor
Extrudability	Good	Good	Poor	Very good	Poor	Poor

[1] Except alloys comprising between 1.8 and 2.2% magnesium.
[2] Except certain alloys: 2219, 7020, 7021.
[3] This is the corrosion resistance of the bare, unprotected metal in natural environments—air, freshwater, seawater—not environments that contain gaseous, liquid or solid chemicals, whether concentrated or as aqueous or organic solutions.

5.2.2. Selecting a metallurgical temper

This criterion is very important when a specific level of mechanical properties and formability is required.

■ Strain-hardenable alloys

The highest formability is offered by tempers O and H111. Among the strain-hardened tempers H1X, formability decreases with the level of strain hardening. Thus, for a given alloy, temper H14 offers less formability than H12, H16 less than H14, and so on.

For a given level of mechanical properties, the partially annealed tempers H2X offer a better formability than their H1X counterparts.

■ Age-hardenable alloys

These alloys are used mainly in structural applications. Their use in humid or aggressive environments should be avoided, unless a suitable protection is provided.

Their formability is maximum in the annealed condition or immediately after quenching.

5.3. PRINCIPAL APPLICATIONS OF ALUMINIUM AND ITS ALLOYS

Since it is not possible to deal with every application of aluminium and its alloys because there are so many, the principal applications are discussed with classic choices of alloys and semi-products (Table A.5.2).

Table A.5.2. Typical applications of aluminium and aluminium alloys

Applications	Main selection criteria of users	Alloys commonly selected by users	Comments
Sheet-metal fabrication	Forming, welding	1200, 1100, 1050A, 3105, 3003, 3004, 5049, 5052, 5454, 5754, 5086, 5083, 6082, 6061	
Mechanical applications	Mechanical properties, machinability	2618A, 2024, 2017A, 2014, 2214, 2030, 2011, 5086, 5083, 6005A, 6082, 6061, 6012, 6262, 7075, 7049A	2030, 2011, 6012 and 6262 are free, machining alloys
Construction of aircraft and spacecraft	Lightness, mechanical properties, formability, machinability, suitability for surface treatments, corrosion resistance	2618A, 2024, 2014, 2214, 2219, 7020, 7075, 7175, 7475, 7050, 7010	
Commercial vehicles	Forming, joining (welding), functionalities of semi-products, appearance, corrosion resistance	3003, 3004, 5052, 5454, 5754, 5086, 5083, 6005A, 6082	
Shipbuilding	Forming, welding, corrosion resistance	5754, 5086, 5083, 6005A, 6082	
Building	Forming, joining, suitability for anodising and painting, corrosion resistance	1050A, 3105, 3003, 3005, 5005, 5052, 6060, 6005A, 6106	Pre-coated coil: 1050A, 3105, 3003, 3005, 5052; Pre-anodised coil: 5005
Outdoor installations, urban amenities	Forming, joining (welding), functionalities of semi-products, appearance, corrosion resistance	3003, 5052, 5086, 5083, 6005A, 6082, 6060, 6106	Tread plate: 3003, 5754, 5086
Heat exchangers	Thermal conductivity, forming, joining (brazing), corrosion resistance	1050A, 1100, 3003, 3005, 6060, 6063, 8011	For brazed exchangers: clad 3003 and 3005
Kitchen utensils	Suitability for deep drawing and surface treatments	1200, 1050A, 8003, 3004, 4006, 4007, 5052, 5754	4006 and 4007 are alloys for enamelling

Part B

The Corrosion of Aluminium

Part B

The Corrosion of Aluminium

Chapter B.1

The Corrosion of Aluminium

Chapter B.1
The Corrosion of Aluminium

The verb "corrode" is derived from the Latin word *rodere* which means "gnaw". This word entered the French language in 1314 to designate the action of gnawing, progressively wearing away by a chemical effect. The noun "corrosion" was derived at the beginning of the 19th century from the Lower Latin *corrosio*, which designates the act of gnawing. It was first introduced in medical vocabulary [1], and later came into use for the phenomenon that forms the subject matter of the present book.

Defining "corrosion" is, however, not an easy task. Corrosion is a slow, progressive or rapid deterioration of a metal's properties such as its appearance, its surface aspect, or its mechanical properties under the influence of the surrounding environment: atmosphere, water, seawater, various solutions, organic environments, etc.

In the past, the term "oxidation" was frequently used to designate what is nowadays commonly called "corrosion". Nevertheless, the former was the right word because corrosion also is an electrochemical reaction during which the metal is oxidised, which usually implies its transformation into an oxide, i.e. into the state in which it existed in the mineral.

1.1. SHORT HISTORICAL INTRODUCTION

The phenomenon of corrosion has been known ever since the discovery of metals. Pliny the Younger was already complaining about the Roman soldiers' weapons getting rusty. Scientific investigation started at the beginning of the 19th century with Nicholson and Carlyle's discovery of the electrolytic decomposition of water by the electric current supplied by a galvanic battery [2].

Humphrey Davy established a relationship between the production of electricity and the oxidation of zinc, in which one of the two metals was copper and the other acted as the generator of electricity. In 1830, the Genevan chemist Auguste de la Rive developed the basis of the electrochemical theory of corrosion. At the beginning of the 20th century, this theory was taken up by Whitney, and completed by Hoar and Evans at the end of the 1920s [3].

During the first quarter of the 20th century, the full economic cost of the corrosion of metals was perceived. The first reported corrosion experiments on aluminium started around 1890, when the metal was available in a quantity sufficient to envisage its use for construction and as kitchen utensils. Its resistance to rainwater and various types of drinks, such as beer, coffee, and tea, was first assessed at the beginning of the 1890s [4].

1.2.   THE ELECTROCHEMICAL BASIS OF METAL CORROSION

Corrosion of metals is caused by the electrochemical reaction between a metal (or an alloy) and an aqueous phase. It proceeds according to a complex electrochemical process that is related to the atomic structure of matter. Matter is built up from elementary particles that carry electrical charges, namely ions and electrons, and from particles that are electrically neutral, namely atoms and molecules. In metals, the electrical environment of atoms is made up of free electrons capable of moving throughout the metal.

   In the aqueous phase, which is a solution, the following species can be found:

-   positive ions (cations) and negative ions (anions),
-   neutral molecules such as water and various undissociated compounds.

   At the interface between metal and water, the transfer of electrical charges leads to electrochemical reactions (Figure B.1.1):

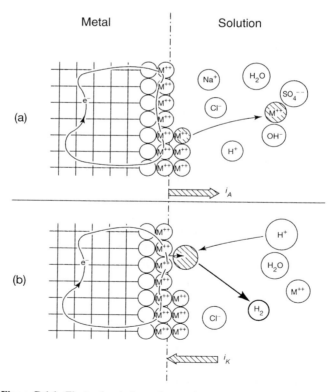

Figure B.1.1. Electrochemical reactions at the metal–solution interface [5].

- The metal atom is oxidised and forms M^{n+} ions that are released in the aqueous phase. This creates a flux of electrons within the metal in the direction solution → metal. The resulting anodic oxidation current i_a flows from the metal to the solution.
- The ions or molecules of the aqueous phase are reduced, which means that they take up electrons from the metal and get transformed into another chemical species. This creates a flux of electrons within the metal in the direction metal → solution. The resulting cathodic current i_k flows from the solution to the metal.

Electrons that interact at the metal–solution interface do not penetrate the solution.

1.2.1. Elementary electrochemical reactions of corrosion

The corrosion of a metal is the result of two simultaneous reactions that are in electrical equilibrium:

- the oxidation of the metal, resulting in a loss of electrons, according to the fundamental reaction

$$M \rightarrow M^{n+} + ne^-$$

It results in an anodic current i_a that flows in the direction

$$metal \rightarrow solution$$

- the reduction of an ion present in the aqueous solution according to the fundamental reaction

$$X^{n-} \rightarrow X + ne^-$$

It results in a cathodic current i_k that flows in the direction

$$solution \rightarrow metal$$

- The reactions of oxidation and reduction proceed at distinct sites of the metal surface. The surface at which oxidation takes place is called the anode. It carries negative charges and is designated by the sign $(-)$; the resulting current is called the anodic current. Reduction takes place on a surface called the cathode, designated by the sign $(+)$; the reducing current is called the cathodic current.

Except when connected to the electrodes of a generator, the metal is electrically neutral, which means that the electron and current fluxes are in equilibrium:

$$\sum i_a = \sum i_k$$

In a given system, all electrochemical reactions result in electrical currents that depend on the differences in potential between the two phases: metal and aqueous liquid. The kinetics of anodic and cathodic electrochemical reactions is represented by the relationship between the potential e and the reaction rate of the corresponding electrical intensity i (Figure B.1.2).

When an electrode is plunged into an aqueous solution, several anodic and cathodic reactions can take place simultaneously, and in principle they need not be related. However, because of the transfer of electrons, interactions between the anodic and cathodic reactions can occur.

Under the conditions of natural corrosion, i.e. without an external source of electrical current, the system formed by the metal and the aqueous solution constitutes an open electrical circuit. No current can be transported from the metal to the solution and vice versa. This means that the anodic current and the cathodic current, flowing in opposite directions, are necessarily equal (point C). This point, which forms the intersection between two polarisation curves, defines the corrosion potential e_{corr} and the intensity of corrosion i_{corr}.

By using Faraday's law

$$m = \frac{1}{96\,500}\frac{A}{n}It$$

where

m is the mass,
A is the atomic mass of the metal (27 for aluminium),
n is the valency (3 for aluminium),
I is the intensity, in amperes, here i_{corr},
t is the time, in seconds,

the mass loss, i.e. the corrosion rate, can be calculated for a given intensity and a given time (month, year, etc.). However, this calculation makes sense only if the corrosion is

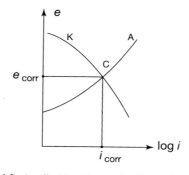

Figure B.1.2. Anodic (*A*) and cathodic (*K*) polarisation curves.

uniform. It does not make any sense when corrosion is localised and proceeds as pitting corrosion, as with aluminium.

These so-called polarisation curves can be determined experimentally, either by varying *i* and measuring *e*, or by varying *e* and measuring *i*. These plots, which are a powerful tool for the study of corrosion phenomena, have been the subject of substantial development since 1950 (see Section B.4.5).

1.2.2. The double layer

These reactions take place at the metal–solution interface consisting in what electrochemists call the double layer, the thickness of which is in the order of 10 nm.

Plunging a metal in an aqueous solution leads to local perturbation of the arrangement of molecules and ions and modifies the distribution of electrical charges throughout the double layer. In order to achieve electrical neutrality, any charges appearing in the liquid close to the interface have to be neutralised by the same number of opposed electrical charges at the surface of the metal.

The double layer is built up from three layers (Figure B.1.3):

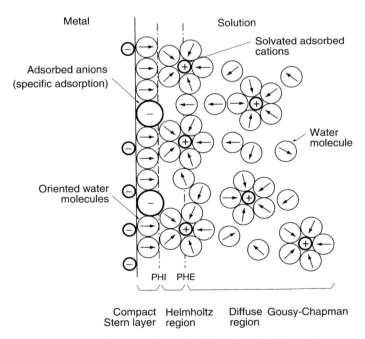

Figure B.1.3. Metal–solution interface: the double layer [6].

–   the compact Stern layer, a few tenths of a nanometre thick, built up from molecules
    and so-called specifically adsorbed species, which are mainly anions of rather small
    size, e.g. chlorides;
–   the Helmholtz region, in which solvatised ions (i.e. ions surrounded by water
    molecules) can be found;
–   the diffuse Gouy–Chapman region, the thickness of which depends on the ionic
    force of the solution.

For electrochemical reactions to take place, the X^- ions to be reduced need to arrive at
the surface of the metal, and vice versa, the cations M^{n+} need to leave the metal's surface.
Several phenomena tend to hinder the movement of ions in the solution (e.g. resistance to
the diffusion of oxygen) or delay the oxidation at the anode (or the reduction at the
cathode), because the electrode surface has undergone modifications for reasons such as
deposition of corrosion products or inhibitors.

This delay in electrochemical reactions, called polarisation, means that the reactions are
no longer reversible (in the thermodynamic sense) and that the corrosion rate decreases.
Polarisation is anodic when delaying reactions at the anode, and cathodic when delaying
reactions at the cathode.

On the other hand, electrode reactions can be accelerated when the equilibrium
reaction

$$M \rightarrow M^{n+} + ne^-$$

is permanently shifted to the right-hand side. This results in continuous dissolution, for
instance under the effect of an external current or powerful etching agents. This is of
course something which corrosion experts try to avoid...if possible.

1.3.   THE ELECTROCHEMICAL REACTIONS IN THE CORROSION OF ALUMINIUM

The fundamental reactions of the corrosion of aluminium in aqueous medium have been
the subject of many studies [7, 8]. In simplified terms, the oxidation of aluminium in water
proceeds according to the equation:

$$Al \rightarrow Al^{3+} + 3e^-$$

Metallic aluminium, in oxidation state 0, goes in solution as trivalent cation Al^{3+} when
losing three electrons.

This reaction is balanced by a simultaneous reduction in ions present in the solution,
which capture the released electrons. In common aqueous media with a pH close to neutral
such as fresh water, seawater, and moisture it can be shown by thermodynamic considerations that only two reduction reactions can occur:

  – reduction of H^+ protons:

$$3H^+ + 3e^- \rightarrow \tfrac{3}{2}H_2$$

H^+ protons result from the dissociation of water molecules:

$$H_2O \rightleftharpoons H^+ + OH^-$$

  – reduction of oxygen dissolved in water:

- in alkaline or neutral media: $O_2 + 2H_2O + 4e^- \rightarrow 4OH^-$
- in acidic media: $O_2 + 4H^+ + 4e^- \rightarrow 2H_2O$.

At 20 °C and under atmospheric pressure, the solubility of oxygen in water is 43.4 mg·kg^{-1}. It decreases with increasing temperature and is no more than 30.8 mg·kg^{-1} at 40 °C, and 13.8 mg·kg^{-1} at 80 °C.

Globally, the corrosion of aluminium in aqueous media is the sum of two electro-chemical reactions, oxidation and reduction:

$$Al \rightarrow Al^{3+} + 3e^-$$

$$3H^+ + 3e^- \rightarrow \tfrac{3}{2}H_2$$

$$\overline{Al + 3H^+ \rightarrow Al^{3+} + \tfrac{3}{2}H_2}$$

or

$$Al + 3H_2O \rightarrow Al(OH)_3 + \tfrac{3}{2}H_2$$

This reaction is accompanied by a change in the oxidation state of aluminium which, from the oxidation state 0 in the metal, is transformed into the oxidation state of alumina (+3), and by an exchange of electrons, since aluminium loses three electrons that are picked up by $3H^+$.

Aluminium corrosion results in the formation of alumina $Al(OH)_3$, which is insoluble in water and precipitates as a white gel, which is found in corrosion pits as white gelatinous flakes. Upon corrosion, 27 g of aluminium will form 33.6 l of hydrogen.

A model calculation shows that in a cube of 1 m^3 filled with water in contact with 5 m^3 of metal surface, in which corrosion develops with a uniform thickness of 1 μm, 5 cm^3 of aluminium will be dissolved (corresponding to 13.5 g), and 16.8 l of hydrogen will be released.

This simple calculation shows that the volume of hydrogen is out of proportion with the intensity of corrosion. It illustrates the hazard resulting from a closed tank in which corrosion takes place: hydrogen will form, and any intervention such as welding without prior degassing can have catastrophic consequences. This type of accident may occur, for instance, when repairing ballast tanks of vessels. This applies, by the way, to any metal: steel, aluminium, etc.

1.4.   THE STANDARD POTENTIAL OF A METAL

The structure of the metal–solution interface can be represented by a series of capacitors (Figure B.1.4) with charges distributed over several planes: metal surface, internal Helmholtz plane, external Helmholtz plane, and in the diffuse zone. This leads to the built-up of a difference in potential between the metal and the solution. This difference is generally called the "absolute potential" of the metal with respect to the solution, or Galvani potential.

In practice, the term "metal potential" is used. By convention,

$$\Delta\phi_{absolute} = \phi_{metal} - \phi_{solution}.$$

1.4.1.   Determination of potentials

The determination of the absolute potential of a metal would require connecting the metal to another solid electrode. However, this electrode would also build up a double layer, whose unknown potential would add to the difference in potential to be measured.

This is why electrochemists carry out this measurement with respect to a reference electrode, with the potential set as constant under standard conditions of temperature and pressure.

By convention, the equilibrium reaction

$$2H^+ + 2e^- \leftrightarrows H_2$$

has been chosen, and the potential 0 V is attributed under standard pressure and temperature conditions (1013 mbar and 20 °C).

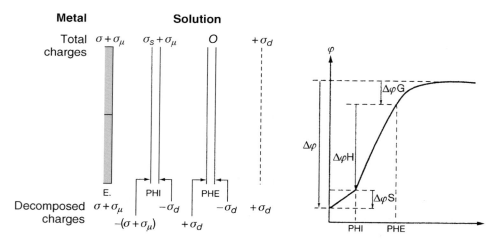

Figure B.1.4. Charges and potentials in the double layer [6].

Under conditions of reversibility, there is a relationship called the Nernst equation between the potential E, the standard electrode potential E^0 and the activity a of the ion M^{n+}:

$$E = E^0 + \frac{RT}{nF} \text{Log } a_{M^{n+}}$$

where

R is the gas constant,
T is the absolute temperature,
F is the Faraday constant,
n is the valency of the ion.

At 20 °C, using decimal logarithms, the Nernst equation becomes:

$$E = E^0 + \frac{0.058}{n} \text{Log } a_{M^{n+}}$$

The potential E is related to the free enthalpy of the reaction

$$M^{n+} + ne^- \rightarrow M$$

by the relation

$$\Delta G = -nFE$$

$$\Delta G = \Delta G_0 + RT \text{ Log } K$$

where (a being the activity):

$$K = \frac{a_M}{a_{M^{n+}}}$$

Conventionally, $a_M = 1$, which leads to

$$E = -\frac{\Delta G}{nF} + \frac{RT}{nF} \text{Log } a_{M^{n+}}$$

With

$$E_0 = -\frac{\Delta G}{nF}$$

this leads to

$$E = E_0 + \frac{RT}{nF} \text{Log } a_{M^{n+}}$$

If the activity a_M equals 1, Log a_M becomes zero, and

$$E = E^0$$

1.4.2. *Values of standard electrode potentials*

In practice, the measurement of standard electrode potentials is carried out with a platinum counter electrode that is dipped into a standard solution of 1 M sulphuric acid with hydrogen bubbling. For the determination of its potential, the metal is dipped into a 1 M standard solution of a salt of this metal (Figure B.1.5).

Based on measurement and calculation, the oxidation reactions of metals can be classified according to their respective potentials, and a scale of potentials can be obtained (Table B.1.1).

Metals having an electronegative potential show a tendency to oxidise and thus to corrode in aqueous media, if the conditions allow. This tendency increases as the potential becomes more electronegative. It is well known that magnesium (standard electrode potential, -2380 mV) degrades much more under the effect of moisture than lead (standard electrode potential, -126 mV).

Aluminium has a highly negative standard electrode potential of -1660 mV, obtained by calculation with respect to the free energy $-\Delta G$, which has one of the highest negative values, -440 Kcal/mol. Therefore, aluminium is expected to be very unstable in the presence of moisture. However, experience shows that this is not the case, because aluminium is covered by a natural oxide layer that modifies its behaviour compared to thermodynamic prediction: aluminium is a passive metal (see Section B.1.8).

On the other hand, metals with a positive potential only have a weak tendency to oxidation; this is the case of metals at the top of Table B.1.1: gold, platinum, palladium

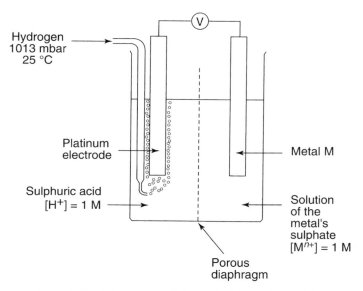

Figure B.1.5. Measurement cell for standard electrode potentials.

Table B.1.1. Standard electrode potentials

Reaction	Potential (V)
$Au \leftrightarrows Au^{3+} + 3e^-$	$+1.42$
$Pt \leftrightarrows Pt^{2+} + 2e^-$	$+1.20$
$Pd \leftrightarrows Pd^{2+} + 2e^-$	$+0.83$
$Ag \leftrightarrows Ag^{2+} + 2e^-$	$+0.80$
$2Hg \leftrightarrows Hg_2^{2+} + 2e^-$	$+0.80$
$Cu \leftrightarrows Cu^{2+} + 2e^-$	$+0.34$
$H_2 \leftrightarrows 2H^+ + 2e^-$	0
$Pb \leftrightarrows Pb^{2+} + 2e^-$	-0.12
$Sn \leftrightarrows Sn^{2+} + 2e^-$	-0.14
$Ni \leftrightarrows Ni^{2+} + 2e^-$	-0.23
$Co \leftrightarrows Co^{2+} + 2e^-$	-0.27
$Cd \leftrightarrows Cd^{2+} + 2e^-$	-0.40
$Fe \leftrightarrows Fe^{2+} + 2e^-$	-0.44
$Cr \leftrightarrows Cr^{3+} + 3e^-$	-0.71
$Zn \leftrightarrows Zn^{2+} + 2e^-$	-0.76
$Ti \leftrightarrows Ti^{2+} + 2e^-$	-1.63
$Al \leftrightarrows Al^{3+} + 3e^-$	-1.66
$Mg \leftrightarrows Mg^{2+} + 2e^-$	-2.38
$Na \leftrightarrows Na^+ + e^-$	-2.71

(it is well know that gold can be found in the metallic state). These metals are also called noble metals.

The standard electrode potential is a thermodynamic parameter and expresses only whether a given reaction is possible or not, and nothing else. It says nothing about the kinetics of such a reaction, that is to say about the speed at which this reaction will proceed. It may happen that the rate of reaction is zero and that the reaction will thus not proceed at all. Passivation of metals is a good example of this phenomenon.

The relative position of two metals on the standard electrode potential scale makes it possible to predict which of the two metals in contact will act as the anode, i.e. which of the two metals will dissolve when the battery so formed starts operating: it is always the more electronegative metal which will dissolve. As an example, when copper ($E = +340$ mV) is coupled with zinc ($E = -760$ mV) in a copper sulphate solution, an electric generator (the so-called Daniel cell) is formed, with zinc acting as the anode. Its electromotive force corresponds to the sum of the absolute values of the standard electrode potentials, i.e. 1.10 V.

1.5. DISSOLUTION POTENTIALS

Standard electrode potentials measured with respect to hydrogen are only of theoretical interest. They are relevant only to pure metals, not to alloys, and do not take into account possible passivation phenomena, as can be seen in the case of aluminium. They are measured in a very peculiar medium, namely a standard solution of the salt of the metal under consideration.

They are only of limited interest for corrosion experts, who prefer dissolution potentials (or corrosion potentials), which are measured with respect to a reference electrode that is easy to use, and using the medium of their choice, such as natural seawater or a standard liquid (Figure B.1.6).

1.5.1. Measurement of dissolution potentials

In the laboratory, the standardised ASTM G69 solution is normally used. It contains 57 g·l^{-1} of sodium chloride and 3 g·l^{-1} of hydrogen peroxide. The latter promotes the reduction of oxygen dissolved in water. Stable and highly reproducible potentials can be obtained in this way. The potentials obtained in these media are mixed, irreversible potentials, in which electrochemical reactions related to both the metal (oxidation) and the electrolyte (reduction of the cation present, in general H^+) are involved.

The dissolution potential of aluminium is measured on a surface which is always covered by a natural oxide layer. This oxide layer consists of three very different elements: anodic pores (about 0.5% of the total surface), the cathodic barrier layer, and thicker areas which are neutral [9]. All parameters that modify the properties of the natural oxide layer will also modify the potential of aluminium.

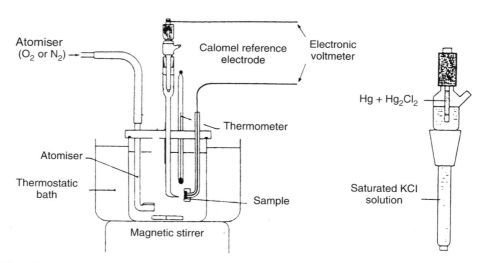

Figure B.1.6. Schematic view of the measurement of dissolution potentials with a saturated calomel reference electrode.

Table B.1.2. Main reference electrodes

Reference electrode	Potential with respect to the hydrogen electrode (mV)
Saturated calomel	+0.244
Calomel in 0.1 M KCl	+0.336
Silver chloride in 0.1N KCl	+0.288
Saturated copper sulphate solution	+0.318
Mercury sulphate, solution saturated in K_2SO_4	+0.615

These measurements vary with the experimental conditions: the nature of the solution, the alloy, and the surface condition of the metal. They can also vary with the duration of immersion, which is chosen according to the alloy and the medium.

The measurement of this potential, which includes everything that happens at the surface of a sample (≈ 1 cm^2), cannot be expected to be as accurate as measurements commonly used in metallurgy (optical microscope, etc.) [10].

For example, in moving seawater, a stationary state is not reached before several days (10–20, depending on the conditions) [11]. This considerably limits the usefulness of these measurements, especially since within several days the sample's surface will undergo changes. Any comparison between alloys or metals will be rather difficult.

Several easy-to-use reference electrodes are known, whose potential has been determined with respect to the standard hydrogen electrode (Table B.1.2) [12]. The most commonly used electrode in aqueous media is the saturated calomel electrode (SCE).

1.5.2. *Dissolution potential scales*

For common metals, dissolution potential scales similar to those in Table B.1.3 are available. It is always necessary to specify the reference electrode and the medium in which the measurement was taken. Dissolution potentials allow classifying metals with respect to each other, which is useful for the prediction of galvanic corrosion in heterogeneous assemblies (see Chapter B.3).

The order of metals and alloys with respect to each other does not vary much in fresh waters, seawater, or brines at ambient temperature. Among the common metals, aluminium is always one of the most electronegative [13].

The potential of a metal immersed in an aqueous solution usually stabilises over several hours or even several days. The potential's evolution over time can give interesting information about the evolution of corrosion or passivation phenomena. The plots given in Figure B.1.7 provide a few classic cases:

Table B.1.3. Dissolution potentials, expressed in mV saturated calomel electrode in natural seawater in motion at 25 °C [14]

Metal or alloy	Dissolution potential (mV)
Graphite	+90
Monel	− 80
Hastelloy C	− 80
Stainless steel	− 100
Silver	− 130
Titanium	− 150
Inconel	− 170
Nickel	− 200
Cupronickel 70–30	− 250
Cupronickel 90–10	− 280
Tin	− 310
Bronze	− 360
Brass	− 360
Copper	− 360
Lead	− 510
Steel	− 610
Cast iron	− 610
Cadmium	− 700
Aluminium (1050A)	− 750
Zinc	− 1130
Magnesium	− 1600

– *curve a*: as the potential becomes more and more noble, the metal (or alloy) becomes passivated,
– *curve b*: the potential becomes less and less noble,
– *curve c*: the potential drops and then increases: attack followed by passivation,
– *curve d*: this is observed when the protective layer of a metal is modified.

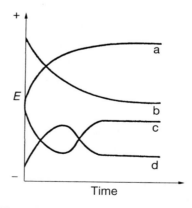

Figure B.1.7. Typical potential-time curves.

1.5.3. Dissolution potential of aluminium

Alloying elements can modify the potential of aluminium in both directions, as can be seen in Figure B.1.8. The dissolution potential of an alloy is determined by the solid solution, the major part of the metal's surface [15].

Zinc strongly decreases the potential. Alloy 7072 at 1% zinc is, therefore, used as a cladding of 3003 (see Section B.5.6). The alloys of the 7000 series have the most electronegative potentials.

Copper alloys of the 2000 series have the least electronegative potentials. Alloy 2017A can, therefore, be protected by a cladding of 1050A (see Section B.5.6).

Dissolution potentials of the most common aluminium alloys are listed in Table B.1.4 [16]. For a given alloy, variations by 50–100 mV from one author to another have been reported.

Although intermetallic phases may have a dissolution potential rather different from that of the solid solution (Table B.1.5), they have no influence on the dissolution potential. However, they may give rise to intercrystalline corrosion, exfoliation corrosion, or stress corrosion if localised at or close to grain boundaries (see Section B.2.3).

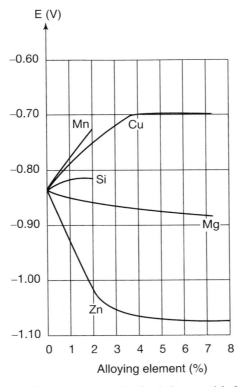

Figure B.1.8. Influence of alloying elements on the dissolution potential of aluminium alloys [17].

Table B.1.4. Dissolution potentials of aluminium alloys (NaCl solution, H_2O_2, ASTM G69)

Alloy	Temper	Potential (mV ECS)	Alloy	Temper	Potential (mV ECS)
1060		−750	5456		−780
1100		−740	6005A		−710
1199		−750	6009	T4	−710
2008	T4	−690	6010	T4	−700
	T6	−700	6013	T6, T8	−730
2014	T4	−600	6053		−740
	T6	−690	6060		−710
2017	T4, T6	−600	6061	T4	−710
2024	T3, T4	−600		T6	−740
	T8	−710	6063		−740
2090	T3, T4	−650	7003		−940
	T8	−750	7005		−840
2091	T3, T8	−670	7039	T6, T63	−840
2219	T3, T4	−550	7049	T7	−750
	T6, T8	−700	7050	T7	−750
3003		−740	7072		−860
3003/7072		−870	7075	T6	−740
3004		−750		T7	−750
5042		−770	7178	T6	−740
5050		−750	7475	T6	−750
5052		−760	8090	T3	−700
5056		−780		T7	−750
5083		−780	42000 (A-S7G03)		−820
5086		−760	45000 (A-S6U3)		−810
5154		−770	51200 (A-G10)		−890
5182		−780	51300 (A-G5)		−870
5454		−770	71000 (A-Z7GU)		−990

In alloys of the series 2000 and 7000, dissolution potentials may also depend on the thermal treatment (Figure B.1.9).

The measurement of dissolution potentials of intermetallic phases is difficult because of their very small size (generally < 100 μm). However an alloy is elaborated, the cooling conditions will have an influence on the size and composition of intermetallics. For this reason, values reported in the literature show some scatter.

1.6. ELECTROCHEMICAL EQUILIBRIUM (POURBAIX) DIAGRAMS

It was Pourbaix's idea to list all the chemical and electrochemical reactions that can take place between a metal and water and to define the domains of stability for each chemical species, as a function of the pH for chemical reactions and as a function of the potential for electrochemical reactions. This approach is based on thermodynamics: the Nernst

Table B.1.5. Dissolution of intermetallic phases (solution NaCl, H_2O_2, ASTM G 69)

Intermetallic phase	Dissolution potential (mV ECS)
Si	− 170
Al_3Ni	− 430
Al_2Cu	− 440 and − 640
Al_3Fe	− 470
1050A	− 750
Al_6Mn	− 760
Al_2CuMg	− 910
$MgZn_2$	− 960
Al_3Mg_2	− 1150
Mg_2Si	− 1190

equation, solubility products, etc. [18]. E–pH diagrams representing three types of equilibrium are used:

– between solid species,
– between two species in solution,
– between a solid species and a species in solution.

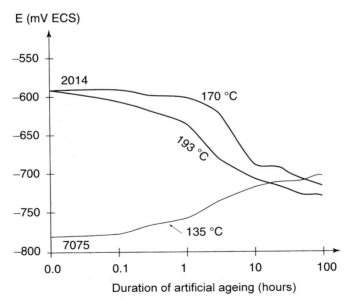

Figure B.1.9. Influence of the duration of ageing treatment on the dissolution potential of alloys 2014 and 7075 [17].

Pourbaix diagrams cover several domains representing three possible situations:

- Corrosion, where a soluble corrosion product exists. Pourbaix has defined a concentration of 10^{-6} M (which amounts to 0.027 mg·l^{-1} for aluminium) as a lower threshold above which it can be assumed that corrosion takes place. On most of these diagrams, upper concentration limits of 10^{-5}, 10^{-4}, 10^{-3}, are also defined.
- Passivation, when an insoluble oxide or hydroxide is formed on the metal's surface.
- Immunity, if the concentration of M^{n+} ions is less than 10^{-6} M.

The E–pH diagram of aluminium (Figure B.1.10) illustrates the amphoteric nature of aluminium: it is attacked both in acidic and alkaline media [19].

These are equilibrium diagrams that determine stable species, their domain of stability and the direction of possible reactions. However, they cannot predict the corrosion rate.

The significance of these diagrams is limited because they refer to an ideal liquid, i.e. chemically pure water at 25 °C, to a metal as pure as possible, and never to an alloy. They do not take into account the possible presence of chloride that plays an important role in pitting corrosion. Furthermore, they do not take into account the nature of the acid and the base that modify the pH value (see Section B.6.1). They do not indicate the risk of cathodic corrosion in the domain of immunity when the potential is highly electronegative (see Section B.5.5). This is due to the method itself, which is based on thermodynamic data without taking into account kinetic data [20].

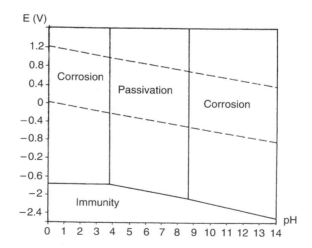

Figure B.1.10. Pourbaix diagram E–pH of aluminium [19].

Taking into account these precautions, it is interesting to plot the E–pH diagram of an aluminium alloy, 5086, in seawater (or more precisely, in a solution of 30 g·l^{-1} of NaCl the pH of which is adjusted [21, 22]).

This diagram is based on the measurement of the following potentials:

- the corrosion potential E_0, which is a mixed metal–water potential;
- the pitting potential E_c, which is the most electronegative potential at which pitting occurs;
- the passivation potential E_p, which is the minimum potential that has to be set in order to passivate existing corrosion pits;
- the potential of uniform anodic attack E_{ga} at which pitting corrosion starts spreading over the entire surface, thus giving rise to a practically uniform attack;
- the cathodic corrosion potential E_{cc} under cathodic polarisation. As for anodic polarisation, there is a potential at which pitting corrosion becomes uniform. There is no reduction of Al^{3+} or AlO_2^- ions, but reduction of water under local alkalinisation (due to consumption of H^+). The natural oxide layer will dissolve in such an alkaline medium.

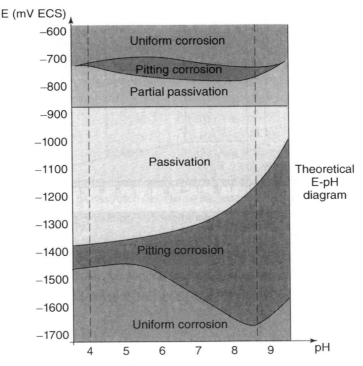

Figure B.1.11. Experimental E–pH diagram of 5086 in the presence of chloride [21].

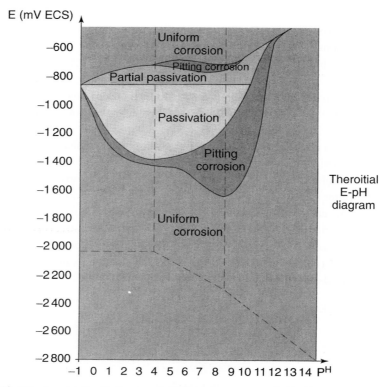

Figure B.1.12. Experimental E–pH diagram of 5086 in the presence of chloride, extrapolated to pH <4 and >9 [21].

The measurement of these potentials leads to the diagrams shown in Figures B.1.11 and B.1.12. The following comments should be made:

– the corrosion potential and the pitting corrosion are very close,
– between pH 4 and 9, the pitting potential does not depend on the pH,
– pitting corrosion takes place only in the range of pH in which the oxide layer is totally insoluble,
– uniform cathodic attack is a catastrophic corrosion that may dissolve up to 10 μm·h^{-1} under cathodic polarisation,
– immunity is theoretically inaccessible at pH values below 9, because at potentials low enough to enter the domain of immunity of aluminium, water is no longer stable, and hydrogen will be released.

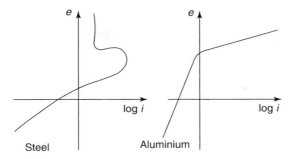

Figure B.1.13. Polarisation curves of aluminium and steel

1.7. THE ELECTROCHEMICAL BEHAVIOUR OF ALUMINIUM

The electrochemical behaviour of aluminium is influenced by the natural oxide layer that governs the corrosion resistance of aluminium.

The potential measured on aluminium does not correspond to that of the metal, but represents a mixed potential between the oxide layer and the metal. The potential of the metal cannot be measured, because in oxidising media such as water, the oxide layer will form immediately, within 1 ms or even less.

For predicting the behaviour of aluminium, the pitting potential does not have the same importance as for steel.

On steel, pitting is controlled by the initiation step (as detected by the pitting potential), and it propagates quickly. On aluminium, there will always be an initiation of a very large number of superficial and very small pits around intermetallic phases that have a cathodic potential with respect to the matrix. However, the propagation of pits cannot be deduced from the measurement of the potential.

When measuring an instantaneous corrosion rate, this value has to be considered as resulting from uniform corrosion. On aluminium, corrosion will progress as pitting in media close to neutral. The measurement of a corrosion current gives some indication of the global corrosion rate, but does not reveal anything about the morphological aspects of corrosion such as diameter, density, and localisation of pits.

Polarisation curves for aluminium do not have the same shape as for steel (Figure B.1.13). There is no domain of passivity, because aluminium is naturally passive.

1.8. ALUMINIUM AS A PASSIVE METAL

Aluminium is naturally passive and, therefore, does not need to be passivated, unlike certain metals such as steel. A metal that can be passivated has undergone a chemical treatment or contains an alloying element, which renders it passive against the medium. It can be depassivated, or may not have been passivated. This does not apply to aluminium, which is of course always covered by its natural oxide layer.

The dissolution potential of aluminium in most aqueous media is in the order of -500 mV with respect to a hydrogen electrode, while its standard electrode potential with respect to this same electrode amounts to -1660 mV. Because of this highly electronegative potential, aluminium is one of the easiest metals to oxidise (Table B.1.1). However, aluminium behaves as a very stable metal, especially in oxidising media (air, water, etc.).

This behaviour is due to the fact that aluminium, like all passive metals, is covered with a continuous and uniform natural oxide film[1] corresponding to the formula Al_2O_3, which is formed spontaneously in oxidising media according to the reaction:

$$2Al + \frac{3}{2}O_2 \rightarrow Al_2O_3$$

The free energy of this oxidation reaction, -1675 kJ, is one of the highest, which explains the very high affinity of aluminium towards oxygen.

The major importance of this oxide film for the corrosion resistance of aluminium was recognised in 1896 by Richards [23]; it has been the subject of a large number of investigations since 1930 [24].

1.8.1. The structure of the natural oxide film

The natural, colourless oxide film is built up from two superimposed layers with a total thickness between 4 and 10 nm (Figure B.1.14):

– the first compact and amorphous layer, in contact with the metal, is called the barrier layer, because of its dielectric properties. It will form at any temperature as soon as the liquid or solid metal comes in contact with air or an oxidising medium; the temperature acts only on the final thickness (Figure B.1.15).

It forms very quickly, within a few milliseconds. The rate of formation is independent of the oxygen partial pressure (Figure B.1.16), as shown by Gulbranssen and Wysong [25]. In practice, this means that the oxide film will reappear immediately after forming or machining operations have destroyed the natural oxide layer locally, even in poorly aerated areas.

Film growth follows a parabolic kinetics up to 350–400 °C [26], and becomes linear at higher temperatures; it proceeds by migration of the Al^{3+} ion across the oxide

---

[1] This expression is used here for the film that forms spontaneously in contact with air or an oxidising medium. It does not cover conversion coatings or anodising coatings. For corrosion products, consisting essentially of hydrated alumina, the term "alumina gel" is used.

Superficial contamination:
Adsorbed water
Residues of rolling oil
Residues of degreasing agents

External film:
Hydrated
boehmite or
bayerite

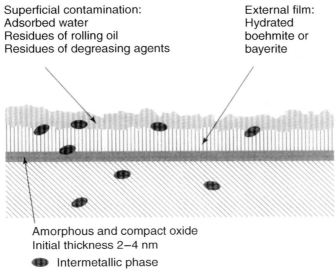

Amorphous and compact oxide
Initial thickness 2–4 nm
Intermetallic phase

Figure B.1.14. Layers and adsorption phenomena on oxide film [32].

layer, according to a mechanism which was described by Cabrera [27]. The maximum thickness of this first layer is in the order of 4 nm [28].

—   the second layer grows on top of the first one, by reaction with the exterior environment, probably by hydratation. Its final thickness will not be reached before several weeks, even months, and depends on the physicochemical conditions [29]

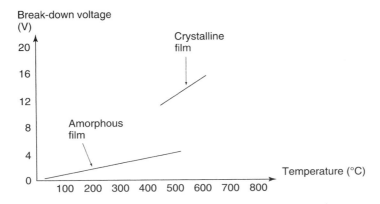

Figure B.1.15. Formation of an oxide film in dry oxygen [29].

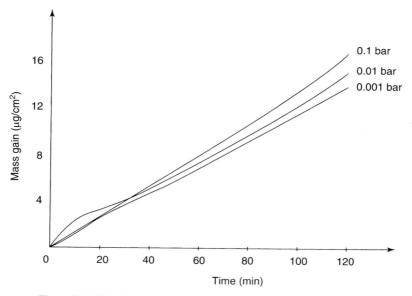

Figure B.1.16. Influence of oxygen pressure on film growth at 500 °C [25].

(relative humidity and temperature) which favour film growth. Godard [30] has monitored the thickness of oxide films as a function of relative humidity over long periods of time (Figure B.1.17). After 4 years, this second layer can reach several tens of nanometres on alloy 3003.

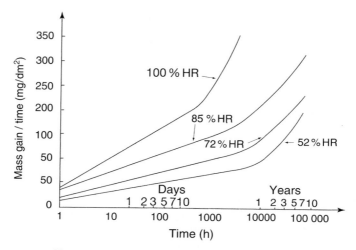

Figure B.1.17. Influence of relative humidity (HR) [30].

Table B.1.6. Allotropic modifications of alumina

Species	Crystal structure	Chemical name	Chemical formula	Temperature range of formation (°C)	Density
Amorphous alumina		Aluminium oxide	Al_2O_3	$<50-60$	3.40
Bayerite	Monoclinic	Aluminium trihydroxide	$\alpha\text{-Al(OH)}_3$	$60-90$	2.53
Boehmite	Orthorhombic	Aluminium oxide hydroxide	$\gamma\text{-AlOOH}$	>90	3.01
Corundum	Hexagonal	Aluminium oxide	$\alpha\text{-Al}_2O_3$	>350	3.98

This second film is less compact than the barrier layer, and it is porous. It reacts with the external environment, either during transformation (rolling, extrusion) by adsorption of residues of rolling oil or forming lubricants, or during service in contact with moisture contained in the atmosphere or with contaminants. As shown in Figure B.1.14, the composition of its surface can be very complex. That is why it is often necessary to regenerate the surface by pickling, before any surface treatments can be applied.

In prolonged contact with water, it tends to grow, especially at high temperatures, and transforms into bayerite and boehmite (see Section B.5.1). The properties of these two forms of aluminium oxide are given in Table B.1.6.

1.8.2. *Influence of alloying elements and additives*

The concentration of alloying elements and additives is generally different in the metal and in the natural oxide film [31]. The rate of formation and the surface properties of the oxide, therefore, depend on its own composition, and not on the composition of the metal underneath.

As an example, the oxidation rate of a metal can be decreased by an alloying element that reacts preferentially with oxygen and forms an oxide layer across which diffusion is slowed down.

Certain elements will strengthen the protective properties of the oxide film by forming mixed oxides, if their structures are compatible. This is the case of magnesium. For this reason, alloys of the 5000 series have excellent corrosion resistance. On the other hand, certain elements such as copper will weaken these protective properties. This explains the poor corrosion resistance of copper-containing alloys of the 2000 and 7000 series.

Table B.1.7. Properties of alumina

Property	Value
Melting point	$2054 \pm 6\,°C$
Boiling point	$3530\,°C$
Linear expansion coefficient at 25 °C	$7.1 \times 10^{-6}\,K$
Thermal conductivity at 25 °C	$0.46\,J \cdot cm^{-1} \cdot s^{-1} \cdot K^{-1}$
Specific heat at 25 °C	$0.753\,J \cdot g^{-1} \cdot K^{-1}$
Dielectric constant at 25 °C	10.6
Electrical resistivity at 14 °C	$10^{19}\,\Omega \cdot cm^{-1}$

1.8.3. Allotropic modifications

Several allotropic modifications of alumina may be present in the natural oxide film, depending on the formation conditions, and especially the temperature of the medium (Table B.1.6).

1.8.4. Properties

The main properties of alumina are listed in Table B.1.7.

Because of the difference in density with respect to the underlying aluminium, the oxide film is in compression: it can sustain deformation without breaking [32]. This explains its excellent resistance during forming operations.

1.8.5. Passivity of aluminium and pH values

The dissolution rate of alumina depends on the pH value, as shown in Figure B.1.18. It is higher at acidic and alkaline pH values, which reflects the amphoteric properties of aluminium oxide.

However, the pH value is not the sole parameter to be considered when predicting the stability of the natural oxide film in aqueous media, and therefore, of aluminium itself: at acidic or alkaline pH, the dissolution rate of aluminium also depends on the nature of the acid or of the base dissolved in water, as shown in Figure B.1.19.

As an illustration, the dissolution rate in a solution of sodium hydroxide at $0.1\,g \cdot l^{-1}$ is 25 times higher than in an ammonia solution at $500\,g \cdot l^{-1}$, although both solutions have very similar pH values, 12.7 and 12.2, respectively.

The same applies in acid solutions: at a given pH value, solutions of hydrochloric acid or hydrofluoric acid are much more aggressive towards aluminium than solutions of acetic acid.

Like all passive metals, aluminium is prone to pitting corrosion in aqueous media close to neutrality. Under these conditions, pitting corrosion depends more on the quantity of anions, such as chlorides, than on variations in the pH value of the aqueous medium.

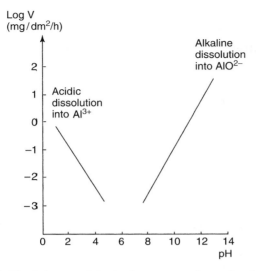

Figure B.1.18. Dissolution rate of alumina in aqueous media as a function of pH [33].

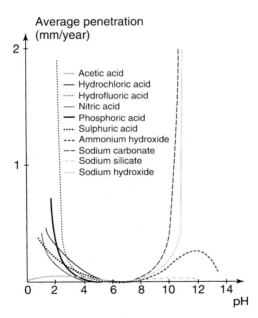

Figure B.1.19. Influence of the nature of acids and bases on the corrosion of 1100 H14 [34].

In organic media, the resistance of aluminium and its alloys also depends on the stability of the natural oxide film. If this film is damaged or destroyed, for example, in contact with certain organic acids at their boiling point, the underlying aluminium may be attacked. This may result, in certain cases, in extremely fast and violent reactions (see Chapters F.2, F.3 and F.6).

REFERENCES

[1] Rey A., (dir.), *Dictionnaire historique de la langue française*, Le Robert, Paris, 1993.

[2] Lynes W., Some historical developments relating to corrosion, *Journal of the Electrochemical Society*, vol. 98, 1951, p. 3c–10c.

[3] Hoar T.P., Evans U.R., Quantitative confirmation of electrochemical theory of corrosion of iron, *Proceeding of Royal Society of London*, vol. 137, 1932, p. 343.

[4] Smith C.A., The corrosion story. Part 3, Zinc, lead and aluminum, *Anti-Corrosion*, vol. 24, 1977, p. 13–15.

[5] Crolet J.L., Présentation des phénomènes de corrosion et des différents moyens de lutte disponibles, *Revue de l'Institut français du pétrole*, vol. 34, 1979, p. 929–946.

[6] Philibert J., Vignes A., Bréchet Y., Combrade P., *Métallurgie: du mineral au matériau*, Masson, Paris, 1998.

[7] Bryan J.M., Mechanism of the corrosion of aluminium, *Chemistry & Industry*, 1948, p. 135–136.

[8] Foley R.T., Nguyen T.H., The chemical nature of aluminium corrosion V. Energy transfer in aluminium dissolution, *Journal of the Electrochemical Society*, vol. 129, 1982, p. 464–467.

[9] Akimov G.W., Electrode potentials, *Corrosion*, vol. 11, 1955, p. 477t–486t, see also p. 515t–534t.

[10] Reboul M., Corrosion galvanique de l'aluminium, *Revue de l'aluminium*, 1977, p. 404–419.

[11] Ailor W.H., *Flowing sea water corrosion potentials of aluminum alloys*, NACE, paper no. 36, Houston, 1970.

[12] Shreir L.L., Jarman R.A., Burnstein G.T., (eds.), *Corrosion Metal/Environment Control*, vol. 2, 3rd edition, tab. 21-7, Butterworth-Heinemann, Oxford, 1994.

[13] Crum J.R., Scarberry R.C., Development of galvanic series in various acid and environments, *Conference of Nickel-base Alloys*, Cincinnati, OH, October 1984, p. 53–57.

[14] Laque F.L., *Marine corrosion, causes and prevention*, Wiley, New York, 1975.

[15] Brown R.H., Fink W.L., Measurement of irreversible potentials as a metallurgical research tool, *Transactions of American Institute of Mining and Metallurgical Engineers*, 1940, p. 1234, technical publication.

[16] Burleigh T.D., Bovard F.S., Rennick R.C., Corrosion potential for aluminium alloys measured by ASTM G. 69, *Corrosion*, vol. 49, 1993, p. 683–685.

[17] Hollingsworth E.H., Hunsicker H.Y., Corrosion of aluminum and alluminum alloys, *Handbook ASM*, vol. 30, 1990, p. 583–609.

[18] Pourbaix M., *Leçons sur la corrosion électrochimique*, Cebelcor, rapport technique, 1957.

[19] Deltombe E., Pourbaix M., *Comportement électrochimique de l'aluminium, diagramme d'équilibre tension pH du système Al–H₂O à 25 °C*, Cebelcor, rapport technique no. 42, décembre 1956.

[20] Scully J.C., *The fundamentals of corrosion*, 2nd edition, vol. 17, Pergamon International, Oxford, International serious of monographs on materials science and technology, 1975.

[21] Gimenez Ph., Rameau J.J., Reboul M., Diagramme experimental potentiel pH de l'aluminium pour l'eau de mer, *Revue de l'aluminium*, 1982, p. 261–272.

[22] Gimenez Ph., Rameau J.J., Reboul M., Experimental pH potential diagram of aluminium for sea water, *Corrosion*, vol. 37, 1981, p. 673–682.

[23] Richards J.W., *Aluminium, its History, Occurrence, Properties, Metallurgy and Application, including its alloys*, Philadelphie, 1896.

[24] Wefers K., Misra C., *Oxides and hydroxides of aluminium*, Alcoa, technical, paper no. 19 revised (450 références).

[25] Gulbransen E.A., Wysong W.S., Thin oxyde film on aluminum, *Journal of Physic and Colloid Chemicals*, vol. 51, 1947, p. 1087–1103.

[26] Mott N.F., Theory of the formation of protective oxides films on metals, *Transactions of the Faraday Society*, vol. 35, 1939, p. 1175–1177.

[27] Cabrera N., Sur l'oxydation de l'aluminium à basse température, *Revue de Métallurgie*, vol. 45, 1948, p. 86–92.

[28] Tolley G., The oxide film on aluminum. Consideration of experimental facts, *Metal Industry*, vol. 77, 1950, p. 255–258.

[29] Hunter M.S., Fowle P., Natural and thermally oxide films on aluminum, *Journal of the Electrochemical Society*, vol. 103, 1956, p. 482–485.

[30] Godard H.P., Oxyde film growth over five years on some aluminum sheet alloys of varying humiditat room temperature, *Journal of the Electrochemical Society*, vol. 114, 1967, p. 354–356.

[31] Smeltzer W.W., Principles applicable to the oxydation and corrosion of metals and alloys, *Corrosion*, vol. 11, 1955, p. 366t–374t.

[32] Dunlop H.M., Benmalek M., Role and caracterization of surfaces in the aluminium industry, *9e Entretiens du Centre Jacques Cartier*, École polytechnique de Montréal, 2 octobre 1996.

[33] Chatalov A.Y., Effet du pH sur le comportement électrochimique des métaux et leur résistance à la corrosion, *Dokl. Akad Nauk, SSSR*, nos 86, 1952, p. 775–777.

[34] Hollingworth E.H., Hunsicker H.Y., Corrosion of aluminium and aluminium alloys, *Metal Hanbook ASM*, vol. 2, 1990, p. 608.

Chapter B.2

Types of Corrosion on Aluminium

Chapter B.2
Types of Corrosion on Aluminium

Different types of corrosion, more or less visible to the naked eye, can occur on aluminium, such as uniform (generalised) corrosion, pitting corrosion, stress corrosion, etc. The predominant type of corrosion will depend on a certain number of factors that are intrinsic to the metal, the medium and the conditions of use.

There is no form of corrosion that is specific to aluminium and its alloys.

2.1. UNIFORM CORROSION

This type of corrosion develops as pits of very small diameter, in the order of a micrometer, and results in a uniform and continuous decrease in thickness over the entire surface area of the metal.

With aluminium, this type of corrosion is observed especially in highly acidic or alkaline media, in which the solubility of the natural oxide film is high (Figure B.1.18). The dissolution rate of the film is greater than its rate of formation; however, the ratio of both rates can change over time. As an example, in sodium hydroxide solutions the dissolution rate has been found to be lower for long-term exposure, in the order of 40 or 80 days, than for tests over 20 days (Figure B.2.1).

The dissolution rate can vary from a few micrometers per year up to a few micrometers per hour, depending on the nature of the acid or base (see Chapters E.4 and E.5). Appropriate inhibitors can reduce it. As an example, sodium silicate greatly reduces the dissolution rate of aluminium in alkaline media.

The rate of uniform corrosion can be easily determined by measuring the mass loss, or the quantity of released hydrogen (see Section B.4.3.2) [2]. This is a useful parameter to evaluate the dissolution rate of aluminium in a pickling bath.

2.2. PITTING CORROSION

This localised form of corrosion is characterised by the formation of irregularly shaped cavities on the surface of the metal. Their diameter and depth depend on several parameters related to the metal, the medium and service conditions.

Aluminium is prone to pitting corrosion in media with a pH close to neutral, which basically covers all natural environments such as surface water, seawater, and moist air.

113

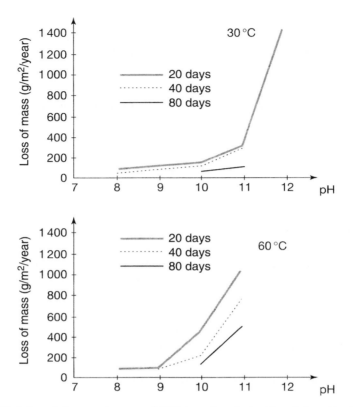

Figure B.2.1. Variation of the corrosion rate of 1050 in sodium hydroxide solutions (the NaOH solution 500 cm³/12 cm² was adjusted every day) [1].

Unlike other metals, corrosion of aluminium is always eye-catching because the corrosion pits are covered with white, voluminous and gelatinous pustules of alumina gel $Al(OH)_3$. These pustules are much bigger than the underlying cavity.

Pitting corrosion occurs when the metal is put into permanent or intermittent contact with aqueous media: water, seawater, rain water, and humidity. Experience shows that when pitting corrosion occurs, it will always develop during the first weeks of exposure.

As an example, a yachtsman who discovers that his boat's aluminium hull structure exhibits superficial pitting corrosion only a few weeks after having been put in service is rightly upset.

Pitting corrosion disconcerts users of aluminium equipment, and sometimes even puzzles corrosion experts who must explain a complex phenomenon and make a forecast on the service lifetime of equipment that seems to be irremediably damaged.

Pitting corrosion is a very complex phenomenon indeed. Even today, its mechanism is not totally understood, in spite of the very large number of studies and publications that have appeared on this subject for 80 years. What is well known are the conditions under which pitting corrosion is initiated and propagates, and how to slow it down. Consequently, nowadays, it is obsolescence of the equipment rather than pitting corrosion that puts an end to the lifetime of most equipment exposed to humid environments for decades.

2.2.1.   *Initiation and propagation of corrosion pits*

Like all passive metals, aluminium is prone to localised corrosion caused by a local rupture of the passive film [3]. This results in a corrosion pit that can propagate, provided that the conditions are favourable.

The electrochemical mechanisms of pitting corrosion are very complex and not totally understood. The practical relevance of these theoretical models is still limited [4]. Pitting corrosion shows two distinct stages: initiation and propagation.

■ Initiation stage

It has long been known that pitting corrosion develops in the presence of chloride. Chloride ions Cl^- are adsorbed on the natural oxide film [5], followed by the rupture of the film at weak points, with formation of microcracks that are a few nanometres wide. Many pits are initiated within a very short time, up to 10^7 cm^{-2}. Their density depends on the alloy: from 10^4 cm^{-2} on 1199 to 10^{10} cm^{-2} on an alloy containing 4% copper.

However, most pits will stop after a few days. Polarisation studies have demonstrated that when pits stop growing, they will be repassivated. When the metal is polarised once again, these passivated pits will not be reinitiated again, but pitting will start on fresh sites.

Oxygen will be reduced slowly in cathodic areas. These areas seem to be intermetallic phases underneath the oxide layer that more or less covers them. Where the film cracks, aluminium will oxidise rapidly, and a complex intermediate chloride $AlCl^{4-}$ will be formed.

■ Propagation stage

Only a minute fraction of initiated pits will continue to propagate according to the two electrochemical reactions [6] (Figure B.2.2):

–   oxidation at the anode formed by the pit's bottom:

$$2Al \rightarrow 2Al^{3+} + 3e^-$$

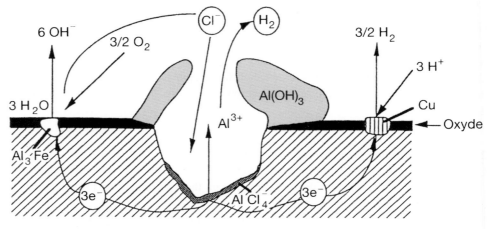

Figure B.2.2. Mechanism of pitting corrosion of aluminium.

– reduction at the cathode outside the cavity:

- of water: $\frac{2}{3}O_2 + 3H_2O + 6e^- \rightarrow 6OH^-$ or
- of H^+: $6H^+ + 6e^- \rightarrow 3H_2$.

If the anode is stable and localised, corrosion will dig a cavity, that is to say a pit. The formation of OH^- ions or the consumption of H^+ ions will locally lead to an excess of OH^- ions, thus to an alkaline pH.

The overall reaction of pitting corrosion on aluminium is

$$2Al + 3H_2O + \frac{3}{2}O_2 \rightarrow 2Al(OH)_3.$$

The dissolution of aluminium by formation of Al^{3+} ions at the bottom of the pit creates an electrical field that shifts Cl^- ions towards the pit's bottom, chemically neutralising the solution and forming aluminium chlorides. Cl^- ions are the most mobile of all ions that participate in these reactions.

The hydrolysis of aluminium chlorides (or of the hydrochlorinated intermediate complex $AlCl^{4-}$) according to

$$Al^{3+} + 3H_2O \rightarrow Al(OH)_3 + 3H^+$$

will lead to the acidification of the pit's bottom up to a pH < 3. The medium becomes very aggressive, leading to autopropagation of the pit.

Al^{3+} ions being highly concentrated at the pit's bottom, they will diffuse towards the pit's opening, where they will meet a medium which is more and more alkaline, especially on the lateral surfaces, where the cathodic reaction leads to alkalinisation.

Al(OH)$_3$ will precipitate. Hydrogen micro-bubbles from the reduction of H$^+$ ions will push the formed aluminium hydroxide to the opening of the pit where it forms a deposit of white pustules.

The reactants, therefore, show a concentration gradient from the bottom up to the rim of the pit. The accumulation of corrosion products at the top of the pit, forming the dome of a volcano, will progressively block the entry of the pit. This will hinder the exchange of ions, especially when chloride Cl$^-$ ions are involved, which explains why pitting slows down or even stops.

Recent studies have shown that pitting corrosion in aluminium is a discontinuous phenomenon.

2.2.2. Rate of pitting corrosion

Experience has shown that in most cases, the rate of deepening of pits formed in natural environments such as freshwater, seawater, and rain water decreases with time. This explains the very long lifetime (several decades) of aluminium used in construction (roof sheet), in naval construction, etc.

Studies performed at the beginning of the 1950s in 25 different waters in Canada [7] showed that the deepening rate of pits on 1100 follows the equation

$$d = kt^{1/3}$$

where

d is the depth of the pit,
t is the time, and
k is a constant depending on the alloy and the service conditions (nature of the alloy, temperature, flow speed of the water, etc.).

This equation can be explained by the fact that for an ideal, hemispherical pit with a radius r, the quantity of dissolved metal during a period of time t is constant and equals

$$\frac{2\pi r^3}{3t} = \text{const.}$$

from which follows

$$d = K\sqrt[3]{t}.$$

This equation was checked by measuring the depth of pits at regular intervals over 13 years in water conveyance installations comprising more than 100 km of tubes (each with a unit

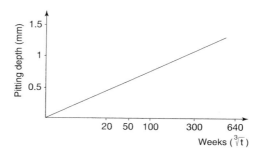

Figure B.2.3. Deepening rate of pits in an installation for water conveyance [7].

length of 15 m) in alloy 5052 (Figure B.2.3), as well as on 1050 alloy in water at 90 °C for 90 days [8] (Figure B.2.4). According to this equation, doubling the thickness of a water pipe increases its lifetime by a factor of 8.

2.2.3. Characterisation of pitting corrosion

Unlike uniform corrosion, the intensity and rate of pitting corrosion can be assessed neither by determining the mass loss, nor by measuring released hydrogen. In fact, these measurements do not make sense because a very deep and isolated pit results only in a small mass loss, whereas a very large number of superficial pits can lead to a larger mass loss.

Pitting corrosion can be assessed using three criteria:

– the density, i.e. the number of pits per unit area,
– the rate of deepening,
– the probability of pitting.

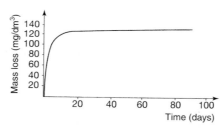

Figure B.2.4. Deepening rate of pits in alloy 1050 exposed to water at 90 °C [8].

■ The density

The measurement of the density does not present any particular difficulty, because one simply counts the number of pits visible on a representative surface area or length. Experience shows that 1 dm^2 of a flat rolled product, and 1 dm of a tube are largely sufficient to obtain a good indication of the density.

Experience further shows that when pits are small in number and disseminated over the surface, they tend to be deeper than if their density is high.

■ The rate of deepening

This is the most important parameter. The rate of pit deepening is far more important than their density, because the service lifetime will depend on the rate of deepening. The pitting depth is unrelated to the thickness of the metal [9].

The pitting depth is measured at the end of the testing period (or after a given period of service). The testing protocol is defined by the statistical treatment of the measurement results at a given time t. On a given surface area, for example, 1 dm^2, the 5 or 10 deepest pits are identified, and their depth is measured.

In practice, this is measured with a microscope at a sufficient magnification. First, the specimens are pickled in order to eliminate any corrosion products. Then the image is focussed first on the uncorroded surface at the pit's rim, and then on the bottom of the pit. This measurement can also be taken by radiography of the samples, followed by image analysis [10].

It has been shown that the deepest pits obey Gumbel's law of distribution of extreme values [11, 12]:

$$\phi(X) = \exp(-\exp(-J)), \qquad \text{wherein } J = \alpha(x - a).$$

Using the Gumbel plot, this law determines the maximum depth of pits. The maximum depth of pits is measured on each specimen of a lot of N identical samples of surface area S. These values are then classified from 1 to N according to increasing depth.

The diagram is set up in the following manner:

–   On the abscissa, on a linear scale, the maximum pitting depth is plotted for each sample,
–   On the ordinate, in a log(log) scale, the cumulative probability $n/(N + 1)$ is plotted, where n is the order of classification of each specimen. If Gumbel's law applies, these points fall on a straight line. This is observed in natural environments (Figure B.2.5).

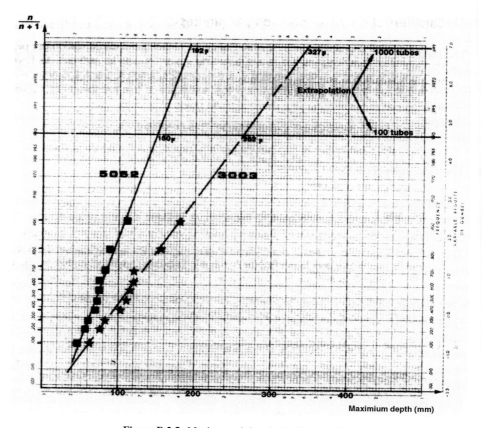

Figure B.2.5. Maximum pitting depth; Gumbel plot.

Repeating these measurements after different durations of exposure makes it possible to check the growth kinetics in $kt^{1/3}$, and to determine the maximum depth of pits at a given time t (Figure B.2.6).

This very interesting approach makes it possible to extrapolate the maximum pitting depth for a structure of any surface area, on the basis of tests using a limited number of identical specimens of surface S and acceptable time intervals: 6, 12, 18 months, etc. In this way, one could determine the corrosion resistance of heat exchangers comprising x thousand tubes at 10, 20, 40 years of service, provided, of course, that tests have been performed according to a rigorous protocol using a representative environment.

■ **Pitting probability**

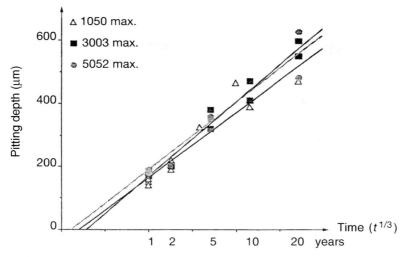

Figure B.2.6. Gumbel plot: pitting depth in marine atmosphere (testing over 2 years).

In addition to the pitting penetration rate, it is also possible to characterise pitting corrosion on aluminium by means of the probability, which is defined as

$$p = 100 \frac{N_\mathrm{p}}{N}$$

where

p is the pitting probability,
N_p is the number of specimens exhibiting pitting,
N is the number of exposed specimens.

A large number of specimens are required for the reliable assessment of this parameter.

Pitting probability and penetration rate are two concepts that characterise the durability of aluminium in water. They are not related to each other: a given aluminium can have a high pitting probability and a low penetration rate, and vice versa. The former is of course preferable.

2.2.4. *Sensitivity of aluminium alloys to pitting corrosion*

All aluminium alloys are prone to pitting corrosion in natural environments. Experience shows that the resistance to pitting corrosion is higher if the density of initiated pits is high.

For thin products with a thickness below 100 μm, the pitting depth can be limited by adding up to 1% iron to the aluminium (alloys 8011A and 8079). The increase in the number of cathodic Al_3Fe intermetallics increases the number of initiated pits.

For the same reasons, a small amount of copper (0.10–0.20%) is added to alloy 3003, in order to multiply the initiation sites of pits on Al_2Cu intermetallics.

The pitting corrosion rate and the pitting depth are independent of metal thickness [13].

2.3. TRANSGRANULAR AND INTERGRANULAR (INTERCRYSTALLINE) CORROSION

Within the metal, at the level of the grain, corrosion may propagate in two different ways:

–   It spreads in all directions (Figure B.2.7): corrosion indifferently affects all the metallurgical constituents; there is no selective corrosion. This is called transgranular or transcrystalline corrosion because it propagates within the grains.
–   It follows preferential paths (Figure B.2.8): corrosion propagates at grain boundaries. Unlike transgranular corrosion, this form of intercrystalline corrosion consumes only a very small amount of metal, which is why mass loss is not a significant parameter for assessment of this type of corrosion. It is not detectable with the naked eye but requires microscopic observation, typically at a magnification of 50. When penetrating into the bulk of the metal, intercrystalline corrosion may lead to a reduction of mechanical properties, especially of elongation, and may even lead to the rupture of components.

Figure B.2.7. Transcrystalline corrosion.

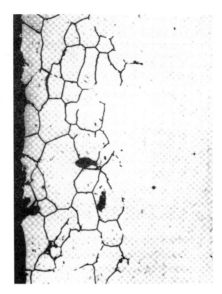

Figure B.2.8. Intercrystalline corrosion.

The propagation of intercrystalline corrosion starts at pits. There is no relationship between the penetration depth of intercrystalline corrosion and the diameter of corrosion pits. This means that intercyrstalline corrosion may also propagate from minute, superficial pits.

2.3.1. Origins of intercrystalline corrosion

Intercrystalline corrosion is caused by a difference in the electrochemical potentials between the bulk of the grain and the grain boundaries where intermetallic phases precipitate. The grain (also called the matrix) comprises a solid solution and dispersed intermetallic compounds. At room temperature, the solubility of iron, nickel, or magnesium in aluminium is so low that the solid solution has a potential very close to that of unalloyed aluminium 1050A. However, when the solid solution is supersaturated or enriched at ambient temperature, the potential depends on the concentration of the alloying element (Figure B.1.8).

The dissolution potential of intermetallic compounds differs from that of aluminium (Table B.2.1). Intermetallic compounds can be

- less electronegative than the solid solution (Al_3Fe, Al_2Cu): they are cathodic with respect to the solid solution, and in the case of intercrystalline corrosion, it is the solid solution that is dissolved;

Table B.2.1. Dissolution potentials of solid solution and intermetallics (solution $NaCl + H_2O_2$ according to ASTM G 69)

Solid solution	Dissolution potential (mV SCE)	Intermetallic phase
	− 170	Si
	− 430	Al_3Ni
	− 470	Al_3Fe
Al−4Cu	− 610	
	− 640	Al_2Cu
Al−1Mn	− 650	
1050A	− 750	
	− 760	Al_6Mn
Al−3Mg	− 780	
Al−5Mg	− 790	
Al−1Zn	− 850	
	− 910	Al_2CuMg
	− 960	$MgZn_2$
Al−5Zn	− 970	
	− 1150	Al_3Mg_2
	− 1190	Mg_2Si

− more electronegative than the solid solution ($MgZn_2$, Al_3Mg_2 and Mg_2Si): they are
   anodic with respect to the solid solution, and in the case of intercrystalline corrosion,
   these intermetallics will be attacked.

Intercrystalline corrosion will develop when three conditions are simultaneously met:

− presence of a corrosive medium,
− difference in potential in the order of 100 mV between intermetallics and the solid
   solution (thus there is no intercrystalline corrosion with the intermetallic phase Al_3Mn),
− continuous precipitation of intermetallics such that intercrystalline corrosion can
   propagate.

For precipitation hardenable alloys, the size and distribution of intermetallic
precipitates depends on the quenching and artificial ageing conditions. The continuous
precipitation of intermetallics at grain boundaries can thus be avoided.

In aluminium alloys, two different situations may occur: precipitation of an anodic
phase, or precipitation of a cathodic phase.

■ Precipitation of an anodic phase

Precipitation of an anodic phase occurs in aluminium alloys of the 5000 series with a
magnesium content above 3.5%. Prolonged heating leads to the precipitation of the phase

β-Al₃Mg₂ at grain boundaries. Its dissolution potential of -1150 mV SCE is highly anodic with respect to the solid solution that has a potential of -780 mV. Under appropriate conditions, namely in presence of a corrosive medium, intercrystalline corrosion can occur. This problem is dealt with in Section B.6.4.

■ Constitution of a depleted, anodic zone

A depleted, anodic zone is constituted in alloys of the 2000 series, such as 2024 (4% copper). When the quenching rate is too low, the hardening phase Al₂Cu, with a potential of -640 mV, will precipitate at grain boundaries. This phase is formed with copper atoms that originate in the vicinity; the zone close to the grain boundaries will, therefore, be depleted in copper. The depleted solid solution has a potential close to -750 mV, which is anodic with respect to grain boundaries (Figure B.2.9).

Under appropriate conditions of intercrystalline corrosion, this depleted zone will dissolve along a path parallel to the grain boundaries.

In general, the corrosion susceptibility of precipitation hardenable alloys depends on their microstructure and on their temper [14]. Tempers T3 and T4 are less susceptible than

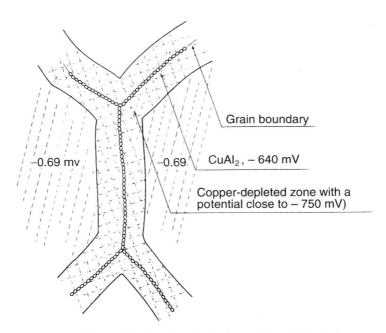

Figure B.2.9. Precipitation of a cathodic phase at the grain boundaries.

T6. On the other hand, the over-aged temper T73 increases the resistance to intercrystalline corrosion (Figure B.2.10).

Intercrystalline corrosion of precipitation hardenable alloys can be controlled by appropriate thermal treatments:

–   high quenching rate,
–   longer ageing for 2000 series alloys,
–   duplex-ageing treatment for 7000 series alloys (temper T73).

■ **Assessment of intercrystalline corrosion**

Usually, the intensity of intercrystalline corrosion is assessed by the number of layers of attacked grains. It is generally admitted that intercrystalline corrosion is only superficial and without danger if limited to three or four layers. This is observed on alloys of the 6000 series that do not contain copper.

The susceptibility to intercrystalline corrosion is tested:

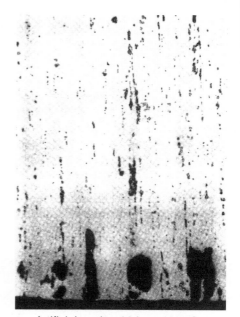

Artificial ageing 16 h at 170 °C   Artificial ageing 20 h at 190 °C

Figure B.2.10. Influence of artificial ageing on the susceptibility to intercrystalline corrosion of 2014.

- for copper-containing alloys (series 2000, 7000 and 6000) according to the standard ASTM G 110 by immersion in a solution of NaCl 58.5 $g \cdot l^{-1}$ + H_2O_2 3 $g \cdot l^{-1}$ at 30 °C for 6 h, followed by microscopic examination of a cross-section,
- for alloys of the 5000 series according to the standard ASTM G 67 by immersion in (NALMT test) in nitric acid at 30°C for 24 h, followed by microscopic examination of a cross-section.

2.3.2. *Unalloyed aluminium (series 1000)*

Aluminium of the series 1000 (including refined aluminium 1199), is also prone to intercrystalline corrosion [15]. In water at a temperature above 60–70 °C, the sensitivity to intercrystalline corrosion increases with increasing purity of the metal. Intercrystalline corrosion is caused by the presence of $AlFe_3$ intermetallics at the grain boundaries [16].

For this reason, AlFeNi alloys were developed at the end of the 1950s, when aluminium was considered for use in contact with hot water in nuclear power plants. The formation of numerous cathodic intermetallics in the matrix reduces the risk of intercrystalline corrosion (see Section D.1.7.3).

2.4. EXFOLIATION CORROSION

Exfoliation corrosion is a type of selective corrosion that propagates along a large number of planes running parallel to the direction of rolling or extrusion [17]. Between these planes are very thin sheets of sound metal that are not attacked, but gradually pushed away by the swelling of corrosion products, peeling off like pages in a book; hence the term "exfoliation corrosion". The metal will swell, which results in the spectacular aspect of this form of corrosion (Figure B.2.11).

This corrosion may be intercrystalline in 2000 and 5000 alloys exhibiting a long-grained structure parallel to the direction of transformation. It can also be found in 7000 alloys with or without copper (which have a low susceptibility to intercrystalline corrosion). It results from the precipitation of parallel stripes of intermetallic phases such as Al_6Mn, $Al_{12}CrMn$, AlFeMn, etc., which are cathodic with respect to the solid solution, and between which there is an anodic zone depleted in Fe and Mn.

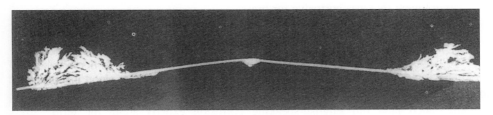

Figure B.2.11. Exfoliation corrosion.

■ **Susceptibility of aluminium alloys**

Welded alloys of the 7000 series without copper, such as 7020, in the T4 temper show a high susceptibility to exfoliation corrosion.

This mainly depends on the conditions of transformation and heat treatments.

■ **Assessment of exfoliation corrosion**

Standardised testing procedures should be used for assessing the susceptibility to exfoliation corrosion [18]. The selection of the test solution and the duration of immersion depend on the alloy series:

–   for 5000 series alloys, the ASSET test (ASTM G 66) is used;
–   for copper-containing alloys of the 2000 and 7000 series, the EXCO test (ASTM G 34) is used.

After the attack, the sample's aspect is compared with control specimens corresponding to various degrees of susceptibility to exfoliation corrosion (see Figure B.4.1).

2.5. STRESS CORROSION

During the rainy season, the British Army in India did not often go out and stored weapons and ammunition in the stables. At the end of the monsoon, when taking up arms again, the soldiers saw that many cartridges in brass 70/30 were unusable because of cracks. Since these problems were related to the season and did not occur during the dry season, corrosion experts in the UK coined the term "season cracking" for this type of corrosion.

It was not until 1921 that an explanation could be given for this phenomenon:

–   deep drawing of the cartridges led to high residual stress,
–   the ammoniated atmosphere due to the decomposition of horse urine created an environment that was corrosive for brass.

Until the early 20th century, steel steam boilers would explode because of rivet rupture. Corrosion experts found anomalous traces of sodium hydroxide in hidden recesses under the rivets. It was only in 1927 that an explanation was given for these cases of rupture due to stress corrosion which had been the cause of many fatal accidents all over the world.

Only much later did English-speaking corrosion experts adopt the term "stress corrosion". In France, until the 1970s, the term "corrosion sous tension" (strain corrosion) was used, which presented a certain ambiguity as to the origin of the tension, mechanical or electrical. Nowadays, the term "stress corrosion" is preferred.

This type of corrosion results from the combined action of a mechanical stress (bending, tension) and a corrosive environment. Each of these parameters alone would not have such a significant effect on the resistance of the metal or would have no effect at all.

The susceptibility of aluminium alloys to stress corrosion was mentioned for the first time in 1922 by Rawdon, who established a relation between susceptibility to intercrystalline corrosion and stress corrosion of Duralumin [19].

Alloys of high mechanical strength of the series 2000, 7000 as well as high-magnesium (≥7%) alloys of the 5000 series can be sensitised to stress corrosion.

2.5.1. Mechanisms of stress corrosion

The mechanisms of stress corrosion have been the subject of numerous studies. They are so complex that it seems to be difficult or even impossible to develop a general theory [20].

In the case of aluminium alloys, propagation of stress corrosion cracks always proceeds along grain boundaries; this form of corrosion is sometimes considered as a special case of intercrystalline corrosion in which mechanical stress accelerates crack propagation along grain boundaries.

However, this view is erroneous because there are aluminium alloys which are prone to intercrystalline corrosion in the absence of mechanical stress, but are not prone to stress corrosion, notably the 6000 series. The hypothesis of propagation by intercrystalline corrosion, therefore, does not adequately explain or predict stress corrosion.

Given the importance of stress corrosion of high-resistance alloys used in aerospace applications [21], this type of corrosion has been the subject of many publications since 1950, and especially in the 1960s, when alloys of the 7000 series (7075, 7010, 7475, etc.) were introduced in the aerospace industry [22–24].

Two possible mechanisms can explain stress corrosion in aluminium alloys:

- electrochemical propagation,
- hydrogen embrittlement.

■ Electrochemical propagation

Brenner in 1932 [25] and Mears, Brown and Dix in 1944 [26] developed the theory of intercrystalline propagation of stress corrosion of alloys of the 2000 and 5000 series.

Propagation proceeds according to an electrochemical mechanism. Once a crack is initiated, the stress resulting from mechanical strain concentrates at the apex of the crack. A zone will be formed which is anodic with respect to the walls of the crack, where the film has formed again, and which constitutes the cathodic zone.

In aqueous media, the crack constitutes a battery, in which the crack's apex is the anode. When the anode is being consumed, it leads to crack propagation. The more the crack propagates, the less the conditions are favourable for a quick reconstitution of the natural oxide film; this accelerates the crack propagation rate (Figure B.2.12).

Stress corrosion of alloys of the 2000 and 5000 series develops only in media containing high amounts of chloride, but not in water having only a low chloride level, or in a humid atmosphere.

A good correlation is found between the sensitivity to intercrystalline corrosion and stress corrosion of these alloys in the marine environment. In practice, an alloy of the 2000 series that is not susceptible to intercrystalline corrosion is not susceptible to stress corrosion either.

■ Hydrogen embrittlement

The electrochemical theory is not applicable to all aluminium alloys [27]. In fact, alloys of the 6000 series are susceptible to intercrystalline corrosion, but not to stress corrosion. On the other hand, alloys of the 7000 series are susceptible to stress corrosion but not to intercrystalline corrosion.

Until the end of the 1960s, the theory of hydrogen embrittlement was considered inapplicable to aluminium alloys, for two reasons:

–   Embrittlement of aluminium by hydrogen does not occur, even at very high pressure. Unlike steel, aluminium alloys are not sensitive to embrittlement in a dry hydrogen atmosphere.
–   The solubility and diffusion rates of hydrogen in aluminium are very low.

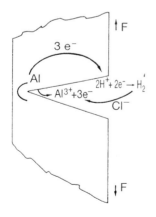
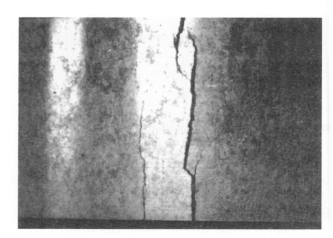

Figure B.2.12. Stress corrosion crack; Electrochemical propagation.

These objections were overcome when it was shown that an AlZnMg alloy prepared from very pure base metals becomes susceptible to stress corrosion in a humid atmosphere, and that this embrittlement is due to the presence of hydrogen in the metal [28, 29].

In fact, during production, hydrogen can penetrate the metal at the metal-oxide film interface by a reduction during corrosion. At the crack's apex, aluminium is not covered by its natural oxide film. It reacts with water and releases hydrogen in statu nascendi, which concentrates at grain boundaries and favours intercrystalline decohesion.

2.5.2. *Susceptibility of aluminium alloys*

Rolling and extrusion leads to a grain orientation parallel to the direction of transformation. Since stress corrosion has an intercrystalline propagation, susceptibility to this type of corrosion is, therefore, not the same in the three directions with respect to the direction of rolling or extrusion (Figure B.2.13):

– long direction,
– long transverse direction,
– short transverse direction.

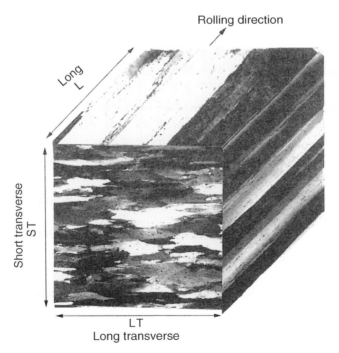

Figure B.2.13. Orientation with respect to the directions of rolling or extrusion.

The resistance to stress corrosion of thick rolled or extruded products depends on the sampling direction of the test pieces. It is always lower in the short transverse direction than in the other two directions.

In general, the resistance to stress corrosion is higher for the following tempers:

–   TX51: semi-products in tempers T451 and T651 have been stretched after quenching (T451) or between quenching and ageing (T651) in order to reduce internal stress. Since the total stress applied is the sum of the internal stresses (caused by the transformation) and the stress stemming from service conditions, it is, therefore, desirable to use semi-products with internal stresses that are as low as possible.
–   T73 that is overaged or duplex aged.

2.5.3.   Detection of susceptibility to stress corrosion

Several standardised tests are available for assessing the susceptibility to stress corrosion; see Table B.4.1. The test pieces are strained by stretching or bending, depending on the type of test. The applied load is typically equal to 75% of $R_{p0.2}$. Testing at a lower load, 50% and 25% of $R_{p0.2}$, makes it possible to determine the threshold stress below which no stress corrosion is observed.

In the laboratory, testing is done in a solution containing 3.5% NaCl with alternating immersion and emersion. The maximum duration, 30–60 days, depends on the alloy and the loading conditions.

■ Stress corrosion criteria

The propagation rate of stress corrosion cracks is very low, between 10^{-6} and 10^{-11} m·s^{-1}. However, propagation, even at this low rate, may lead to catastrophic mechanical failure, due to the weakening of the structures.

By analogy with failure mechanics, two criteria can be determined for assessing the susceptibility to stress corrosion of aluminium alloys:

–   the limit of susceptibility to stress corrosion, which depends both on the alloy and the sampling direction (Figure B.2.14),
–   the threshold of non-cracking k_{1SCC}, which is always lower than k_{1C}.

Three stages [30] can be distinguished (Figure B.2.15):

–   *(A) initiation*: k_1 being slightly higher than k_{1SCC}, the crack propagation rate increases very rapidly,
–   *(B) propagation*: the crack propagation rate is independent of k_1 because of corrosion, and the rate does not increase,
–   *(C) rupture*: the material becomes unstable and fails.

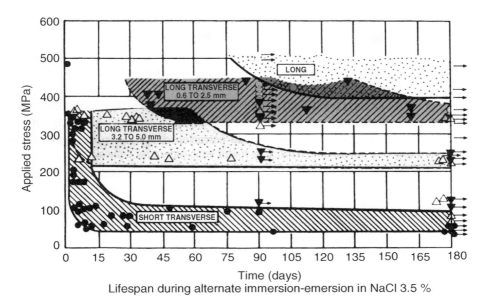

Figure B.2.14. Resistance to stress corrosion of 7075 T6 and sampling direction [31].

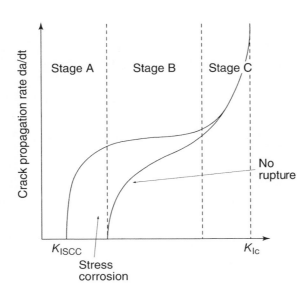

Figure B.2.15. Propagation of stress corrosion.

2.6. FILIFORM CORROSION

Filiform corrosion is specific to lacquered metal. This is mainly an alteration of surface appearance. The underlying metal only suffers a very superficial attack, not exceeding a depth of a few tens of microns.

It develops as narrow filaments, about 0.1–0.5 mm wide and a few millimetres long, which propagate at the metal–lacquer interface. Swelling of corrosion products deforms the lacquer film and appears as very narrow wires which progress like mole tunnels underneath the lacquer film (Figure B.2.16).

Filiform corrosion always starts at coating defects, such as scratches, and weak points: beards, cut edges or holes. It can be seen after several years of service.

This type of corrosion, described for the first time in 1944 [32], was observed on lacquered steel first and then on high-resistance aluminium alloys for aerospace applications [33].

Coil-coated strip in aluminium alloys is not prone to this type of corrosion. After the introduction of new lacquering techniques by electrostatic deposition of powders on profiles or sheets in aluminium alloys of the 3000, 5000 and 6000 series, filiform corrosion appeared in the 1980s, mainly in the coastal areas of Northern Europe.

2.6.1. *Mechanisms of filiform corrosion*
Several mechanisms have been proposed in order to explain this special type of corrosion.

Figure B.2.16. Filiform corrosion.

■ **Cathodic delamination**

Filiform corrosion results from the formation of a small hygroscopic disk having a relative humidity in equilibrium with the ambient atmosphere. This lentil is the active head of the filament's point where the cathodic reduction of water occurs Figure B.2.17:

$$O_2 + 2H_2O + 4e^- \rightarrow 4OH^-$$

Local alkanisation leads to delamination of lacquer layers, which is promoted by the edge effect resulting from the swelling of corrosion products [34].

Since the daily rate of progression is constant and approximately equal to the diameter of the lentil, the underlying metal is only very superficially attacked, with a depth of a few tens of micrometres.

■ **Anodic dissolution**

The filiform corrosion "cell" is made up of the active head and an inert tail where corrosion products accumulate. The head, where chlorides concentrate and acidification occurs, forms the anode, where aluminium oxidation takes place:

$$Al \rightarrow Al^{3+} + 3e^-$$

$$Al^{3+} + 3H_2O \rightarrow Al(OH)_3 + 3H^+.$$

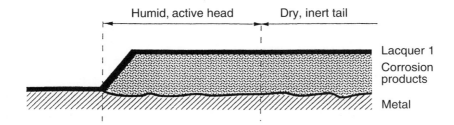

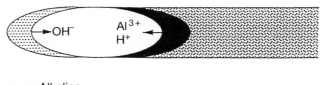

Figure B.2.17. Filiform corrosion; cathodic delamination.

The downstream area forms the cathodic zone, where oxygen is reduced:

$$O_2 + 2H_2O + 4e^- \rightarrow 4OH^-.$$

This cell works according to the classic reaction of aluminium corrosion in the presence of chloride (Figure B.2.18).

Water and oxygen diffuse across the desiccated corrosion products of the filament's tail, and water migrates by osmosis towards the head [39].

2.6.2. Factors of filiform corrosion

■ The nature of the coating

Filiform corrosion occurs with all types of paints: acrylic lacquers, epoxy-polyamides, epoxy-amines and polyurethanes, and whatever the classic mode of application, whether with liquid paint or electrostatic powdering. It does not occur under sealed coatings such as electrician's tape.

■ The surface preparation

This is a very important factor [36]. Filiform corrosion develops on metal that has received no surface preparation, or poor preparation, or on a metal whose surface has been contaminated before lacquering.

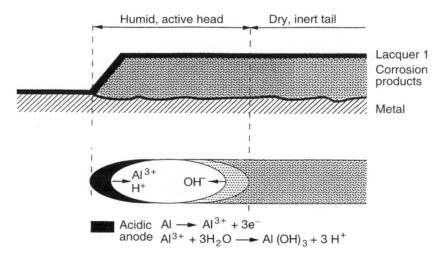

Figure B.2.18. Filiform corrosion; anodic dissolution [35].

■ **The nature of the alloy**

The nature of the alloy is not an important factor, because filiform corrosion may affect all aluminium alloys. A recent collaborative study conducted by three European transforming companies, Alusuisse, Hydro Aluminium and Pechiney, showed that for the most commonly used alloys in the construction industry, 6060 and 6063, the alloy composition has no influence, except when the copper concentration exceeds 0.1%.

2.7. WATER LINE CORROSION

Experience has shown that when metallic structures are partly immersed in water, a localised and more intense corrosion very often occurs on the immersed portion just below the air–water interface (Figure B.2.19). This is caused by the difference in aeration between the surface of the liquid and the zone immediately underneath.

This differential aeration was shown by Evans in the following experiment: a container with an alkaline solution is subdivided into two parts of equal volume by means of a porous ceramic wall, across which the ions present in the solution can migrate. In each compartment, a steel electrode is plunged. The two electrodes have identical dimensions in terms of metal composition and surface state.

As soon as air is introduced in one of the two compartments, a current forms between the two electrodes, the anode being the electrode of the compartment that is not aerated.

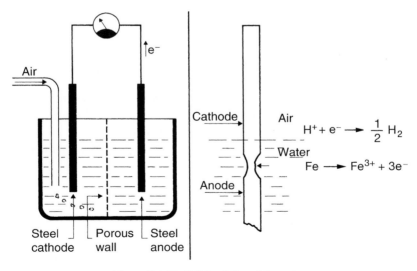

Figure B.2.19. Cell for differential aeration.

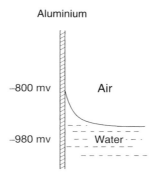

Figure B.2.20. Waterline corrosion of aluminium [37].

Experience has shown that on an aluminium structure that is partly immersed in seawater, water line corrosion may occur as rather scattered, superficial pitting, with a depth not exceeding a few tenths of a millimetre. This can be observed on the uncoated aluminium hull of barges.

The electrochemical behaviour of aluminium at the water line differs from that of iron, because aluminium is a passive metal. Corrosion develops in the meniscus, where the water film is very thin. It has been shown that in the case of aluminium, it is not the difference in aeration that is important, but the difference in the concentration of chloride in the thinnest part of the water film, where evaporation is fast. The chloride concentration is higher in the thin part of the film. Moreover, the more electronegative the dissolution potentials, the thinner the film is. The upper part of the meniscus is, therefore, the anodic zone, which corrodes preferentially (Figure B.2.20) [37].

Experience shows that the zone of air–water separation is not very prone to pitting corrosion. This has been shown on the testing raft for corrosion studies that has been immersed for 25 years in the Mediterranean Sea at Pechiney's outdoor corrosion testing station at Salindre-de-Giraud, France [38].

Uncoated barges in 5083 and 5086 series alloys as well as floats in 5754 at landing stages of marinas (tens of thousands of these floats are used in France, the oldest for 25 years) exhibit no preferential corrosion in the tidal range.

The air emerging zone of structures embedded in concrete or plaster shows no preferential corrosion either (see Chapter G.4).

2.8.   CREVICE CORROSION

Crevice corrosion is a localised corrosion in recesses: overlapping zones for riveting, bolting or welding, zones under joints, and under various deposits (sand, slag, precipitates, etc.). These zones, also called crevices, are very tiny and difficult to access for the aqueous liquid that is covering the rest of the readily accessible surfaces (Figure B.2.21). This type of corrosion is also known as deposit attack.

As soon as an electrochemical reaction occurs in this confined volume, the composition of the contained liquid will change. As a consequence, the dissolution potential becomes more electronegative, and the surface in the recess becomes anodic with respect to the rest of the structure [39]. In the crevice, aluminium is oxidised according to the reaction

$$Al \rightarrow Al^{3+} + 3e^-$$

whereas at the rim of the crevice, oxygen is reduced:

$$O_2 + 2H_2O + 4e^- \rightarrow 4OH^-.$$

While the metal in the recess is being corroded, oxygen dissolved in the liquid will be consumed. Since the local geometry limits diffusion, the recess will be depleted in oxygen, with an excess of Al^{3+} ions. This leads to an inflow of Cl^- chloride ions. Aluminium chloride will hydrolyse:

$$AlCl_3 + 3H_2O \rightarrow Al(OH)_3 + 3H^+ + 3Cl^-$$

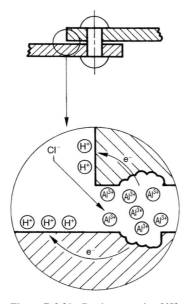

Figure B.2.21. Crevice corrosion [40].

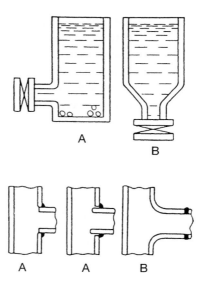

Figure B.2.22. Designs that avoid crevice corrosion [39] (A: Corrosion is favouried; B: Corrosion is avoided).

Because of the crevice's small volume, its medium will very quickly become acidic, with a pH value of 2–3.

Unlike certain alloys such as stainless steel, aluminium exhibits a rather low susceptibility to crevice corrosion. When a bolted or riveted assembly that has been exposed to a liquid such as seawater for a long period is dismantled, it appears that crevices are usually sealed by corrosion products of aluminium. This is certainly one of the reasons why aluminium has a rather low susceptibility to crevice corrosion.

Nevertheless, it is recommended that as far as possible, irregular or intermittent welding beads be avoided in reservoirs and tanks, where deposits can accumulate (Figure B.2.22).

Crevice corrosion associated with galvanic corrosion may develop under steel washers (ordinary or stainless) used for mounting aluminium wall panels exposed to the seashore [41]. For this reason, assemblies must be mounted with great care in such an aggressive environment in order to avoid the formation of zones that retain humidity, which would favour galvanic and crevice corrosion.

2.9. CAVITATION

Cavitation occurs when the hydrodynamic pressure exceeds the vapour pressure of a moving liquid. Gas bubbles form within the liquid, which thus becomes a two-phase system. These bubbles will be crushed against the metal surface at high speed, an attack that leads to cavities with rounded contours. This degradation is caused by the combination of a mechanical effect and corrosion of the metal. The natural oxide film is destroyed and the aluminium is attacked; there is a competition between tearing off the film and reforming it. It is not possible to estimate the individual contributions of the mechanical effect and corrosion.

Cavitation develops quickly: within a few days, and even within a few hours, a tube in 3003 several millimetres thick may become heavily perforated, all the more so the higher the liquid's speed and temperature.

Since this is partly a mechanical phenomenon, a high superficial hardness of the metal will favour resistance to cavitation. A direct relationship between hardness and resistance to cavitation has been found. Aluminium is less resistant than other metals (cast iron, steel, etc.) and needs to be replaced by a harder metal such as cast iron if it is not possible to adapt the flow regime of the fluid.

Cavitation is not a stable regime and can be avoided by increasing the diameter of the pipes that form the circuit, appropriate dimensioning of accessories such as pumps, and avoiding sudden changes in flow direction. Prevention of cavitation in aluminium equipment is based on the same problems and the same solution as for other metals.

2.10. EROSION

Corrosion by erosion occurs in moving media. This type of corrosion is related to the flow speed of the fluid. It leads to local thinning of the metal, which results in scratches, gullies, and undulations, which are always oriented in the same direction, namely the flow direction.

Experience with tubular heat exchangers in aluminium alloys of the series 3000, 5000 and 6000 used for desalination of seawater has shown that aluminium can withstand a flow speed in the order of $2.5-3$ m·s^{-1} at temperatures up to 130 °C with no erosion corrosion. This range corresponds to the usual flow speed in industrial installations. Tests with distilled water at 100 °C have shown that the erosion of aluminium starts at a flow speed in the order of $12-15$ m·s^{-1} [42].

Avoiding erosion corrosion on aluminium does not require any specific precautions, only those commonly applied to other metals.

2.11. MICROBIOLOGICAL CORROSION

Microbiological corrosion of a metal may be caused by

-   heterotrophic bacteria that require an organic carbon source in order to develop. They can assimilate a wide range of organic substances, transforming them into carbon dioxide, metabolites and organic acids, which are released in the nearby environment. It is this modification of the environment (not the bacteria themselves) which leads to corrosion;

—   autotrophic bacteria, which develop only in the presence of inorganic elements, while taking up energy from chemical or photochemical reactions [43]. No autotrophic bacteria for aluminium and its alloys are known.

The only well-documented case of microbiological corrosion of aluminium is related to fuel tanks of jet aircraft. Around 1950, the US Air Force detected the first case of corrosion of such a fuel tank [44].

Fuel is never totally anhydrous; it always contains traces of humidity (75 ppm at 25 °C) that will separate from kerosene and accumulate in zones of the tank that are difficult to drain and in which corrosion may occur. Fuel contains microorganisms that will find food there.

It has been shown that corrosion in jet aircraft tanks is mainly caused by the growth of the bacterium *Cladosporium Resinae* in kerosene [45].

Corrosion develops as pitting at the fuel–water interface. Like any other type of corrosion, it follows an electrochemical mechanism. Oxidation of kerosene by bacteria releases organic acids that modify the pH of the medium. Microbial deposits form anodic sites by local acidification. The oxidation reaction consumes the oxygen dissolved in kerosene and in water.

Enzymes contained in bacteria preferentially attack zinc and magnesium present in certain aluminium alloys such as 7075 [46].

The prevention of microbiological corrosion in tanks of jet aircraft is based on bactericides that are either water-soluble such as strontium chromate, or kerosene-soluble such as mono-ethylene glycol.

There is no specific microbiological corrosion of aluminium in aqueous media, especially in seawater, which contains a wide range of bacteria.

2.12.   CORROSION PRODUCTS

According to the basic corrosion reaction of aluminium:

$$2Al + 6H_2O \rightarrow 2Al(OH)_3 + 3H_2$$

the corrosion product is aluminium hydroxide $Al(OH)_3$, commonly called alumina. Freshly formed alumina contains a great deal of water. After calcination at 1000 °C, the mass loss can reach 60% [47].

Alumina is a gelatinous white gel that covers corrosion pits. After several weeks of exposure to air, part of the water will evaporate and the alumina will look like a white powder. Whether dry or not, the alumina adheres well to the metal surface.

Alumina that is formed during corrosion at room temperature is amorphous [48], whatever the type of corrosion from which it originates. On the metal surface, its structure changes slowly (and more quickly in solution) as follows [49]:

$$\text{Amorphous alumina} \rightarrow \text{Boehmite} \rightarrow \text{Bayerite} \rightarrow \text{Hydrargilite.}$$

■ Elimination of corrosion products

It may be necessary to eliminate corrosion products for several reasons:

- measurement of the variation in the mass before and after corrosion testing,
- cleaning of corroded surfaces.

For laboratory tests, the surface and its irregularities caused by pitting can be cleaned by pickling for a few minutes with concentrated nitric acid at room temperature, together with vigorous brushing with a hard polymer wire brush.

For the elimination of corrosion products on a construction (ship, tank, etc.), in order to refresh the surface, for example, as a preparation for lacquering, sand blasting is the best approach. Solutions of corrosive products should never be introduced in a tank, ship, etc., because it is often difficult to properly rinse surfaces of intricate shapes such as those of tanks, ships, etc. The retention of pickling chemicals may eventually lead to severe corrosion.

■ Analysis of corrosion products

Often, analysis of corrosion products is requested because there is a tendency, rightly, to consider that they may allow tracing the origin of corrosion in service.

Most qualitative analyses show the presence of several constituents, including

- chlorides,
- sulphates,
- carbonates.

However, experience shows that the knowledge of the concentration of these constituents rarely provides useful information for the determination of the causes of corrosion in service. Chlorides are ubiquitous, their origin is marine (see part D).

The presence of sulphates, especially at high concentrations, can hardly tell more than the atmospheric origin of corrosion (see Section C.2.6), and the presence of carbonates indicates corrosion in water [50].

In general, the qualitative and quantitative analysis of corrosion products does not yield information that could be useful for determining the origin of corrosion under service conditions.

This is not surprising as such, because in most cases, corrosion is due to the unfavourable design of a construction and to anomalous service conditions. Therefore, it cannot be expected that the detailed analysis of corrosion products can reveal the source of corrosion under service conditions. However, there are two exceptions: traces of copper and mercury, whatever their origin, may be considered to be a likely source of corrosion in service. Since the use of mercury is increasingly restricted, it is rarely found in corrosion products.

REFERENCES

[1] Tabrzi M.R., Lyon S.B., Thompson G.E., Ferguson J., The long term corrosion of aluminium in alkaline media, *Corrosion Science*, vol. 32, 1991, p. 733–742.

[2] Reboul M., Méthodes de mesure de la corrosion généralisée de l'aluminium et de ses alliages, *Revue de l'Aluminium*, nos 419, 1973, p. 62–73.

[3] Kaesche H., Mécanisme de la corrosion par piqûres, *Corrosion Traitements Protection Finition*, vol. 17, 1969, p. 389–396.

[4] Reboul M., Warner T., Mayet H., Baroux B., A ten step mechanism for the pitting corrosion of aluminium alloys, *Corrosion Reviews*, vol. 15, nos 3–4, 1997, p. 471–496.

[5] Bogar F.D., Foley F.D., The influence of chloride ion on the pitting of aluminium, *Journal of the Electrochemical Society*, vol. 119, 1973, p. 462–464.

[6] Brown R.H., Mears R.B., The electrochemistry of corrosion, *Transactions of the Electrochemical Society*, vol. 74, 1938, p. 495.

[7] Godard H.P., The corrosion behavior of aluminium in natural waters, *The Canadian Journal of Chemical Engineering*, vol. 38, 1960, p. 167–173.

[8] Groot C., Troutner V.H., *Corrosion of aluminium in tap water*, USAEC, report HW 43085, June 1956.

[9] Champion F.A., Metal thickness and corrosion effects, *Metal Industry*, vol. 74, 1949, p. 7–9.

[10] Reboul M., Odièvre T., Warner T.J., *Pit depth measurement on aluminium alloys by image analysis of radiographic films*, Eurocor, Trondheim, 1997, p. 271–278.

[11] Aziz P.M., Application of the statistical theory of extreme values to the analysis of maximum pit depth data for aluminum, *Corrosion*, vol. 12, 1956, p. 495t–506t.

[12] Gumbel E.J., *Statistical theory of extreme values and some practical applications*, US Department of Commerce, Applied Mathematics Series 33, 1954.

[13] Champion F.A., Metal thickness and corrosion effects, *Metal Industry*, vol. 74, 1949, p. 7–9.

[14] Lifka B.W., Sprowls D.O., *Significance of intergranular corrosion in high-strength aluminum alloy products*, ASTM, STP, vol. 516, 1972, p. 120–144.

[15] Rohrmann F., *Transactions of the Electrochemical Society*, vol. 66, 1934, p. 229.

[16] Peryman E.C.W., Grain-boundary general corrosion of high-purity aluminium and hydro-chloric acid, *Journal of Metals*, vol. 5, 1953, p. 911–917.

[17] Ketcham S.J., Shaffer I.S., *Exfoliation corrosion of aluminum alloys, ASTM, STP*, vol. 516, 1972, p. 3–16.

[18] Sprowls D.O., Walsh J.D., Shumaker M.B., *Simplified exfoliation testing of aluminum, alloys, ASTM, STP*, vol. 516, 1972, p. 38–65.

[19] Rawdon H.R., Krynitski A.I., Berliner J.F., Brittleness developed in aluminium and duralumin by stress and corrosion, *Chemical Metallurgy Engineering*, vol. 26, 1922, p. 154–160.

[20] Desjardins D., Oltra R., Introduction à la corrosion sous contrainte, in Desjardins D., Oltra R., *Corrosion sous contrainte, phénoménologie et mécanismes*, Bombannes, Éditions de Physique, Les Ulis, 1990, p. 1–17.

[21] Bodu J.J., Expérience pratique de la corrosion sous contrainte dans l'industrie aéronautique, in Desjardins D., Oltra R., *Corrosion sous contrainte, phénoménologie et mécanismes*, Bombannes, Éditions de Physique, Les Ulis, 1990, p. 797–817.

[22] Sprowls D.O., Brown R.H., Stress-corrosion mechanisms for aluminum alloys, *Conference Fundamentals Aspects Stress Corrosion Cracking, NACE, Houston*, 1967, p. 466–512.

[23] Brown R.H., Sprowls D.O., Shumaker M.B., The resistance of wrought high strength aluminum alloys to stress-corrosion cracking, *Conference Stress Corrosion Cracking. A state of the art, ASTM, STP*, vol. 518, 1972, p. 87–118.

[24] Speidel M.O., Hyatt M.V., *Stress-corrosion cracking of high strength aluminum alloys*, vol. 2, Plenum Press, New York, Advances in corrosion science and technology, 1972, p. 115–335.

[25] Brenner P., Studies on stress-corrosion cracks in light metals, *Z. Metallkunde*, vol. 24, 1932, p. 145–151.

[26] Mears R.B., Brown R.H., Dix E.H., A generalized theory of stress-corrosion of alloys, *Symposium on Stress-Corrosion Cracking of metals ASTM and AIME*, 1944, p. 323–344.

[27] Reboul M., La corrosion sous contrainte des alliages d'aluminium, in Desjardins D., Oltra R., *Corrosion sous contrainte, phénoménologie et mécanismes*, Bombannes, Éditions de Physique, Les Ulis, 1990, p. 623–643.

[28] Scamans G.M., Alani R., Swann P.R., Pre-exposure embrittlement and stress-corrosion failure in AlMgZn alloys, *Corrosion Science*, vol. 16, 1976, p. 443–459.

[29] Ambat R., Dwaarkadasa E.S., Effect of hydrogen in aluminium and aluminum alloys: A review, *Bulletin of Materials Science*, vol. 19, 1996, p. 103–114.

[30] Lemaître C., Influence de l'environnement mécanique sur les propriétés de tenue à la corrosion des alliages métalliques, *Métaux Corrosion-Industrie*, vol. 61, 1986, p. 1–16.

[31] Sprowls D.O., Brown R.H., What every engineer should know about stress-corrosion of aluminium, *Metal Progress*, vol. 81, 1962, p. 77–83.

[32] Sharman C.F., Filiform under film corrosion of laquered steel surface, *Nature*, nos 3890, 1944, p. 621.

[33] Rique J.P., La corrosion filiforme dans les peintures pour l'aéronautique, *Surfaces*, vol. 117, 1984, p. 55–66.

[34] Kaesche H., Études relatives à la corrosion filiforme des tôles d'acier laquées, *Werkstoffe und Korrosion*, vol. 11, 1959, p. 668–681.

[35] Steele G.D., Filiform corrosion on architectural aluminium. A review, *Anti-Corosion Materials & Methods*, vol. 41, 1994, p. 8–12.

[36] Pietschmann J., Filiform corrosion of organically coated aluminium. Part 1, *Aluminium*, vol. 69, 1993, p. 1019–1023, Part 2, p. 1081–1084.

[37] Fiaud C., Azerad G., Grosgogeat E., Corrosion de l'aluminium sous couche mince électrolytique, corrosion à la ligne d'eau et corrosion au stockage, *Journées Durabilité de l'aluminium et de ses alliages dans les industries électriques, Cefracor*, 1986, p. 33–37.

[38] Reboul M., Touche M., *Examen de deux radeaux en aluminium après 8 et 35 ans en mer*, rapport Pechiney CRV, 1983.

[39] Rozenfeld I.L., *Crevice corrosion of metals and alloys*, NACE, 3, Localized Corrosion, 1974, p. 373–398.

[40] Fontana M.G., Greene N.D., *Corrosion engineering*, McGraw-Hill, New York, 1998.

[41] Betts A.J., Boulton L.H., Crevice corrosion: Review of mechanism, modeling and mitigation, *British Corrosion Journal*, vol. 28, 1993, p. 279–295.

[42] Dillon R.L., Hope R.S., *Erosion–corrosion of aluminum alloys*, REV, rapport HW-74359, April 1953.

[43] Gatelier C.R., Pollution microbienne des carburants et corrosion des réservoirs, *Corrosion Traitements Protections et Finitions*, vol. 21, 1973, p. 103–109.

[44] Engel W.B., Role of metallic ion concentration in the corrosion of metals, *25th Conference NACE*, 1970, p. 588–596.

[45] Hedrick H.G., Crum M.G., Reynolds R.J., Culver S.C., Mechanism of microbiological corrosion of aluminum alloys, *Electrochemical Technology*, 1967, p. 75–77.

[46] Hedrick H.G., Microbiological corrosion of aluminum, *25th Conference NACE*, 1970, p. 609–619.

[47] Godard H.P., Cooke W.E., The analysis and composition of aluminium corrosion products, *Corrosion*, vol. 16, 1960, p. 181t–187t.

[48] Takabeya R., Aluminium corrosion products formed in atmospheric environments, *Corrosion Engineering*, vol. 36, 1987, p. 279–286.

[49] Lepina L., Ose Z., Colloidal-chemical effects at surfaces of metals and inhibition of corrosion in salt solution. Composition and structure of the corrosion products of aluminium in potassium chloride solutions, *Latvijas PSR Zinntnu Akad Vestis*, nos 6, 1950, p. 35–46.

[50] Heine R.A., Observations on the corrosion of aluminium and its alloys in relation to chemical examination of the corrosion products, *Journal of Applied Chemicals*, vol. 9, 1959, p. 43–49.

Chapter B.3

Galvanic Corrosion

Chapter B.3
Galvanic Corrosion

Galvanic corrosion has appeared ever since two different metals were put together in a liquid medium, that is to say since the Iron Age, when iron was put in contact with brass and copper. Marine archaeology shows such cases of galvanic corrosion in wrecks of ancient ships.

The first technical report on galvanic corrosion was addressed to the British Admiralty in 1763. The hull of the English frigate *Alarm* had been lined with copper plate in order to prevent the devastating effect of shipworms on the wood of the ship and prevent barnacles from attaching themselves to the hull, resulting in a reduction in the vessels' speed. The plates had been fixed with iron nails.

After 2 years of navigation in the Caribbean Sea, the ship was careened in the dry dock, and it was discovered that many copper plates had been lost during navigation over the sea because the nail heads were corroded. A detailed inspection showed that those nail heads that had resisted corrosion were insulated from the copper plate by pieces of brown wrapping paper that had been used as a packaging material for the plates [1].

From this misadventure, the engineers of the time drew the conclusion that any direct contact between iron and copper in seawater should be avoided.

3.1. ALUMINIUM AND GALVANIC CORROSION

From its ranking on the scale of potentials (Table B.1.1), aluminium is more electro-negative than most common elements: steel, stainless steel, cuprous alloys, etc. Whether in mechanical applications, building, electrical engineering, ship building, heat exchangers or circuits of aqueous liquids (fresh water or seawater), it is common to find heterogeneous assemblies in which there is contact between aluminium alloy components and components made of other metals.

It is not possible to use aluminium only in homogeneous systems consisting in a single aluminium alloy, for both technical and economic reasons. Here are two examples:

−   The first example is a ship. The propeller shaft is very often made of stainless steel, because for reasons related to mechanical resistance, it is impossible to manufacture it in aluminium alloys, except perhaps in 7075. The propeller is very often made in brass, and sometimes in cast aluminium, for small crafts. This has not been a hindrance for the construction of ships in aluminium, even very large ones, because it is possible to control galvanic coupling in the hull.

–   The second example concerns open or closed circuits of aqueous liquids. Accessories such as valves, taps, joints, bends, pump bodies and volutes are almost never made in aluminium, because they cannot be made as easily in aluminium as in copper. Hence, when heat exchangers and pipes are made in aluminium, insulation must be provided between aluminium and other parts of the circuit, and corrosion inhibitors added to the liquid, if it circulates in a closed loop.

For a very long time, galvanic corrosion of aluminium in contact with other metals was a major concern to users, to such an extent that it slowed down the development of applications of aluminium whenever the issue of its behaviour in contact with other metals was raised. Nowadays, it is surprising to discover the extensive precautions that were taken in former times with heterogeneous assemblies exposed to ambient atmosphere.

Decades of experience in construction, and especially with metal fittings joined with screws in stainless steel, as well as in civil engineering, ship building and other uses, have made it possible to better evaluate the risks of galvanic corrosion depending on the metals and alloys in contact with aluminium, their application and their environment.

It was thus necessary to re-evaluate the risk of galvanic corrosion and to abandon certain common assumptions, most of which had been based on laboratory tests that were not representative of the reality of the applications, especially in the maritime environment. Experience, however, can provide valuable information [2].

3.2. DEFINITION OF GALVANIC CORROSION

When two dissimilar metals are in direct contact in a conducting liquid, experience shows that one of the two may corrode. This is called galvanic corrosion.[1] The other metal will not corrode; it may even be protected in this way. This corrosion is different in its kind and intensity from the one that would occur if they were placed separately in the same liquid.

Unlike other types of structural corrosion, galvanic corrosion does not depend on the metal's texture, temper, etc.

Galvanic corrosion may occur with any metal, as soon as two are in contact in a conductive liquid. It works like a battery; we will recall its principles below.

3.3. FORM OF GALVANIC CORROSION

The appearance of galvanic corrosion is very characteristic. It is not dispersed like pitting corrosion, but highly localised in the contact zone with the other metal.

---

[1] In memory of Luigi Galvani, an Italian physician who in 1786 discovered the electrical effect of the coupling of two dissimilar metals.

The attack of aluminium is regular and progresses in depth as craters that are more or less rounded. The zone affected by galvanic corrosion often has a shinier aspect than the rest of the surface.

All aluminium alloys undergo identical galvanic corrosion. In addition, intercrystalline and exfoliation corrosion may occur in those alloys of the 2000, 5000, 6000 and 7000 series that are susceptible to these forms of corrosion.

3.4. PRINCIPLE OF A BATTERY

Galvanic corrosion works like a battery[2] built from two electrodes:

–   the cathode, where reduction takes place,
–   the anode, where oxidation takes place.

These two electrodes are plunged into a conductive liquid called an electrolyte, which is normally a diluted acid solution, such as sulphuric acid N/10, or a salt solution, such as copper sulphate. The two electrodes are externally linked by an electrical circuit that ensures a circulation of electrons. Within the liquid, the transport of electrical current proceeds by the movement of ions, i.e. by ionic transport. The liquid thus provides an ionic junction (Figure B.3.1).

Figure B.3.1 shows a cell in which the electrolyte is a solution of sulphuric acid H_2SO_4 which totally dissociates in water (since it is a strong acid) by forming H^+ ions that are responsible for the acidity of the medium. The following electrochemical reactions occur:

–   the zinc anode oxidises:

$$Zn \rightarrow Zn^{2+} + 2e^-$$

–   on the copper cathode, protons H^+ are reduced (this is the only cation present which can be reduced):

$$2H^+ + 2e^- \rightarrow H_2.$$

---

[2] Volta, a professor at the University of Pavia in Italy, invented the first electrical battery. In 1800, he built a device consisting of a stack of sequences of three washers always disposed in the same order: a copper disk, a felt disk soaked with a salt solution, and a zinc disk. Up to 20 sequences were put in a series (in the electrical sense) and stacked one on top of the other. (see J. Talbot, "Les éléments chimiques et les hommes," SIRPE, 1965, p. 182.)

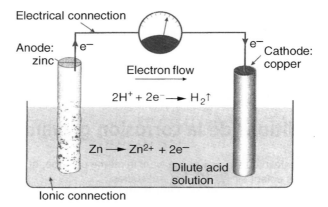

Figure B.3.1. Principle of a galvanic cell.

The net reaction is

$$Zn + 2H_2O \rightarrow Zn(OH)_2 + H_2.$$

This cell produces electricity while consuming zinc that precipitates as zinc hydroxide $Zn(OH)_2$.

Three conditions must be met simultaneously for a cell to function:

- two metals of a different nature which form the two electrodes,
- presence of an electrolyte,
- electrical continuity.

In galvanic corrosion, the cathodic reactions (on the metal which does not undergo corrosion) depend on the medium:

- if it is acid:

$$O_2 + 4H^+ + 4e^- \rightarrow 2H_2O$$

  or

$$2H^+ + 2e^- \rightarrow H_2$$

- if it is an alkali

$$O_2 + 2H_2O + 4e^- \rightarrow 4OH^-$$

If one of these three conditions is not met, for example, if the electrical connection is interrupted, the cell will not produce a current, and the oxidation at the anode will not take place (nor the reduction at the cathode).

3.5. CONDITIONS OF GALVANIC CORROSION

Galvanic corrosion is based on the same principle, and the three following conditions must be simultaneously met:

– different types of metals,
– presence of an electrolyte,
– electrical continuity between the two metals.

3.5.1. Different types of metals
Whenever two different types of metals are in contact, galvanic corrosion is possible. The metal with an electronegative potential (or the more electronegative metal, if both are electronegative) acts as the anode.

In order to evaluate which of the metals will undergo galvanic corrosion, the dissolution potentials of the most common metals and alloys must be compared.

Experience shows that galvanic corrosion only occurs when the two metals in contact have a difference in potential of at least 100 mV. The intensity of galvanic corrosion is not related to the difference in potential between the two metals.

According to the ranking of aluminium and its alloys on the potential scale (Table B.1.4), in virtually any assembly with other common metals, aluminium will be the anode of the resulting cell and hence likely to suffer galvanic corrosion, if the conditions are favourable.

3.5.2. Presence of an electrolyte
The contact area must be wetted by an aqueous liquid in order to ensure ionic conduction. Otherwise there will be no possibility of galvanic corrosion. This is the case of the tank of an electrical transformer, in which possible junctions between aluminium conductors and copper conductors, immersed in the oil bath of the transformer, are not a source of galvanic corrosion of aluminium.

Galvanic corrosion between metals may also occur in certain ionic media such as liquid fluorine, liquid ammonia [3], and concentrated nitric acid.

3.5.3. Electrical continuity between the metals
Electrical continuity between the metals may be achieved either by direct contact between the two metals or by a connection such as bolts.

3.6.   GALVANIC CORROSION PARAMETERS

Since galvanic corrosion works like a battery, Faraday's law applies. In theory, this makes it possible to calculate the quantity of metal that is consumed at the anode:

$$m = \frac{1}{96\,500}\frac{A}{n}It$$

wherein

> m is the mass,
> A is the metal's atomic mass (27 in the case of aluminium),
> n is the valency (3 in the case of aluminium),
> I is the intensity (in amperes) of the coupling current (which can be measured by classic electrochemical methods),
> and t is the time in seconds.

The intensity of the coupling current, and thus the amount of galvanic corrosion, depends on several factors [4], including the nature of the electrolyte, polarisation phenomena, the ratio between the surfaces of the two metals, their distance, and temperature.

3.6.1.   *The nature of the electrolyte*

The more conductive the electrolyte the greater the galvanic corrosion. It will thus be more intense in brine and in seawater whose resistance is in the order of several ohms per centimetre, than in freshwater or rainwater whose resistance attains 2000–3000 Ω, depending on the origin of the waters. In exposure to weathering, galvanic corrosion will be greater at the seaside than in the countryside, where its effect is very limited (Figure C.4.1).

3.6.2.   *Polarisation phenomena*

Galvanic corrosion depends on both anode and cathode reactions. It can be slowed down by the polarisation of the anode or cathode surfaces. Polarisation is caused in particular by the accumulation of insoluble corrosion products on the anode surface. This is the case with alumina.

On bolted aluminium–steel assemblies with unprotected contacts, an important accumulation of alumina is observed in the contact area after a few years of immersion in seawater. This accumulation is capable of deforming the aluminium sheets. This "cataplasm" may stop galvanic corrosion, at the price of a certain deformation of the aluminium sheets.

It is obviously not appropriate to rely on possible anodic or cathodic polarisation phenomena in order to limit the damage caused by galvanic corrosion. Prevention is necessary at the assembly design stage.

Among parts recovered from the wreck of the *Titanic*, a megaphone in the form of a sawn-off cone was found, crimped at each end with a wire of—probably originally galvanised—steel. After 70 years in seawater at great depth, only part of the aluminium sheet in contact with the wire had been consumed.

3.6.3. *The surface ratio between the two metals*

The current density at the anode, which governs the dissolution rate of the metal, depends on the ratio [5]:

$$K = \frac{\text{cathodic surface area}}{\text{anodic surface area}}$$

The most favourable case is a very large anodic surface and a small cathodic surface, for example, an aluminium sheet of large surface area mounted with a few bolts made of stainless steel. Experience shows that if this assembly is immersed, severe galvanic corrosion will occur in the contact area with the stainless steel bolts.

In practice this parameter serves no purpose because galvanic corrosion always develops next to the zone of heterogeneous contacts.

3.6.4. *The distance*

Galvanic corrosion is local corrosion. It is limited to the contact zone. The intensity of corrosion decreases rapidly with increasing distance, even by a few centimetres, from the point of contact between the two metals. This decrease is greater when the electrolyte is a poor conductor.

It is for electrical reasons that this type of corrosion is so localised. Electrical current flows according to a path as linear as possible. Since galvanic corrosion often tends to develop in depth, it is not uncommon that galvanic corrosion perforates pieces several millimetres thick.

3.6.5. *Temperature*

An increase in temperature may modify the dissolution potentials and accelerate galvanic corrosion.

3.7. PRACTICAL ASPECTS OF GALVANIC CORROSION

Two very important factors need to be taken into account when analysing the risk of galvanic corrosion of aluminium:

– whether the heterogeneous assembly is immersed,
– the nature of the metal in contact with aluminium.

3.7.1. Submerged (or embedded) structures

Some structures are permanently immersed in water or seawater such as a hull in aluminium with water points made in stainless steel, heat exchanger circuits in copper–nickel, scraper bridges in settling basins of water-treatment plants, etc.

Galvanic corrosion also develops at unprotected heterogeneous contacts that are embedded in the soil (see Chapter G.1).

Zones that permanently retain humidity such as low points, basins, i.e. the hidden recesses where wet dust or mud accumulates, are sites where galvanic corrosion may develop. This happens especially in hidden recesses of vehicles at bolted or riveted assemblies between steel and aluminium where mud projected from the road may accumulate.

All galvanic corrosion couples will be effective, including couples with stainless steel. Whatever the metal in contact with aluminium, it will be necessary to neutralise the coupling in order to avoid galvanic corrosion, which otherwise will be unavoidable.

As mentioned above, in the case of an assembly with steel, the accumulation of corrosion products (alumina) may slow down or even stop galvanic corrosion. However, sometimes this happens only after serious damage to the aluminium parts has already occurred.

3.7.2. Emerged structures

Emerged assemblies only occasionally get wet by rain and bad weather, condensation, etc., and are not located in a zone that retains humidity. This is the most frequent case in the building sector, in civil engineering, industrial vehicles, on board vessels, in marina equipment, etc.

Several aspects need to be taken into consideration:

–   Galvanic corrosion is irregular, depending on the weather conditions, because moisture is needed in order to keep it going.
–   Galvanic corrosion occurs only in the vicinity of the contact areas and requires humidity. Experience shows that severe galvanic corrosion may develop on aluminium cladding panels mounted on steel rails without intermediate insulation; condensation or retention of humidity may occur in the confined space between the wall and the cladding panel.
–   Galvanic corrosion in bad weather conditions has only a low intensity, if it develops at all, due to the very low conductivity of rainwater. In the maritime environment, galvanic corrosion is more intense, as can be seen in Figure C.4.1.
–   The influence of the nature of the metal in contact with aluminium, which will be discussed below.

Landing stages of marinas are assembled with bolts made in ordinary or stainless steel. The inspection of this equipment shows that no galvanic corrosion can be detected with the naked eye at heterogeneous contacts located several centimetres above the water level and exposed to permanent humidity. Many other similar examples are known.

3.8. THE INFLUENCE OF THE TYPE OF METAL IN CONTACT WITH ALUMINIUM IN EMERGED ASSEMBLIES

The relative position of the two metals or alloys on the scale of dissolution potentials (Table B.1.4) only indicates the *possibility* of galvanic coupling when the difference in potential is sufficiently high. It says no more than that, and especially nothing about the rate (or intensity) of galvanic corrosion, which may be zero or insignificant, or even undetectable. Its intensity depends on the type of the metals involved [6, 7].

3.8.1. Unalloyed steel

In mechanical applications, aluminium parts exposed to bad weather may be joined with screws made in ordinary steel. Experience shows that aluminium suffers only very superficial corrosion in the contact areas with steel screws. Rust discolouration, which has no effect on aluminium, deeply impregnates the alumina layer and stains the surface.

In fact, contact with unprotected steel will affect the general appearance and esthetical attractiveness of a structure made in aluminium alloy more than its ability to resist corrosion.

Several explanations can be put forward: the formation of a film with corrosion products on the contact surfaces, i.e. rust on the steel and alumina on aluminium, will slow down electrochemical reactions.

3.8.2. Galvanised or cadmium-plated steel

On the scale of potentials, zinc is more electronegative than aluminium, while cadmium has a potential very close to that of aluminium. Galvanised or cadmium-coated steel fasteners can, therefore, be used to join and assemble structures made from aluminium alloys. It should just be remembered that when these coatings become too worn to protect the steel and the aluminium, the previous scenario applies in which there is contact between the aluminium alloy and bare steel. Chromium-plated steel does not lead to galvanic corrosion with aluminium, as long as chromium covers the nickel underlayer; however, the contact between aluminium and the nickel underlayer would lead to galvanic corrosion of aluminium.

3.8.3. Stainless steel

Although there is a great difference in potential between stainless steel and aluminium alloys, in the order of 650 mV, it is unusual to see galvanic corrosion on aluminium in contact with stainless steel. Aluminium alloy components are often assembled using stainless steel bolts, e.g. in metallic fittings.

3.8.4. *Copper and copper alloys*

While contact between aluminium alloys and copper and cuprous alloys (bronze, brass) causes no appreciable galvanic corrosion of aluminium under atmospheric conditions, it is nevertheless advisable to provide insulation between the two metals to localise surface corrosion of the aluminium.

It should be recalled that the corrosion product of copper, verdigris, attacks aluminium and may be reduced under the formation of small copper particles. These particles in turn cause localised pitting corrosion of aluminium.

3.8.5. *Mercury*

Like copper salts, mercury salts lead to severe pitting corrosion of aluminium.

Due to its volatility, mercury can easily be transported by moving fluids such as natural gas that contains minute traces [8]. Experience shows that mercury may be concentrated in plants where natural gas undergoes liquefaction and regasification; it may damage aluminium heat exchangers [9]. For this reason, mercury is trapped in specific devices.

Mercury itself leads to severe corrosion of aluminium, which appears as very narrow white lines, possibly thicker than 1 cm. It may also lead to intercrystalline corrosion and rupture at cracks [10].

The mechanism of the attack of aluminium by mercury is rather complex. A spontaneous reaction between the mercury film, aluminium, humidity and oxygen from air occurs (Figure B.3.2) [11]. While mercury is insoluble in aluminium, aluminium is slightly soluble in mercury (0.002% at room temperature). When mercury is wetting the aluminium surface, it keeps it activated, because no oxide layer can form. Aluminium will dissolve in the mercury and become oxidised in contact with air.

Eventually, the liquid mercury will transport aluminium to the outside, where it becomes oxidised in contact with air and humidity. There is no consumption of mercury during this reaction, which, once started, will never stop [12].

Decontamination of aluminium is a difficult task [13]. It is, therefore, not advisable under any circumstances to carry mercury on an aircraft or a vessel. More generally, mercury should never be used in the presence of aluminium.

3.8.6. *Other metals*

Only lead and tin will be discussed here. Contact with these metals should be avoided, because in humid media, they may lead to severe galvanic corrosion of aluminium.

Red lead-based paints or paints containing mercury salts (which are now forbidden by regulations) or copper salts should never be used on aluminium structures.

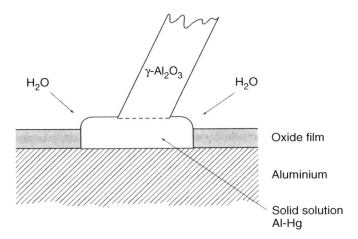

Figure B.3.2. Action of mercury on aluminium [11].

3.8.7. Graphite and graphitic products (seals, grease, etc.)

Although not a metal, graphite is an electrical conductor and is porous. With aluminium, it forms a battery that works quite well! In a humid environment, the contact with graphite results in severe galvanic corrosion of aluminium. That is why it is recommended not to use natural or synthetic rubber joints loaded with graphite. Graphitic greases should also be avoided on aluminium and its alloys.

3.8.8. Carbon fibre composites

The contact between carbon-fibre reinforced plastics (CFRP) or graphite epoxy composite materials (GECM) and aluminium leads to galvanic corrosion of aluminium. Testing with salt sprays or natural atmospheres shows severe attack in the contact zone [14]. The local change in pH by release of OH^- ions during galvanic corrosion of aluminium may damage certain composite materials [15].

3.9. CONTACT BETWEEN ALUMINIUM ALLOYS

The dissolution potentials of alloys of the 5000 and 6000 series as well as those of magnesium and silicon-containing casting alloys of the series 40000 and 50000 are very close to each other, and very similar to that of unalloyed aluminium of the 1000 series (Table B.1.4). Therefore, there is no risk of galvanic corrosion between these materials.

In aluminium ship building, planking and bulkheads are always in alloys of the 5000 series (5083, 5086, 5754, etc.), while stiffeners, handrails and floors are in extruded profiles in 6005A, 6082, etc., welded with 4043A or 5356. Such welded assemblies (possibly with A-S7G (42100)) are also widely used in coastal equipment: landing stages of marinas, buoys, etc. The filler metal, if selected according to the principles of good engineering practice, is not a contributory factor to galvanic corrosion or any other type of corrosion.

However, galvanic corrosion may occur between alloys of the 1000 and 2000 series, as well as between 7072 and alloys of the 3000 series. The difference in potential between these alloys can be used for cathodic protection by cladding layers (see Section B.1.5): 2024 may thus be protected by 1050A, and 3003 by 7072.

3.10. PREDICTING GALVANIC CORROSION

Predicting galvanic corrosion is a serious problem. For contacts between common metals, in particular between ordinary steel and stainless steel, experience shows that laboratory testing always leads to more severe results than what is actually observed under conditions of weathering.

Most laboratory tests, including electrochemical tests, say nothing about the intensity of possible galvanic corrosion of aluminium in contact with other metals.

In practice, for constructions exposed to bad weather, laboratory tests can predict a possible risk of galvanic corrosion and determine the efficiency of the planned protection.

However, for common situations of heterogeneous contacts, the risk of galvanic corrosion can only be assessed by professional experience. Over the years, prescriptions concerning the protection of heterogeneous contacts have been progressively adapted on the basis of accumulated experience.

Of course, this is not an issue for immersed assemblies that must be protected in any case.

Several methods are available for the measurement of galvanic corrosion between aluminium and other metals or alloys.

3.10.1. Dissolution potentials

If the dissolution potential of an alloy cannot be found in numerical tables, or when a measurement in a specific medium is required, one simply selects the most appropriate reference electrode (in general a saturated calomel electrode) and carries out the measurement (see Section B.1.5).

The comparison of potentials yields nothing more than their difference. It says nothing about the intensity of possible galvanic corrosion, which is not necessarily proportional to this difference.

3.10.2. Polarisation curves

Polarisation curves can be used to determine the intensity of possible galvanic corrosion. However, for practical reasons, these measurements are carried out in conductive electrolytes. They, therefore, greatly overestimate the intensity of possible galvanic corrosion. Experience has shown that all galvanic couples tested by this method work under experimental conditions, but not necessarily in practice.

3.10.3.   *Manufacture of assemblies and salt spray or immersion testing*

The advantage of salt spray or immersion tests is that they test rather realistic assemblies that are close to that of an actual structure, for example, profiles or sheet joined with bolts in stainless steel.

They show possible risk of galvanic corrosion and make it possible to study how to protect aluminium.

However, it should be noted that heterogeneous assemblies between aluminium and other metals exposed to salt sprays would always exhibit galvanic corrosion and generally very severe corrosion. This intensity is totally unrelated to what is observed during long-term exposure in outdoor corrosion testing stations, even in coastal areas or after an extended time of service.

In fact, under salt-spray conditions, the galvanic coupling with stainless steel works very well, while no visible galvanic corrosion is observed on contacts between stainless steel and aluminium under service conditions, even in coastal areas.

3.10.4.   *Preparation of assemblies and testing in natural atmospheres*

There are several possibilities for preparing and testing assemblies: prototype assemblies can be produced and exposed to natural atmosphere, or test pieces consisting of threaded bars made in steel, copper, etc. can be provided, around which aluminium wire, weighed before and after the test, is coiled (wire-on-bolt test) [16].

The interpretation of the results can be tricky, because if the test lasts too long, galvanic corrosion is slowed down due to the accumulation of corrosion products between the spirals of the wire. The duration of these trials should be rather short and not exceed a few months. Taking into account this short duration, the season in which the test is started does indeed matter.

3.11.   PREVENTION OF GALVANIC CORROSION

Galvanic corrosion can be prevented in several ways.

3.11.1.   *Insulation of the two metals present*

One practical way of preventing possible galvanic corrosion is to insulate the two metals in contact from one another as carefully as possible [17]. This is achieved by placing a high ohmic resistance, i.e. an insulating material (Figure B.3.3) such as neoprene or any other suitable polymer, between them.

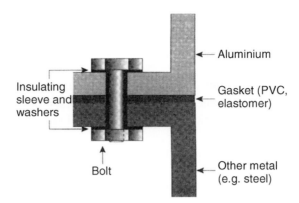

Figure B.3.3. Insulation between aluminium and another metal.

The electrical resistance, measured on a dry assembly, should be as high as possible, typically several hundred thousand ohms. In order to achieve this, the bolts, whatever their material (steel, stainless steel or aluminium alloy), must be carefully insulated from the two metals in presence, by using insulating sleeves in the bores and insulating washers under the bolt heads and nuts.

Such a luxury insulation is necessary only if the heterogeneous assembly is immersed (or located in an area that will retain water), and provided that safety rules do not oblige the use of a common grounding. In this case, the common grounding will transform the immersed aluminium structures into sacrificial anodes for the cathodic protection of steel constructions.

The insulation of flanges in aluminium and steel in a circuit can be achieved by using an elastomer joint maintaining a gap between the metals that is sufficient to avoid galvanic corrosion.

It should be recalled that rubber charged with carbon or graphite may lead to severe galvanic corrosion by its simple presence as a charge.

Assemblies between steel and aluminium can be simplified by using transition joints that are weldable on both sides (Figure B.3.4) [18]. Their use has become widespread in shipbuilding for welding joints between steel and aluminium (such as aluminium superstructures on a steel deck) and could be used also in the transportation sector [19].

3.11.2. *Painting of the surfaces in contact*

The surfaces in contact can be painted with effective results if the paint resists well. The cathodic surface needs to be painted preferentially.

Anodising does provides no effective protection against galvanic corrosion.

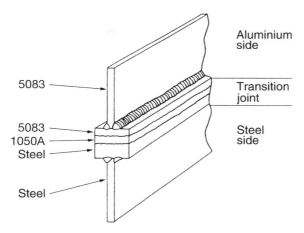

Figure B.3.4. Transition joint between steel and aluminium.

3.11.3. *Use of aluminium bolts*

An elegant way to prevent galvanic corrosion, and one which may turn out to be economical, is to use bolts in an aluminium alloy that is adapted to the service conditions (level of mechanical properties, adequate protections) when assembling structures made of aluminium alloys. This eliminates any need for insulation.

3.11.4. *Neutralisation of galvanic coupling*

This method is widely used in ship building for the neutralisation of galvanic coupling with a sacrificial anode, generally made in zinc or in a specific aluminium alloy (Figure B.5.4). In order for this protection to be effective, all metallic parts must be at the same potential and the anodes must not be painted; they should all be inspected regularly to check whether they are in perfect working order and to replace them as necessary. In seawater, such anodes are effective up to a distance of 10 m.

This type of protection is effective only for constructions that are simple in shape, such as hull structures, and in a very conductive medium, such as seawater. Satisfactory results cannot be expected in freshwater.

Remark

Metallic coatings deposited by electrolytic processes on aluminium such as nickel can provoke galvanic corrosion of the underlying aluminium. These products should not be exposed to aggressive environments.

On alloys of the 2000 and 7000 series, heterogeneous contacts may lead to inter-crystalline corrosion or exfoliation corrosion, even when exposed to bad weather.

REFERENCES

[1] Trethewey K.R., Sargeant D.A., The galvanic effect: a continuing corrosion problem, *Metals and Materials*, vol. 8, 1992, p. 378–382.

[2] Vargel C., La corrosion des bateaux en aluminium, *Loisirs Nautiques*, nos 241, 1991, p. 16–24.

[3] Toy S.M., English W.D., Crane W.E., Studies of galvanic corrosion couples in liquid fluorine, *Corrosion*, vol. 24, 1968, p. 418–421.

[4] Oldfield J.W., *Electrochemical theory of galvanic corrosion*, ASTM, STP 978, 1988, p. 5–22.

[5] Mansfeld F., Henkel J.V., Galvanic corrosion of aluminum alloys. III. Effect of area ratio, *Corrosion Science*, vol. 15, 1975, p. 239–250.

[6] Godard H.P., Galvanic corrosion of aluminium in the atmosphere, *Materials Protection*, 1963, p. 38–47.

[7] Caruthers W.H., Aluminium and its alloys, *Process Industries Corrosion*, NACE, Houston, 1975, p. 201–206.

[8] Leeper J.E., Mercury corrosion in liquefied natural gas plants, *Energy Processing Canada*, vol. 73, 1981, p. 46–51.

[9] Bodle W.W., Attari A., Serauskas R., Considerations for mercury in LNG operations, *International Conference Liquefied Natural Gas*, session II, Chicago, 1980.

[10] English J.J., Kobrin G., Liquid mercury embrittlement of aluminum, *Materials Selection and Design*, 1989, p. 62–63.

[11] Pinel M.R., Bennett J.E., Voluminous oxidation of aluminium by continuous dissolution on a wetting mercury film, *Journal of Materials Science*, vol. 7, 1972, p. 1016–1026.

[12] Bennett J.E., Pinel M.R., Reactions between mercury wetted aluminium and water, *Journal of Materials Science*, vol. 8, 1973, p. 1189–1193.

[13] Allsopp H.J., *A chemical treatment for mercury accidentally spilled in aircraft*, RAF, technical report 77014 AD A O43160, 1977.

[14] Brown A.R.G., Corrosion of carbon fiber reinforced plastic to metal couple in saline water, *Conference Plastics and Polymers*, paper 35, 1974, p. 230–241.

[15] Boyd J., Chang G., Webb W., Speak S., Gerth D., Reck B., Galvanic Corrosion Effects on Carbon Fiber Composites, *36th International SAMPE Symposium*, 1, vol. 36, San Diego, 1991, p. 1217–1231.

[16] Compton K.G., Mendizza A., *Galvanic couple corrosion studies by means of the threated bolt and wire test*, ASTM, STP 175, 1956, p. 116–125.

[17] Rowe L.C., *Preventing galvanic corrosion*, Society of Automotive Engineers, 1974, p. 40–45.

[18] Transitions spacers developed for dissimilar metal joints to minimize corrosion and simplify joining, *Welding Journal*, vol. 56, 1977, p. 51–53.

[19] Baboian R., Bardner H., Reducing corrosion in aluminium–steel joints, *Automotive Engineering*, vol. 102, 1994, p. 103–105.

Chapter B.4

Testing Methods

Chapter B.4
Testing Methods

Corrosion resistance of a metal or alloy is a part of its use properties, to the same extent as its mechanical properties, fatigue resistance, etc. It, therefore, needs to be assessed.

However, corrosion is a complex phenomenon. Different types of corrosion can occur, which do not necessarily develop at the same time and under the same service conditions or in the same media. Corrosion resistance, therefore, needs to be evaluated with respect to several parameters concerning the alloy, the environment, the service conditions, etc.

Certain corrosion testing methods are common to several metals: determination of the mass loss or the volume of released hydrogen, salt spray testing, etc. The testing solutions, however, are in most cases specific to a given metal and a given form of corrosion. For example, stress corrosion testing of stainless steel is carried out at 154 °C in boiling solutions containing 44% magnesium chloride $MgCl_2$ (ASTM G 36). On aluminium alloys, this test is performed at room temperature in a brine solution containing 3% sodium chloride (ASTM G 103).

Corrosion is not an instantaneous phenomenon. It proceeds more or less rapidly, depending on the metal, the medium and the form of corrosion. In practice, a potential user cannot wait for real-time corrosion testing results in order to select a material for a given use such as construction. The same applies to a supplier who wishes to grant a warranty for the material offered for sale.

That is why corrosion experts use accelerated corrosion tests in their laboratories in order to greatly reduce this waiting period. Often, media rich in sodium chloride are used. However, while these tests should be sufficiently severe in order to accelerate the corrosion phenomena, they must also be discriminating, and therefore, their severity needs to be adapted accordingly.

4.1. TYPES OF TESTS

Three types of tests can be distinguished according to where they are carried out [1, 2]: in the laboratory, in outdoor exposure stations, or on prototypes.

4.1.1. Laboratory testing
Laboratory tests are generally done in synthetic media that are as aggressive as possible (while still being selective) in order to assess as quickly as possible (the time scale is in

the order of 1 day) the corrosion resistance of a metal or alloy. The testing protocols are normally very strict in order to avoid the risk of dispersion that is inherent to all short-term testing. Most often (though not exclusively), brine solutions containing between 3 and 5% NaCl are used.

4.1.2.   Testing in outdoor exposure stations
Outdoor tests aim at assessing the behaviour of metals and alloys in natural environments (as compared to artificial media used in laboratory tests). These are generally long-term tests, which can last for several years. It is not uncommon to find reports on test results carried out over 20 years and longer in outdoor exposure stations.

4.1.3.   Tests on prototypes in feed-back loops
These tests are carried out when it is not possible to rely on tests on small test pieces in contact with media that often are only loosely related to the real-life medium.

4.2.   GOAL OF CORROSION TESTING METHODS

For assessing the resistance to corrosion, corrosion experts can choose among several testing methods that have different goals.

4.2.1.   Routine testing
The goal is to test the quality of a coating, such as an anodising coating, or the susceptibility to a given form of corrosion. Laboratories rely on specific methods, most of them standardised, such as

–   Salt spray testing of coatings, in which the test pieces are exposed for a predetermined period of time: 100, 200, 400 h. These are typically reception tests which suppliers have to perform according to specifications.
–   The ASSET test (ASTM G 66) to detect the susceptibility of alloys of the 5000 series to exfoliation corrosion.

The corrosive media are usually only somewhat related to service conditions. This does not matter; what is required here is that the testing method be selective with respect to flawed manufacturing or a specific form of corrosion. It is, however, impossible to deduce the lifetime of the product from these tests. This debate has been going on for a long time, and there is no answer to this question [3].

As an example, take the EXCO test used to assess susceptibility to exfoliation corrosion of alloys of the 2000 and 7000 series. There is obviously no relationship between a solution of pH 0.4 that contains

–   sodium chloride: 4.0 M
–   potassium nitrate: 0.5 M
–   nitric acid: 0.1 M

on the one hand, and a natural atmosphere, even the most aggressive one, at the sea shore, on the other hand. The same question arises with stress corrosion testing of high-resistance alloys of the 2000 and 7000 series.

Many comparative studies have been carried out in order to correlate laboratory testing based on specific media with testing in natural atmospheres in outdoor testing stations [4–7]. This can validate accelerated testing methods and ensure that stress corrosion testing methods are sufficiently severe in order to be both selective and discriminating.

4.2.2. Assessment of compatibility

Compatibility assessment tests aim to evaluate the resistance of a metal or alloy in a given medium such as hypochlorite solution or phenol solution or any other chemical product or media of more complex composition. The method, long in use, is quite simple: samples are immersed in the relevant solutions for a defined period of time (hours, days), depending on the medium's aggressiveness.

The type of corrosion, if there is corrosion, is detected visually. If the corrosion is uniform, the dissolution rate is determined by measuring the mass loss. This kind of test has evaluated the resistance of aluminium in many media, and the reported results are summarised in Parts E and F.

4.2.3. Performance of an alloy in a given medium

Performance tests try to answer users' classic questions: "What will the lifetime of my roof be? of my heat exchanger ?", etc. These tests are generally performed under conditions as close as possible to reality:

–   *in outdoor corrosion testing stations* for assessing resistance to atmospheric corrosion;
–   *in closed loops or prototypes*, when more complex processes are tested. The behaviour of aluminium in seawater desalination plants has been the subject of many studies using small prototypes operating under the same conditions as industrial desalination plants. The same approach has been used for solar energy conversion systems. Corrosion resistance testing of aluminium alloys in motor engine cooling liquids are also carried out in closed loops;

–   *on equipment in service*, conditions where it is common to replace a component with a component made in the alloy to be tested. This can be carried out with vehicles [8], in water treatment plants, chemical reactors, water pipes [9], etc. This method has the advantage of taking into account real-life service conditions, but only if the tested components are used under representative and well-defined conditions. For example, how can results obtained on vehicles operating in a region where roads are snow-covered in winter (and therefore, salted) be transposed to those operating in tropical zones?

In order to ensure that the results of these tests (which can be very expensive) can actually be exploited, they have to be calibrated with respect to industrial use conditions. Certain media such as seawater are notoriously difficult to reproduce under laboratory conditions [10]. Whenever possible, natural seawater should be preferred for determining the resistance to corrosion in seawater.

4.2.4. Characterisation of new products

When modifying the composition of an alloy or its heat treatment conditions (this is especially important for age-hardenable alloys), it may be necessary to study the influence of these modifications on corrosion resistance. These tests must be validated by comparison with a reference sample, which is usually the product that is to be replaced.

4.2.5. Corrosion mechanisms

Corrosion mechanism tests are most often fundamental studies based on specific testing procedures and specific investigation techniques. This topic is beyond the scope of the present book.

4.3. EXPRESSION OF RESULTS

Corrosion testing procedures yield qualitative and quantitative results that can assess the resistance to corrosion and possibly establish a classification.

4.3.1. Qualitative results

Qualitative results are visual observations made with the naked eye or under the optical microscope: general appearance, type of corrosion, etc. For certain methods, standardised test methods include standard images for the classification of samples, as shown in Figure B.4.1.

Figure B.4.1. Standard images for the ASSET test (according to ASTM G66): degree of exfoliation corrosion.

4.3.2. *Quantitative results*

■ **Volume of released hydrogen**

Hydrogen release measured during a test in a reagent for selective attack (Figure B.4.2) corresponds to the anodic dissolution reaction of aluminium.

$$Al \rightarrow Al^{3+} + 3e^-$$

$$3H^+ + 3e^- \rightarrow \tfrac{3}{2} H_2$$

This method is valid only for uniform corrosion.

■ **Mass loss**

Mass loss is the oldest test method for corrosion. Test pieces are weighed before and after testing. Like the volume of released hydrogen, these results are significant only for uniform corrosion.

■ **Changes in mechanical properties**

The mechanical properties are measured on machined specimens in reference sheet and in sheet after exposure to a corrosion test. This method has long been used for testing in

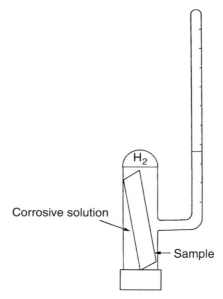

Figure B.4.2. Measurement of hydrogen release [16].

outdoor testing stations. It gives a fair indication of the intensity of corrosion with medium-gauged sheet, up to a thickness of a few millimetres. For a given mean pitting depth, the relative drop in mechanical properties decreases with increasing thickness.

■ **Depth of pitting**

For pitting corrosion, pitting depth is the most significant measurement; the results can be analysed by statistical methods (see Section B.2.2).

4.4. STANDARDISATION OF CORROSION TESTING METHODS

Corrosion testing methods are like systems of units: comparable testing results from different laboratories can be obtained only on the basis of common, standardised testing methods. The first systematic tests of atmospheric corrosion of steel were launched in the United States by ASTM [11].

The first ASTM standard on salt spray testing goes back to the 1930s.

In the early 1920s, when corrosion studies of common metals became important, working groups of corrosion experts and metallurgists from several aluminium-producing companies were set up in order to develop corrosion testing methods common to all laboratories. Most of this standardisation work on atmospheric corrosion testing was carried out in the United States under the umbrella of the ASTM [12].

Standardisation may focus on the testing equipment, such as chambers for salt spray testing, and on the testing protocol. In order to limit the dispersion of the results, the testing conditions define as precisely as possible the sample preparation, the chemicals, the test duration, the expression of results, etc.

The development of a testing method often requires a considerable amount of work done in several laboratories, until the test truly reflects the susceptibility to a given form of corrosion such as exfoliation corrosion.

The most common standards for aluminium are given in Tables B.4.1 and B.4.2.

4.5. ELECTROCHEMICAL TEST METHODS

Electrochemical methods are popular because they can be carried out in a few hours. They are based on the measurement of the current i as a function of the voltage applied to the test piece [13]. The polarisation curve in an aqueous medium is the sum of two electrochemical reactions (Figure B.4.3) [14]:

– on the cathodic curve, reduction of H^+ ions:

$$3H^+ + 3e^- \rightarrow \tfrac{3}{2}H_2$$

– on the anodic curve, oxidation of aluminium:

$$Al \rightarrow Al^{3+} + 3e^-$$

In fact, there is little advantage to plot these curves completely, because only the area around the corrosion potential (the dissolution potential) is useful for determining the corrosion current i_{corr}.

The corrosion current can be measured in two ways:

– by the Tafel slope,
– by the polarisation resistance.

4.5.1. Tafel slopes

Tafel slopes are determined from the tangents of anodic and cathodic curves in the vicinity of the corrosion potential on a semi-logarithmic plot $E = \log i$ (Figure B.4.4) [15]. In the case of aluminium, the anodic polarisation curve has no linear portion, and the cathodic

Table B.4.1. Common standards for corrosion testing of aluminium

Type of corrosion	Standard	Title of the standard	Subject matter	Medium
Immersion in a reagent	ASTM G 31	Standard practice for laboratory immersion corrosion testing of metals	Conditions of immersion testing in a reagent	
Pitting corrosion	ASTM G 46	Standard guide for examination and evaluation of pitting corrosion	Assessment of pitting corrosion (form, density, depth)	
Intercrystalline corrosion	ASTM G 67	Standard test method for determining the susceptibility to intergranular corrosion of 5XXX series aluminium alloys by mass loss after exposure to nitric acid	Determination of the susceptibility to intercrystalline corrosion of 5000 series alloys	Concentrated HNO_3
	ASTM G 110	Standard practice for evaluating intergranular corrosion resistance of heat treatable aluminium alloys by immersion in sodium chloride + hydrogen peroxide solution	Intercrystalline corrosion of alloys of the 2000 and 7000 series	$NaCl$ 57 g$\cdot$l^{-1}, H_2O_2 10 ml of a 30% solution
Exfoliation corrosion	ASTM G 112	Standard guide for conducting exfoliation corrosion tests in aluminium alloys	Testing corrosion of exfoliation corrosion of aluminium alloys	
	ASTM G 34	Standard test method for exfoliation corrosion susceptibility in 2XXX and 7XXX series aluminium alloys (EXCO test)	Determination of susceptibility to exfoliation corrosion of alloys of the 2000 and 7000 series alloys (EXCO)	$NaCl$ 4.0 M, KNO_3 0.5 M, HNO_3 0.1 M
	ASTM G 66	Standard test method for visual assessment of exfoliation corrosion susceptibility of 5XXX series aluminium alloys (ASSET test)	Exfoliation corrosion of 5000 alloys	NH_4Cl 1.0 M, NH_4NO_3 0.25 M, $(NH_4)_2C_4H_2O$ 0.01 M, H_2O_2 0.09 M
Stress corrosion	ASTM G 49	Standard practice for preparation and use of direct tension stress-corrosion test specimens	Preparation of test specimens for stress corrosion testing	
	ASTM G 58	Standard practice for preparation of stress-corrosion test specimens for weldments	Preparation of test pieces of welded metal for stress corrosion testing	

ASTM G 103	Standard test method for performing a stress corrosion cracking test of low copper containing Al–Zn–Mg alloys in boiling 6% sodium chloride solution	Conditions for stress corrosion testing of alloys of the 7000 family with less than 0.25% copper	NaCl 6%
ASTM G 44	Standard practice for alternate immersion stress-corrosion testing in 3–5% sodium chloride solutions.	Stress corrosion test with alternate immersion–emersion	NaCl 4.5%
Stress corrosion ASTM G 47	Standard test method for determining susceptibility to stress-corrosion cracking of high-strength aluminium alloy products.	Conditions of stress corrosion testing for alloys of the series 2000 (1.8–8% copper) and 7000 (0.4–2.8% of copper)	NaCl 3.5%
ASTM G 64	Standard classification of resistance to stress-corrosion cracking of heat-treatable aluminium alloys	Stress corrosion susceptibility index for alloys of the 2000, 6000 and 7000 series	
Filiform corrosion ASTM D 2803	Standard guide for testing filiform corrosion resistance of organic coatings on metal	Determination of the susceptibility to filiform corrosion	
Measurement of dissolution potentials ASTM G 69	Standard practice for measurement of corrosion potentials of aluminium alloys		NaCl: 58.5 g·l^{-1}, H$_2$O$_2$: 9 ml of a solution at 30%
Cavitation erosion ASTM D 2809	Standard test method for cavitation erosion characteristics of aluminium pumps with engine coolants	Resistance to erosion and cavitation	Engine cooling liquids
Atmospheric corrosion ASTM G 50	Standard practice for conducting atmospheric corrosion tests on metals	Conditions for exposure of test pieces	Natural atmospheres
Galvanic corrosion with atmospheres ASTM G 116	Standard practice for conducting wire-on-bolt test for atmospheric galvanic corrosion		Natural atmospheres

EXCO: EXfoliation COrrosion; ASSET: Ammonium-Salt Solution Exfoliation Test.

Table B.4.2. Common standards for corrosion testing test equipment and various standards

Type of corrosion	Standard	Title of the standard	Subject matter
Immersion into a reagent	ASTM G 31	Standard practice for laboratory immersion corrosion testing of metals	Conditions of immersion testing in a reagent
	ASTM G 1	Recommended practice for preparing, cleaning and evaluating corrosion test specimens	Preparation of test specimens for corrosion testing and evaluation of results
Salt spray	ASTM B 117	Standard practice for operating salt spray (fog) apparatus	Description of equipment for salt spray testing and conditions of exposure
	ASTM G 85	Standard practice for modified salt-spray (fog) testing	Special acetic salt spray test
	ASTM B 368	Standard method for copper–acetic acid–salt spray (fog) testing (CASS test)	Testing conditions for copper–acetic spray
	ISO 9227	Corrosion testing in artificial atmospheres	Salt spray testing
	ASTM B 680	Standard test method for seal quality of anodic coatings on aluminium by acid dissolution	Measurement of the thickness of anodic coatings
Cleaning solutions	ASTM D 930	Standard test method of total immersion corrosion test of water-soluble aluminium cleaners	Measurement of the aggressiveness of cleaning solutions
	NF A 91-451	Anodised aluminium and aluminium alloys—Qualification, of cleaning products	Resistance to cleaning products
Artificial seawater	ASTM D 1141	Standard specification for substitute ocean water	Artificial seawater
Solar energy	ASTM E 712	Standard practice for laboratory screening of metallic containment materials for use with liquid in solar heating and cooling systems	Resistance of metal in thermal solar energy systems
English glossary of terms relating to corrosion	ASTM G 15	Standard definitions of terms relating to corrosion and corrosion testing	Definition of English terms relating to the corrosion of metals

CASS test: Copper Accelerated acetic–Salt Spray.

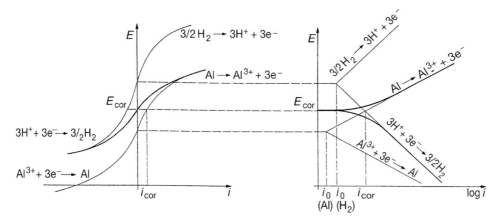

Figure B.4.3. Polarisation curves of aluminium [16].

part must be used, with the condition that a reduction of oxygen does not occur:

$$\tfrac{1}{2}O_2 + H_2O + 2e^- \rightarrow 2OH^-$$

In order to avoid this reaction, these curves are determined in a solution that is deaerated by bubbling with an inert gas, most often hydrogen [16].

4.5.2. Polarisation resistance

Electrochemists have shown that the polarisation curves $E = f(i)$ are linear in the vicinity of the corrosion potential, within a zone of about 10 mV in the anodic and cathodic directions, and that their slope is inversely proportional to the corrosion rate and bears the dimension of a resistance (Figure B.4.5) [17].

This method can be used if the potential is stable and if the polarisation curve in the vicinity of the polarisation potential is linear, which is not always the case.

4.5.3. Limits of electrochemical methods

Electrochemical methods have been very fashionable in many laboratories since the 1950s, including for studying the corrosion of aluminium. However, theoretical and practical reasons considerably limit the usefulness of these methods for aluminium. (The following remarks apply only to aluminium).

■ Theoretical reasons

The corrosion current is related to mass loss through Faraday's law (Figure B.4.6). It can be related to a decrease in thickness only if corrosion is uniform. On aluminium,

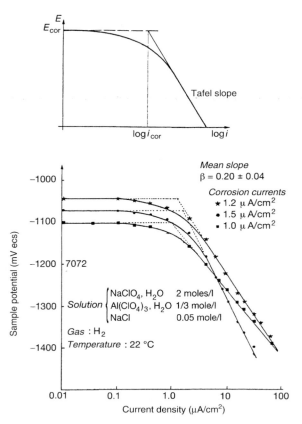

Figure B.4.4. Determination of the corrosion current from the Tafel slopes. Cathodic curve [16].

this type of corrosion only occurs in very acidic or very alkaline media (see Section B.2.1).

The most common type of corrosion on aluminium in media close to neutral, i.e. natural media, is pitting corrosion. In this case, the measurement of the corrosion current makes no sense, and it will not give any information on the type of corrosion.

The electrochemical behaviour of aluminium is strongly influenced by the permanent presence of a natural oxide film on its surface. Therefore, a mixed potential corresponding to the pitting potential is measured on aluminium (see Section B.1.7); this potential represents a threshold below which pitting corrosion can be prevented.

However, this pitting potential does not have the same meaning as in the case of steel, where initiation is followed by rapid propagation. On aluminium, the initiation of a pit may

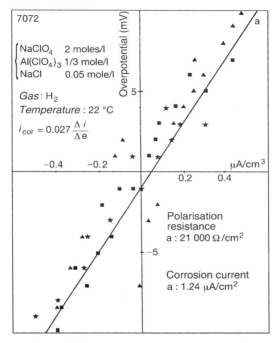

Figure B.4.5. Polarisation resistance of 7072 [16].

occur without propagation (see Section B.2.2). The pitting potential gives no information on the propagation of pitting.

The comparison of the dissolution potentials of aluminium alloys may reach absurdity, for example, leading to a preference for alloys of the 2000 series, which have a dissolution potential far less negative, about -650 mV, over those of the 5000 series, which have a more electronegative potential, on the order of -800 mV (Table B.1.3). And yet the latter show excellent corrosion resistance, while alloys of the 2000 series are highly susceptible to pitting corrosion in natural environments.

■ Practical reasons

The seductive advantage of electrochemical measurement is the rapidity and the capability of programming modern measurement equipment in such a way that the variations of potential or intensity occur at a very low rate, for example, to remain under operating conditions close to a certain reversibility. On the other hand, a corrosion current of 1 mA·cm^{-2} (which is a reasonable order of magnitude for aluminium, easily measured)

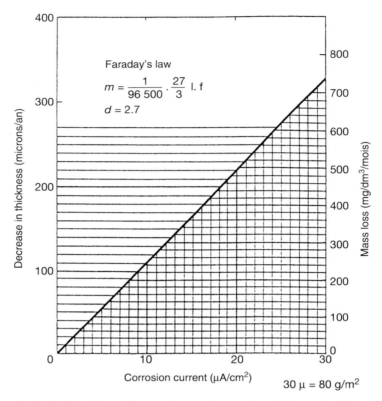

Figure B.4.6. Relation between the corrosion current, mass loss and decrease in thickness [16].

corresponds to a mass loss of 25 mg·dm^{-2} per month. This is the minimum duration for the test if good weighing accuracy is required.

Electrochemical tests, whatever the metal to be tested, must be carried out following very strict protocols that must be based on basic electrochemical considerations. The measurement of the current corresponding to secondary electrochemical reactions, especially oxygen reduction at the cathode, should be avoided. That is why controlling the medium is so important: aerated, deaerated, under nitrogen, under hydrogen, etc., all conditions that need to be specified when analysing the validity of the results.

While measuring currents is easy with modern equipment designed for plotting polarisation curves, the situation is different for measuring potentials. Often, it will be necessary to wait for several hours, or even several days, until the potential is reasonably stable, as shown in Figure B.4.7.

In fact, electrochemical testing methods can only be used under very well-controlled conditions and in very special cases of corrosion, or for fundamental studies. Because of

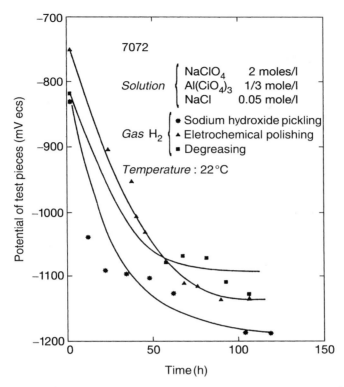

Figure B.4.7. The potential of an aluminium alloy as a function of time [16].

the peculiar electrochemical behaviour of aluminium, these are of little practical importance for experts in the field of aluminium corrosion.

REFERENCES

[1] Champion F.A., *Corrosion testing procedures*, Chapman & Hall, Londres, 1952.
[2] Thompson D.H., General test and principles, *Handbook on corrosion testing and evaluation*, Ailor W.H. (ed.), Wiley, New York, 1971, p. 115–141.
[3] Ketcham S.J., Jankowski E.J., *Developing an accelerated test: Problems of pitfalls*, ASTM, STP 866, 1985, p. 14–23.
[4] Sprowls D.O., Summerson T.J., Lofting F.E., *Exfoliation corrosion testing of 7075 and 7178 aluminium alloys*, ASTM, STP 558, 1974, p. 99–113.
[5] Sprowls D.O., Walsh J.D., Shumaker M.B., *Simplified exfolation testing of aluminum alloys*, ASTM, STP 516, 1972, p. 38–65.
[6] Humphries T.S., Nelson E.E., *Seacost stress-corrosion cracking of aluminium alloys*, NASA, technical memorandum 82393, January 1981.

[7] Lifka B.W., Corrosion resistance of aluminium alloys plate in rural, industrial and sea-cost atmospheres, *Aluminium*, vol. 63, 1987, p. 1256–1261.

[8] Haynes G.S., Baboian R., *Simulating automotive exposure for corrosion testing of trim material*, ASTM, STP 970, 1988, p. 18–26.

[9] Pathak B.R., *Testing in fresh waters. Handbook on corrosion testing and evaluation*, Wiley, New York, 1971, p. 553–574.

[10] Dexter S.C., *Laboratory solutions for studying corrosion of aluminium alloys in sea water*, ASTM, STP 970, 1988, p. 217–234.

[11] Coburn S.K., Atmospheric tests, *Handbook on Corrosion testing and evaluation*, Ailor W.H. (ed.), Wiley, New York, 1971.

[12] Borgmann C.W., Mears R.B., The principles of corrosion testing, *ASTM Symposium of Corrosion Testing*, 1937, p. 3–35.

[13] Liening E.L., Electrochemical corrosion testing techniques. The theory and the practice, *Process Industries Corrosion,* NACE, Houston, 1986, p. 85–122.

[14] Brown R.H., Mears R.B., The electrochemistry of corrosion, *Transactions of the Electrochemical Society*, vol. 74, 1938, p. 495–517.

[15] Stern M., Geary A.I., Electrochemical polarization. A theorical analysis of the shape of polarization curves, *Journal of the Electrochemical Society*, vol. 104, 1957, p. 56–63.

[16] Reboul M., Méthodes de mesure de la corrosion généralisée de l'aluminium et de ses alliages, *Revue de l'Aluminium*, no. 419, 1973, p. 399–412.

[17] Stern M., A method for determining corrosion rate from linear polarisation data, *Corrosion*, vol. 14, 1958, p. 60–64.

Chapter B.5

Protection Against Corrosion

Chapter B.5
Protection Against Corrosion

Corrosion experts proclaim, and they are right in doing so, that aluminium resists corrosion well. Why is it nevertheless necessary to protect it against corrosion? The answer to this question depends on the alloy, applications and use.

It is well known that copper-containing alloys of the 2000 and 7000 series are not sufficiently corrosion-resistant to be used without protection in humid media such as ambient atmosphere, and all the more in aggressive media such as marine environments.

Incidentally, it was for the protection of aircraft in Duralumin that chemists looked for surface treatments in the 1920s. Anodising, invented by Bengough and Stuart in 1923, was a response to this need.

On the other hand, strain-hardenable alloys of the 1000, 3000 and especially 5000 series, as well as age-hardenable alloys of the 6000 series have good corrosion resistance. They can, therefore, be used without protection in many environments such as maritime environments. Many vessels in aluminium are painted on the visible parts and for decorative purposes only. Many small crafts in aluminium are not painted at all. The same is true for landing stages of marinas, road sign supports, etc. Water staining and superficial pitting corrosion do not affect the solidity of the component.

For certain applications, no pitting corrosion or water staining is acceptable. In most applications in the construction industry, metallic fittings, metallic shutters, interiors, etc., appearance is essential. Aluminium, therefore, needs to be protected for durability of the surface aspect and against pitting corrosion. Therefore, most rolled and extruded semi-products for use in buildings are protected by anodising or lacquering.

But can the user rely on a coating in order to use aluminium in a very hostile environment? Is it possible, for example, to transport or to store highly aggressive chemicals in tanks protected by a specific coating? In most cases, the answer is clearly "no". One cannot rely on the resistance and durability of any coating (paint, polymer, etc.). The slightest defect in the coating would result in localised corrosion, which can reach very considerable proportions. Therefore, equipment in aluminium for the transportation and storage of chemicals almost never has an interior paint coating.

There are several ways to prevent corrosion of a metal:

- modifying its surface properties by application of boehmite coatings, chemical conversion coatings, anodising, or cladding;
- protecting it against the environment by a continuous coating: paint, lacquer;
- modifying the properties of the medium by inhibitors.

5.1. BOEHMITE COATINGS

Boehmite (γ-AlOOH) coatings are natural oxide coatings formed by the reaction with water at a temperature of at least 75 °C according to the reaction:

$$Al + 2H_2O \rightarrow AlOOH + \frac{3}{2}H_2$$

Boehmite formation starts at grain boundaries and then advances to the surface. The conditions of boehmite film growth in water or vapour have been studied by Altenpohl [1]. Boehmite coatings are grown by a prolonged contact, on the order of a few hours, with vapour of deionised water at a temperature between 110 and 160 °C, depending on the purity of the metal. The deionised water must have a high resistivity, above 0.4×10^6 Ω·cm, as well as a pH between 4 and 11. It must not contain more than 1 ppm of silica SiO_2. The presence of oxygen in the water has no influence on the boehmite film growth rate [2]. The thickness of the boehmite film reaches a maximum after a treatment time of 15 h, and will be higher in water vapour at 120 °C than in boiling water (Figure B.5.1).

If the temperature is too high, the formation of boehmite will be localised at grain boundaries and may degenerate to intercrystalline corrosion. The purer the metal, the lower the temperature limit. In practice, it is not recommended to exceed the following temperatures:

–   105 °C with 1199,
–   150 °C with 1080, and
–   180 °C with 1050A and aluminium alloys.

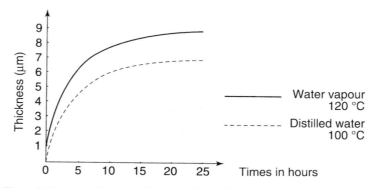

Figure B.5.1. Rate of boehmite film growth in boiling water and in water vapour [1].

Before growing a boehmite film, the surface should be cleaned by pickling at room temperature for 30–40 s in a bath of the following composition:

– nitric acid HNO_3: 25%
– sodium fluoride NaF: 4%
– deionised water: 71%,

followed by rinsing with distilled water.

The thickest boehmite coatings are obtained in alkaline media; these coatings present the best resistance to corrosion. However, they have a milky, white colour, while coatings obtained with water vapour are very clear.

Boehmite coatings between 1 and 2 µm thick (depending on the alloys) can be obtained after immersion for 8 h in a solution containing 3 g triethanolamine $(C_2H_5)_3N$. Ammoniac solutions at 80 °C at a concentration of $2 \ g \cdot l^{-1}$ are also a good medium for growing boehmite coatings.

Experience shows that aluminium may naturally develop a boehmite film when put in prolonged contact with weakly mineralised water (less than $1 \ g \cdot l^{-1}$ of salinity), and even with unpolluted seawater, at a temperature above 60 °C [3]. Coatings obtained in this way do not have exactly the same composition as boehmite, but when they form and when they are stable, they provide an excellent protection against corrosion in hot water; in fact, aluminium is very stable under these conditions. Of course, the temperature limit above which the metal becomes prone to intercrystalline corrosion (see Section B.2.3) must be respected.

Boehmite coatings have been known for many years, but have not been used as widely as one could have expected. This is mainly because of the long duration of the treatment and its cost. Moreover, solutions for growing boehmite coatings need to be regenerated frequently.

5.2. CHEMICAL CONVERSION TREATMENTS

Chemical conversion treatments form a thin, complex oxide layer which may reach 0.05–0.15 µm. The formation of these complex oxides may be obtained in acidic or alkaline phosphate or chromate baths [4]. As in the case of boehmite formation, the chromate treatment starts at grain boundaries [5].

These treatments are mainly used for forming a base layer for paints, lacquers and adhesives. They slightly increase the corrosion resistance of aluminium.

There are two types of chemical conversion treatments:

– chromate treatment,
– chromate–phosphate treatment.

5.2.1. Chromate treatment

The first chemical conversion coating process was developed by O. Bauer and O. Vogel in 1915. It consists of treating aluminium in a solution of potassium carbonate, sodium bicarbonate and potassium dichromate for 2 h at 90 °C. A dark grey layer is formed by this process.

A variant of this process was patented in 1923, known under the abbreviation MBV (modified Bauer–Vogel process). It led to many patents and modifications before 1940 [6]: EW, LW, Pylumin, Alrok, etc.

Aluminium is immersed for 5–10 min in a bath at 90–95 °C containing

- sodium carbonate Na_2CO_3: 50 $g \cdot l^{-1}$
- sodium dichromate $Na_2Cr_2O_7$ (or potassium dichromate $K_2Cr_2O_7$): 15 $g \cdot l^{-1}$.

Nowadays, these treatments are only rarely used, since a wide range of recently developed methods are faster and more efficient.

5.2.2. Chromate–phosphate treatment

Many methods have been developed since 1945. They are better known by their trademark names:

- Alodine by the American Paint Company [7],
- Bonderite by Continental Parker,
- Iridite by Allied Kelite Product [8].

These solutions comprise phosphates, fluorides and chromates [9].
The following methods can be distinguished [10]:

- *Conversion treatments based on phosphates of trivalent chromium*, which yield a deep emerald green layer. The baths comprise chromium anhydride, phosphoric acid and hydrofluoric acid. This type of conversion is widely used for products that are lacquered by the coil-coating process and used in the building industry. The weight of the layer ranges from 400 to 800 $mg \cdot m^{-2}$, and its thickness is roughly 400 nm for a deposition of 1 $g \cdot m^{-2}$. These layers combine good corrosion resistance with excellent adherence for lacquer.
- *Conversion treatments based on chromium chromate.* The colour of the layer depends on its thickness; it may be more or less yellow, even dark brown. These solutions contain chromic acid, hydrofluoric acid, and possibly potassium ferricyanide. Because of its excellent corrosion resistance with no paint, this type of conversion coating is used in the aerospace industry. The weight of the layer ranges from 250 to 400 $mg \cdot m^{-2}$, and its thickness is on the order of 600 nm for a deposition of 1 $g \cdot m^{-2}$.

Table B.5.1. Main anodising treatments

Alloy series	Typical treatment	Properties of the anodic coating	Typical application
1000	Sulphuric acid anodising (continuous or static)	Transparency, reflectivity, brilliance. The lower the Fe and Si content, the more transparent the layer	Decoration, optical reflectors
3000	Sulphuric acid anodising	Alloys easy to anodise. Protection, decoration, good corrosion resistance, resistance to scoring	Architecture, building, kitchen utensils
5000	Chromic acid anodising (continuous or static), colourless or electrolytically coloured	Durable aspect, corrosion protection. Most alloys easy to anodise	Architecture, building
2000	Chromic acid anodising, sealed with dichromate. Sulphuric acid anodising, sealed with dichromate	Surface preparation for adhesive bonding. Limited film thickness, more or less porous layer due to copper dissolution. Provides limited protection against corrosion	Aeronautical and mechanical applications
7000	Chromic acid anodising on 7075. Sulphuric acid anodising on 7020. Hard anodising on 7075, 7049A	Most alloys easy to anodise, copper-containing alloys less so. Protection, decoration, durability	Aeronautical and mechanical applications. Sporting goods
6000	Sulphuric acid anodising, colourless or electrolytically coloured	Most alloys easy to anodise. Decoration, good durability of surface aspect, good corrosion protection	Metallic fittings, buildings, boat masts
40000	Sulphuric acid anodising	Good durability of surface aspect, corrosion protection. Grey colour due to presence of silicon	Mechanical and electrical applications 42100 (A-S7G03), 44200 (A-S13)
50000	Colourless sulphuric acid anodising	Good durability of surface aspect, corrosion protection. Most alloys easy to anodise	Fittings for buildings, 51100 (A-G3T). Upperworks, 51300 (A-G6)

5.2.3.   Conversion treatments other than chromic

Chromates have been known to be inhibitors since 1907, and the use of zinc chromate in primer coats used to prepare aluminium for painting dates back to 1931. Given their high efficiency, it was not until the end of the 1980s that replacing them began to be considered.

However, the toxicity of hexavalent chromium and the cost of its elimination from surface treatment baths have generated a tendency to replace chromate-based conversion treatments. Several products are presently being evaluated as a replacement for chromates [11]:

- cerium salts [12, 13],
- permanganate [14],
- titanium or zirconium oxides [15],
- lithium salts [16],
- molybdates.

5.3.   ANODISING

The idea to thicken the natural oxide layer was first mentioned by Buff in 1857 [17]. He had observed that aluminium became covered with an oxide film when placed in the anodic position in electrolysis cells.

This idea was taken up at the beginning of the 20th century, and a first patent was filed in 1911 by the Frenchman de Saint-Martin, who proposed an anodising process in a sulphuric medium containing iron sulphate.

Several processes and variants thereof have been developed since 1923:

- chromic anodising, patented by Bengough and Stewart in 1923, aiming at the protection of aeroplanes in Duralumin;
- sulphuric anodising, patented by Gower and O'Brien in 1937;
- oxalic anodising, developed during the 1930s in Japan and later in Europe under the name Eloxal (Elektrolytisch oxidiertes Aluminium).

Chromic and sulphuric anodising, as well as their variants, have been widely used. These surface treatments are specific to aluminium. They consist of growing oxide layers whose structure and properties differ from those of natural oxide layers on aluminium. Their thickness ranges from a few micrometers up to 100 μm, i.e. 1000–10 000 times the thickness of the natural oxide layer, which is of the order of 5–10 nm [18].

Anodising is carried out for a wide variety of reasons:

–   to decorate aluminium products,
–   to protect against weathering,
–   to improve resistance to abrasion and increase superficial hardness,
–   to improve the adhesion of organic coatings (adhesives, lacquer, paint),
–   to modify its dielectric properties (insulation), and
–   to modify optical properties (reflectivity).

Several types of anodising are used, each of which may have several variants:

–   sulphuric acid anodising, mainly used for protection against atmospheric corrosion and durability of the surface appearance;
–   chromic acid anodising, mainly used in the aerospace industry with 2000 and 7000 series alloys;
–   integral colour anodising, used especially in architecture;
–   phosphoric acid anodising, used as a surface preparation prior to adhesive bonding;
–   barrier anodising, used for refined metal 1199 for electrical applications (capacitors);
–   hard anodising, which increases the surface hardness of the metal; and
–   electrolytic brightening.

In general, the suitability for anodising depends on the composition and the microstructure of the aluminium, and may vary from one alloy series to another. The resulting anodic coatings will not necessarily have the same properties (Table B.5.1).

Like natural oxides, anodic coatings have a poor resistance to acidic or alkaline media (Figure B.1.18). Consequently, specific cleaning agents must be used for the maintenance of anodised surfaces. It should also be remembered that anodic coatings do not eliminate the risk of galvanic corrosion.

5.3.1. Sulphuric acid anodising

This is the most common anodising process. It is used for decorative purposes, for hard coatings and for protection against weathering. The classic treatment parameters (architectural quality) are

–   concentration of H_2SO_4: 200 ± 20 $g \cdot l^{-1}$,
–   concentration of dissolved aluminium in the bath: 15 $g \cdot l^{-1}$ maximum,
–   stirred bath, in order to avoid overheating (not more than 20 °C),
–   direct current input: 1.5 ± 0.2 $A \cdot dm^{-2}$.

This treatment is performed discontinuously on profiles, castings or sheet, or continuously on coiled strip. The oxide thickness depends on the intended use:

–   reflectors: 2–4 μm,
–   decorative use (furniture, vehicles): 5–8 μm,
–   architectural use: 15–25 μm,
–   hard anodising: 50–100 μm.

The structure of anodic coatings depends on the composition of the bath and the process parameters. It consists of hexagonal cells pierced by micropores, whose diameter, say for a coating thickness of 15 μm, is about 1000 times less than their film thickness.

The cell diameter depends on the electrolysis voltage, while the diameter of the pores depends on the electrolyte's anions (Figure B.5.2). These layers are not formed directly on the metal, but on a barrier layer.

These porous coatings lend themselves to adsorption coloration, either by immersion in a bath of special dye or by electrolytic coloration treatments. The latter are performed with alternating current, directly after anodising.

Whether coloured or not, anodic coatings need to be sealed in boiling demineralised water in order to achieve good resistance against weathering. Hydration of the coating closes the micropores by swelling. The duration of a sealing treatment is the same as for the anodising treatment [19].

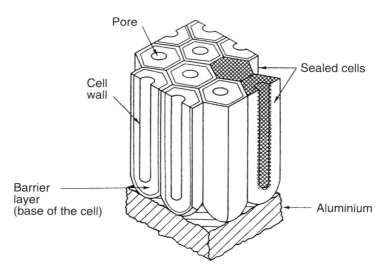

Figure B.5.2. Schematic representation of a sulphuric acid anodic oxide coating.

The weathering resistance of certain alloys, in particular the 2000 and 7000 series, can be enhanced by adding potassium dichromate to the sealing bath. This imparts a slightly green–yellow tinge to anodic coatings.

5.3.2. Chromic acid anodising

This process is widely used on aeronautical alloys of the 2000 and 7000 series, in order to improve their corrosion resistance in service. The thin anodic layer (5 μm), which is normally grey on 1050A, can assume more or less dark shades depending on the alloys. It provides a good hold for paint and adhesives.

5.3.3. Hard anodising

Hard anodising refers to a number of low-temperature processes that produce thick (50–100 μm) and dense oxide layers. They have better abrasion resistance than the best treated steels, and their electrical insulation properties are on a par with those of porcelain.

Hard anodising is used in the electrical and mechanical industries. The friction coefficient is reduced by a number of impregnation products: lanolin, teflon, molybdenum disulphide, etc. Given the thickness of these layers, it may be necessary in some cases to allow for dimensional variations after anodising.

5.3.4. Phosphoric acid anodising

This recently developed process creates oxide films that are highly porous and ideal for preparing surfaces for adhesive bonding.

The treatment can be carried out

– with direct current (the Boeing process), for aircraft bonding, lasting about 10 min,
– with alternating current.

The surface properties last for several months after treatment. These methods are used industrially with baths that have a long lifetime and are free from toxic components.

5.4. PAINT AND OTHER ORGANIC COATINGS

Painting aluminium most often serves a decorative purpose. This is true especially in the construction industry and in shipbuilding (with the exception of hulls in contact with seawater). Since aluminium is nontoxic, it needs to be coated with an antifouling paint compatible with aluminium (i.e. without copper and mercury salts) in order to prevent fouling.

Most organic coatings such as paints, varnishes, etc., can be applied on aluminium and its alloys using either traditional techniques or electrostatic power application. It is impossible to obtain a proper adherence of an organic coating on aluminium without an appropriate surface preparation. It has long been accepted that the durability of coatings on aluminium is determined first by the preparation of the base metal surface [20]. A suitable preparation of the surface usually starts with degreasing, followed by eliminating existing oxides, forming a base layer, and applying a primer.

Since the corrosion resistance of the base metal itself is very good, the resistance of organic coatings is remarkable even after many years of exposure [21].

When the protecting effect of the coating eventually fades, possible corrosion of the aluminium base metal will not degrade the surface appearance, as observed in the case of rust runoff on steel. The renovation of paint coatings on aluminium will be less frequent than on other metals.

5.5. CATHODIC PROTECTION

Cathodic protection of steel by zinc was discovered in 1742 by Malouin. The industrial production of galvanised steel began in 1837 [22].

Cathodic protection consists in putting the metal to be protected in the situation of a cathode. This is achieved by linking it to a metal that has, in the medium being considered, a more electronegative dissolution potential. The difference in potential between the two metals should be at least 100 mV. Two different methods of cathodic protection are used (Figure B.5.3):

– *sacrificial anodes*: this is the case of zinc protecting steel by dissolving; this method is used most often for aluminium;
– *forced currents*: a current is applied between the (non-consumable) anode and the cathode (i.e. the metal to be protected) so that the cathode potential is sufficiently decreased.

5.5.1.   *Cathodic protection of aluminium*

Cathodic protection for aluminium is more difficult than for steel, because it takes place at a potential that is significantly more electronegative, below -800 mV SCE. At this level, the reduction of ions H^+ is possible according to the reaction:

$$2H^+ + 2e^- \rightarrow H_2$$

If the medium, such as soil, is not replenished easily, H^+ ions that have been reduced are not replaced quickly, and the medium becomes alkaline by an excess of OH^-. As the

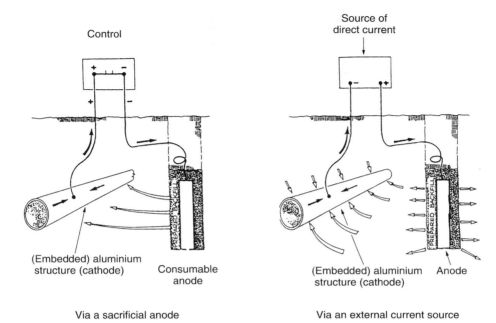

Figure B.5.3. Cathodic protection [52].

potential gets even more electronegative, several other reactions are possible:

$$2H_2O + O_2 + 4e^- \rightarrow 4OH^-$$

or

$$2H_2O + 2e^- \rightarrow 2OH^- + H_2$$

In all these cases, OH^- ions will be produced. They give rise to so-called cathodic corrosion of aluminium:

$$2Al + 2OH^- + 2H_2O \rightarrow 2AlO_2^- + 3H_2$$

which usually develops as bright craters.

The intensity of cathodic corrosion of aluminium does not depend on the nature of the anions or on the concentration of cations [23]. According to the Pourbaix diagram (Figure B.1.10), cathodic protection of aluminium must be carried out at highly electronegative potentials, on the order of -1750 mV SCE, which is well below the stability domain of water. At such a potential, there will always be cathodic corrosion.

The goal of cathodic protection of aluminium is to avoid pitting corrosion. It operates at a potential very close to the dissolution potential; a very low or even minute level of uniform corrosion can be accepted [24].

It is generally accepted that cathodic protection can be obtained in the range of potentials from -830 to -1130 mV SCE with no risk of cathodic corrosion [25]. In this range, the risk of cathodic corrosion in seawater is very low.

Accumulation of OH^- ions close to the aluminium surface leads to the precipitation of insoluble magnesium and calcium hydroxides. They precipitate as a very thin film that will slow down cathodic corrosion of aluminium [26]:

$$Ca^{2+} + HCO_3^- + OH^- \rightarrow CaCO_3 + H_2O$$

and

$$Mg^{2+} + 2OH^- \rightarrow Mg(OH)_2.$$

Any movement of the water that replenishes the electrolyte in contact with the metal will reduce the risk of corrosion because it dilutes the OH^- ions.

5.5.2. Set-up of cathodic protection of aluminium

Taking into account the dissolution potential of aluminium, anodes can be made either in zinc or in a special aluminium alloy called Hydral®, which contains indium (0.015–0.025%) or tin (0.10–0.20%). Magnesium anodes must not be used, because they lower the potential too much and will thus lead to severe cathodic corrosion of aluminium.

It is very important to control both the potential (in order to avoid cathodic corrosion) and the current density. The current density should be distributed as uniformly as possible in order to prevent certain zones of the structure from developing a positive or negative polarity.

Cathodic protection of unprotected structures is thus easy to set up in seawater. It is more difficult with carefully insulated and embedded structures.

As in the case of other metals, the location of the anodes must be adjusted such that the current lines take linear paths. That is why cathodic protection is ineffective in dissymmetric volumes that have complex shapes and are possibly subdivided, such as tanks or ballast tanks.

5.5.3. Use of cathodic protection of aluminium

Aluminium alloys of the 5000 and 6000 series have an excellent resistance against corrosion in seawater. Cathodic protection of these alloys is thus unnecessary and never used in practice. Cathodic protection of aluminium hull structures is, however, needed to neutralise galvanic coupling between aluminium and other metals. This is achieved by distributing sacrificial anodes in zinc or special aluminium alloys such as Hydral® over the hull structure close to the heterogeneous contacts, in order to neutralise galvanic coupling (Figure B.5.4).

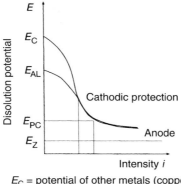

E_C = potential of other metals (copper,
  stainless steel, etc.)
E_{AL} = potential of aluminium
E_{PC} = potential of cathodic protection
E_Z = anode potential

Figure B.5.4. Neutralisation of galvanic coupling by cathodic protection.

When the conductivity of the medium is weak, which is the case in fresh water, cathodic protection is provided by nonsacrificial anodes, made for example in platinum-coated titanium, which operate with an external direct current source.

5.6. CLADDING

Cladding is another form of cathodic protection. A difference in potential of at least 100 mV can be found between certain wrought aluminium alloys. This makes cathodic protection of a core alloy possible by cladding with a more electronegative alloy (Table B.5.2).

In order to increase the resistance to atmospheric corrosion of 2000 series alloys used in the aerospace industry, in 1927 Dix from Alcoa proposed [27] protecting them by a cladding in 1050, which he called "Alclad".[1] In Europe, this product was called Vedal [28].

This development made it possible to use highly resistant alloys such as 2024 in environments where previously they could not be used without protection.

Clad products are produced by hot rolling an ingot of the core alloy on which, on one or both sides, plates of the selected cladding alloy are clad; the cladding thickness on each face generally ranges between 2 and 5% of the total thickness. The most common couples of rolled semi-products are listed in Table B.5.3.

---

[1] US patent 1 865 089 of June 28, 1932.

Table B.5.2. Dissolution potentials, measured in a solution NaCl (57 g·l^{-1}) + H$_2$O$_2$ (3 g·l^{-1})

Cladding	Potential	Core	Potential
1050A	− 770	2024 T4	− 610
		2014 T6	− 680
7072	− 860	3003	− 750
		6061	− 720
		7075 T4	− 740

Tubes that are clad inside and outside can be produced by co-extrusion. The most economical approach is to manufacture seam-welded tubes from sheet that is clad on one or both sides. This is commonly done for irrigation tubes in 3003/7072, and for heat exchanger tubes used in cars.

Tests in outdoor testing stations have shown that cladding provides an effective protection for many years, whether exposed to weathering or immersed in seawater [29]. The consumption of the cladding avoids corrosion of the underlying core (Figure B.5.5).

The solubility of copper and magnesium at high temperatures in aluminium is high enough to generate a risk of diffusion of these elements in the 1050 cladding layer during heat treatments, and especially during solution heat treatment. Copper diffusion may lead to a reduction in the difference in potential between the cladding such that the cladding will no longer protect the core [30].

It has been shown that even a discontinuous cladding may still protect the core, up to a distance that increases with the conductivity of the medium. This protection extends to several centimetres when immersed in seawater [31].

This explains why zones where the cladding has locally been eliminated by welding, machining, riveting or corrosion will nevertheless be effectively protected. Of course, when the cladding is entirely consumed over large areas, pitting corrosion may initiate.

Table B.5.3. Clad rolled semi-products

Couple		Cladding		Average cladding thickness per	
Core	Cladding	On one face	On both faces	face (%)	Applications
2024	1050A		×	2–5	Aircraft
2618A	7072		×	2–5	Aircraft
2214	1050A		×	2–5	Aircraft
3003	7072	×	×	5–10	Deep forming, irrigation tubes, etc.
7075	7072		×	2–5	Aircraft
7475	7072		×	2–5	Aircraft

Figure B.5.5. Protection by cladding.

5.7. ZINC COATING

It is well known that 7072 effectively protects alloys such as 3003 against pitting corrosion in water, seawater, etc. However, the use of clad semi-products is limited to simple shapes manufactured from sheet, which is the only form of clad semi-products available.

In order to protect intricate shapes like profiles, fin tubes, etc., two solutions are available:

– chemical zinc deposition followed by hot diffusion,
– thermal projection or plasma projection of zinc.

5.7.1. Chemical zinc deposition

Zinc can be deposited on aluminium from alkaline zinc solutions [32]. Up to 60 mg·dm^{-2} can be deposited after a few minutes of immersion in the deposition bath at 25 °C. It is then necessary to heat for several hours at 400–500 °C, in order to allow the zinc to diffuse into the aluminium. A mean zinc content of 1% over a depth of 50 μm is obtained by this diffusion treatment, which corresponds to an in situ cladding with 7072 (Figure B.5.6).

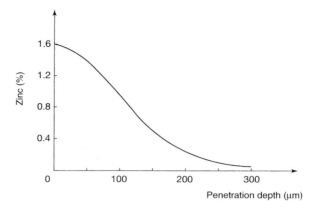

Figure B.5.6. Diffusion of zinc in 6061 [33].

The diffusion heat treatment lowers the mechanical properties. The resistance of this protection against weathering or water has not been reported. This process is not widely used, probably because anodising is easier to carry out.

5.7.2. *Zinc metallisation*

Up to 120 μm of zinc can be deposited by plasma metallisation. Experience has shown that this coating is not very effective and has a poor resistance to seawater. The coating being porous, the attack of zinc by seawater results in very voluminous corrosion products that lead to the formation of blisters at the interface between the metal and the coating [34]. The difference in potential between aluminium and zinc, approximately 400 mV, is too high.

Zinc coating of 5000 and 6000 series alloys for classic marine applications seems to be unnecessary and superfluous, because these alloys show an excellent corrosion resistance in marine environment. Zinc coating has no advantage, and moreover, its corrosion resistance is precarious.

On the other hand, industrial equipment such as regasification plants for liquid natural gas operating by seawater spraying and located in closed and polluted sites are usually protected by thermal or plasma projection coatings with aluminium–zinc alloys, and are additionally protected with specific paints.

5.8. INHIBITORS

The word "inhibit" is derived from the Latin word "inhibere", which means to stop, to refrain. The word "inhibition" has been used in chemistry and in corrosion science since 1907.

The idea of using inhibitors is very old, going back to the Middle Ages: armourers used to add flour or yeast in order to prevent fragilisation of the weapons during acid pickling [35]. The role of chromates was discovered around 1910; sodium silicate has been used as an inhibitor on aluminium since 1929 [36]. Inhibitors are substances (or a combination of substances) that, added in a low concentration to an environment, will cancel or reduce its aggressiveness towards a metal.

With aluminium, inhibition aims at reducing or cancelling pitting corrosion in contact with water or aqueous media, mainly in closed circuits of heat exchangers. There are also other uses of inhibitors, for example in pickling baths.

The number of inorganic and, mostly, organic substances that have been proposed as inhibitors of metals and alloys is considerable[2] and has led to many patents.

5.8.1. Mechanism of inhibition

The mechanism of inhibition has been the subject of many hypotheses and is still rather poorly understood [37]. Inhibition is unlikely to be a simple phenomenon, but rather a sequence of processes that are more or less well identified.

Inhibitors can act on the medium by modifying its properties, thus reducing its aggressiveness towards the metal. Their action is probably mainly within the volume of the Helmholtz double layer.

Inhibitors can act on the metal by modifying its surface state, or they can modify the anodic or cathodic reactions, or even both at the same time. They form a monomolecular layer on the metal and thus create a barrier to the environment.

It is convenient to classify inhibitors according to their physico-chemical mechanism:

–   *products that adjust the pH value*, such as buffer solutions which can be a mixture of weak acids and their corresponding salts, for example a mixture of acetic acid and sodium acetate, which sets the pH at 4;
–   *products forming insoluble aluminium compounds*;
–   *products forming protective adsorbed layers*, such as chromates, molybdates, vanadates, arsenates and phosphates;
–   *products forming organic films*, such as amides and sulfamates. They contain a polar molecule that creates a covalent bonding with the metal.

Inhibitors can also be classified as to their electrochemical mechanism [38]:

–   *passivating inhibitors*, which are oxidants that are easy to reduce, such as chromates, nitrites or even oxygen. They form a mixed oxide film on the metal;

---

[2] For aluminium, more than 800 abstracts can be found in the *Chemical Abstracts* since its first volume (1907) to volume 124 (1996)

–   *anodic inhibitors*: silicates, tungstates, permanganates, phosphates, chromates,
    nitrites. They modify the anodic reactions. Their concentration must be sufficiently
    high to cover all of the anodic surface;
–   *cathodic inhibitors* such as magnesium or calcium salts modifying the cathodic
    reactions. They slow down corrosion by increasing the overpotential of hydrogen
    (H^+) reduction on the cathodic zones.

The effect of inhibitors depends on conditions such as [39]

–   temperature: adsorption decreases with increasing temperature;
–   pH: chromates are effective at a pH above 3;
–   the surface/volume ratio;
–   the inhibitor's concentration.

In order to be effective, inhibitors need to be adsorbed at the metal surface. This is the
first stage of the mechanism of organic inhibitors. Two types of adsorption can be
distinguished:

–   *Physical adsorption* (physisorption): this can either be of the Van der Waals type,
    involving weak attraction forces and spreading out over the whole surface, or of the
    electrostatic type if the inhibitor is dissociated in the solution. The cations resulting
    from that dissociation fix on cathodic sites. Such an inhibitor is called a cathodic
    inhibitor.
–   *Chemical adsorption* (chemisorption), by formation of a covalent bonding using
    electrons common to the inhibitor and the metal. This type of adsorption is
    irreversible and normally leads to inhibition of the anodic reaction.

5.8.2.   *Use of inhibitors*

First of all, it should be noted that there is no such thing as a universal inhibitor capable of
protecting all metals in a single medium. Some inhibitors may protect several metals, while
others are highly specific to one metal or alloy (Table B.5.4).

Each metal or alloy in a multi-metal circuit must be protected by its own inhibitor(s).
This explains the complexity of certain formulations of liquid inhibitors used for certain
circuits.

Inhibitors are commonly used in the following situations:

–   *In closed circuits of heat exchangers*, such as cooling systems of car engines,
    solar heating, etc.

Table B.5.4. Specificity of inhibitors

Inhibitor	Steel	Aluminium	Copper
Silicate	×	×	×
Phosphate	×	×	
Nitrite	×		
Nitrate		×	
Borate	×		
Sodium benzoate	×	×	×
Mercaptobenzothiazole			×
Triethylamine	×		

These are heterogeneous circuits comprising several metals and alloys: steel or cast iron, aluminium, cuprous alloys. Each metal, therefore, needs to be protected against corrosion by the coolant, usually water that contains antifreeze. For that reason, there has been a significant tendency since the end of the 1960s to specify specifically designed liquids for this kind of circuit. Moreover, the corrosion of components in cuprous alloys may lead to corrosion of the heat exchangers or other aluminium components. For this reason, these liquids should contain effective inhibitors for copper alloys [40]. The development of aluminium heat exchangers in cooling systems of thermal engines and in heat recovery systems (solar, etc.) has been made possible by the general use of coolants specifically adapted to aluminium [41, 42].

- *During acidic or alkaline pickling treatments* generally applied at the beginning of surface treatments. The goal of using inhibitors is to minimise the attack of the metal during its short immersion in the pickling bath.
- *When cleaning aluminium equipment*, especially in alkaline medium, which is widely used for degreasing equipment used in the food industry. Blackening can be reduced by silicates or fluorosilicates, but it cannot be totally prevented [43, 44].

Inhibitors can also be used in very specific cases such as in protecting aircraft against corrosion during cleaning operations [45]. Inhibitors are very rarely used in open systems, because of their cost and environmental problems.

5.8.3. *Choice of inhibitors*
It is impossible to list all the inhibitors that have been proposed for protecting aluminium. Specialised publications should be studied for this purpose [46]. The inhibitors most commonly mentioned for aluminium [47–49] are listed in Table B.5.5. It should be noted that inhibitors may be subject to ageing, and that their effectiveness is not always total.

Table B.5.5. Inhibitors for aluminium

Medium	Inhibitor
Sodium acetate	Sodium silicate
Hydrochloric acid < 5%	Nicotine, thiourea, acridine
Diluted nitric acid < 10%	0.10% hexamethylene tetramine
Diluted nitric acid < 10%	0.1% alkali chromate, 0.50% hexamethylene tetramine
Diluted nitric acid 2–5%	0.50% hexamethylene tetramine
Phosphoric acid < 2%	0.5–1% chromic acid
Sulphuric acid < 5%	Potassium chlorate, iodate or bromate
Ethyl alcohol	0.03% sodium carbonate, lactates, acetates, borates
Methyl alcohol	Sodium chlorate and sodium nitrate
Dehydrated alcohols	Traces of water
Higher alcohols	0.1–0.5% benzotriazole
Potassium carbonate < 5%	0.2–1% sodium silicate, 0.05% sodium fluorosilicate
Sodium carbonate < 5%	0.2–1% sodium silicate, 0.05% sodium fluorosilicate
Calcium chloride, concentrated solutions	2% sodium chromate
Aromatic chlorine derivatives	0.1–2% nitrochlorobenzene
Water in closed circuits	0.2% sodium silicate or 0.2% sodium chromate
Ethylene glycols, antifreezing liquids	Borates, sodium phosphate, sodium nitrate, sodium tungstate, sodium molybdate, amines, sodium nitrate, sodium benzoate
Hypochlorites < 1%	Sodium silicate
Hydrogen peroxide < 30%	$0.1–0.5 \text{ g·l}^{-1}$ sodium silicate, ammonium nitrite, potassium nitrite
Dehydrated phenols	Traces of water
Alkali phosphates < 1%	0.2% sodium silicate
Sodium hydroxide < 1%	Sodium silicate, sodium chromate, potassium permanganate 3–4%
Sodium sulphate	1% sodium silicate
Carbon tetrachloride	0.02–0.05% formamide
Trichloroethylene	Trimethylamine
Triethanolamine	1% sodium silicate

The choice of inhibitors must conform to applicable standards and regulations concerning toxicity and environmental protection. These also apply to closed circuits. This limits the use of certain inhibitors, in particular sodium and potassium chromate, which were widely used because they are the most effective inhibitors for aluminium. The same applies to the use of volatile inhibitors.

Their replacement by other inorganic salts such as vanadates, molybdates, etc. has been investigated for several years [50, 51].

REFERENCES

[1] Altenpohl D.G., Use of boehmite films for corrosion protection of aluminium, *Corrosion*, vol. 18, 1962, p. 143t–153t.

[2] Hart R.K., The formation of films on aluminium immersed in water, *Transaction of the Faraday Society*, vol. 53, 1956, p. 1020–1027.

[3] Uchiyama T., Hasegawa M., Matsumoto H., A study on chemical conversion coating formed on aluminium in boiling sea water, *Journal of Metals Finishing Society Japan*, vol. 37, 1986, p. 28–33.

[4] Koudelkhova M., Augusynski J., Berthou H., On the composition of the passivating films formed on aluminium in chromate solutions, *Journal of the Electrochemical Society*, vol. 124, 1977, p. 1165–1168.

[5] Brown G.M., Shimoizu K., Kobayashi K., Thompson G.E., Wood G.C., The development of chemical conversion coating on aluminium, *Corrosion Science*, vol. 35, 1993, p. 253–256.

[6] Helling W., Neunzig H., Das Färben von Aluminium und seinen Legierungen mit anorganischen Stoffen auf MBV-Grundlage, *Aluminium*, vol. 18, 1936, p. 608–617.

[7] Hess C., Le procédé Alodine. Protection galvanique de l'aluminium, *Revue de l'Aluminium*, vol. 174, 1951, p. 44–50.

[8] Sticklen R., New protective treatment for aluminium simplifies processing at reduced cost, *Materials & Methods*, vol. 35, 1952, p. 91–95.

[9] Spencer L.F., Conversion coatings. Chromates films, *Metal Finishing*, vol. 58, 1960, p. 58–65.

[10] Masson J.C., *Les traitements de conversion des alliages de l'aluminium*, CFPI, Gennevilliers, 1990.

[11] Hinton B.R., Corrosion prevention and chromates, the end of an era, *Metal Finishing*, vol. 89, nos 9, 1991, p. 55–61, see also vol. 10, pp. 15–20.

[12] Mansfeld F., Wang Y., Corrosion protection of high copper aluminium alloys by surface modification, *British Corrosion Journal*, vol. 29, 1994, p. 194–200.

[13] Schmith M., Aljinovic L.J., Radosevic J., Examination of efficiency of cerium passivation coating on a aluminium alloys, *Eighth European Symposium on Corrosion Inhibitors*, Ferrara, 1995, p. 851–860.

[14] Biber J.W., A chrome-free conversion coating for aluminium with the corrosion of chrome, *Conference Lightweight Automotive Applications*, Orlando, 1995, p. 392/1–392/13.

[15] Snodgrass J.S., Weir J.R., Corrosion test selection for aluminium autobody sheet with chromoum-free pretreatments, *Conference Lightweight Automotive Applications*, Orlando, 1995, p. 380/1–380/11.

[16] Buscheit R.G., Drevien C.A., Martinez M.A., Stoner G.E., Processing and properties of chrome-free conversion coatings on aluminium, *Conference Lightweight Automotive Applications*, Orlando, 1995, p. 390/1–390/14.

[17] Buff H., *Liebigs Ann. Chem.*, vol. 102, 1857, p. 265–284.

[18] Patrie J., Le mécanisme de l'oxydation électrolytique et la formation des couches d'oxyde d'aluminium, *Revue de l'Aluminium*, 1949, p. 397–403, see also 1950, p. 3–7.

[19] Whitby L., Anodic oxyde coating. Influence of sealing treatments on protective value, *Metal Industry*, vol. 72, 1948, p. 400–403.

[20] Guilhaudis A., Bourbon R., La protection de l'aluminium et de ses alliages par peinture, *Revue de l'Aluminium*, nos 206, 1954, p. 7–10, see also vol. 207, p. 47–51.

[21] Lundberg C.V., *Long-term weathering of organic and inorganic coatings on steel and on aluminium*, ASTM, STP 1086, 1990, p. 122–155.

[22] Lynes W., Some historical developments relating to corrosion, *Journal of the Electrochemical Society*, vol. 98, 1951, p. 3c–10c.

[23] Van de Ven X., Koelmans H., The cathodic corrosion of aluminum, *Journal of the Electrochemical Society*, vol. 123, 1976, p. 143–144.

[24] Cerny M., Present state of knowledge about cathodic protection of aluminum, *Zashchita Metallov*, vol. 11, 1975, p. 687–698.

[25] Whiting J.F., Wright T.E., Cathodic protection for an uncoated aluminium pipeline, *Corrosion*, vol. 17, 1961, p. 8.

[26] Watkins K.G., Davies D.E., Cathodic protection of 6351 aluminium alloys in sea water: protection potential and surface pH effects, *British Corrosion Journal*, vol. 22, 1987, p. 157-161.

[27] Dix E.H., Alclad, a new corrosion resistant aluminium product, National Advisory Committee for Aeronautics, technical note no. 259, August 1927.

[28] Pubellier M. et al., Bimetal: le Vedal, Aciers spéciaux, *Métaux et Alliages*, vol. 9, 1934, p. 535–536.

[29] Walton C.J., Sprowls D.O., Nock J.A., Resistance of aluminum alloys to Weathering, *Corrosion*, vol. 9, 1953, p. 345–358.

[30] Keller F., Brown R.H., *The heat treatment of 24S and Alclad 24S alloy products*, Alcoa, technical paper no. 9, 1943.

[31] Gauthier G., Lacunes de couches d'aluminium de placage de l'alliage A-U4G (Duralumin) compatibles avec la protection contre la corrosion, *Revue de métallurgie*, vol. 50, 1953, p. 551–557.

[32] Wernick S., Pinner R., Surface treatment and finishing of light metals. Part 10, The zinc immersion process, *Sheet Metals Industry*, vol. 32, 1955, p. 189–197.

[33] Bothwell M.R., New technique enhances corrosion resistance of aluminium, *Metal Progress*, vol. 87, 1965, p. 81–83.

[34] Kweon Y.G., Coddet C., Behavior in sea water of zinc-base coatings on aluminium alloy 5086, *Corrosion*, vol. 48, 1992, p. 97–102.

[35] Putilova I.N., Balezin S.A., Barranik V.P., *Metallic Corrosion Inhibitors*, Pergamon Press, Oxford, 1960.

[36] Röhrig H., Additions which reduce the attack of solutions on aluminium, *Aluminium*, vol. 17, 1935, p. 559–562.

[37] Oakes B.D., Historical review inhibitors mechanisms, *NACE Corrosion*, 1981, paper No. 248.

[38] Riggs O.L., Theoretical aspects of corrosion inhibitors and inhibition, *Corrosion Inhibitors*, Nathan C.C. (ed.), NACE, Houston, 1973, p. 7–27.

[39] Frasch J., Problèmes actuels d'inhibition, *Corrosion et Anticorrosion*, vol. 14, 1966, p. 204–209, see also p. 261–269.

[40] Jackson J.D., Miller, P.D. Fink F.W., Boyd, W.K., *Corrosion of materials by ethylene glycol–water*, Battelle, report DMIC 216, May 1965.

[41] Sigurdsson H., Tsuvik R., Corrosion of aluminium in water based heat exchangers systems, *10th Scandinavian Corrosion Congress*, June 1986, p. 65–72.

[42] Humphries T.S., *Corrosion inhibitors for solar heatings and cooling systems*, US Department of Energy, report DOE/NASA TM-78180, 1978.

[43] Ouvrier-Buffet J., Pagetti J., Recherche de quelques inhibiteurs de corrosion de l'alliage aluminium–silicium–magnésium dans des solutions de soude 0.1 N, *Métaux Corrosion et Industrie*, vol. 50, 1975, p. 180–188.

[44] Dauphin G., Labbé J.P., Michel F., Pagetti J., Triki E., Inhibition de la corrosion de quelques alliages d'aluminium au cours des opérations de nettoyage et de désinfection pratiquées dans l'industrie alimentaire, *Métaux Corrosion et Industrie*, vol. 52, 1977, p. 253–261.

[45] Khobaib M., Wahldiek F.W., Lynch C.T., New concepts in multifonctionnal corrosion inhibition for aircraft and other systems, *Conference AGARD*, vol. 315, 1981.

[46] Desai M.N., Desai S.M., Gandhi M.H., Shah C.B., Corrosion inhibitors for aluminium and aluminium based alloys, *Anti-Corrosion*, vol. 18, nos 4, 1971, p. 8–13.

[47] Desai M.N., Desai S.M., Gandhi M.H., Shah C.B., Corrosion inhibitors for aluminium and aluminium based alloys. Part I, *Anti-Corrosion*, vol. 18, nos 5, 1971, p. 4–10.

[48] Roebuck A.H., *Inhibition of aluminium. Corrosion Inhibitors*, Nathan C.C. (ed.), NACE, Houston, 1973, p. 240–244.

[49] Junière P., Sigwalt M., *Les applications de l'aluminium dans les industries chimiques et alimentaires*, Masson, Paris, 1961.

[50] Wilcox G.D., Gabe D.R., Warwick M.E., The role of molybdates in corrosion prevention, *Corrosion Reviews*, vol. 6, 1986, p. 327–365.

[51] Monticelli C., Brunoro G., Zucchi F., Fagioli F., Inhibition of localized attack on the aluminium alloys AA6351 in glycol/water solutions, *Werkstoffe und Korrosion*, vol. 40, 1989, p. 393–398.

[52] Ellis W.J., *Fundamentals of cathodic protection*, NACE, Houston, Fourth Western States Corrosion Seminar, 1970, p. 6/1–6/10.

Chapter B.6

The Corrosion Behaviour
of Aluminium Alloys

Chapter B.6
The Corrosion Behaviour
of Aluminium Alloys

Corrosion is a complex phenomenon that depends on many parameters that are related to the environment or to the metal. The occurrence of corrosion is sometimes difficult to explain because the cause of corrosion has not been identified, or because the theoretical foundations are sometimes insufficient to give a satisfactory answer.

For users as well as for corrosion experts, corrosion has a relative aspect. In principle, corrosion is defined as the degradation of a metal by its environment. At which point is this then called corrosion? This question is neither simple nor innocent. Appropriate electrochemical methods measure corrosion currents as low as 1 μA, which corresponds to a dissolution rate of less than 1 μm per year.

In practice, corrosion needs to be associated with a risk that is deemed either acceptable or not acceptable. As an example, for roofing sheet 0.80 mm thick, corrosion means perforation rather than a uniform decrease in thickness of a few micrometers per year. Water staining is not acceptable for luxury goods, is only slightly acceptable for kitchen utensils, but will not even be taken into consideration for many components for which it is a part of the natural surface aspect.

It is the task of corrosion experts to predict what will happen, to highlight possible limitations in lifetime or in acceptable stress for the material under consideration, to the extent allowed by theory and experience. Many grey and uncertain areas still exist. These will incite corrosion experts to adopt a cautious attitude, but also promote a certain boldness, because in many cases, predictions can be made based on experience with existing applications, without taking an inconsiderate risk.

Corrosion is a complex phenomenon that is sometimes difficult to explain; it is not just corrosion but several possible types of corrosion for a given metal. No metal or alloy is capable of resisting all possible aqueous media, even at room temperature. The corrosion resistance of a metal or alloy depends on many factors that are inherent to the metal itself, the environment in which it is placed, and the conditions of use.

6.1. FACTORS RELATED TO THE ENVIRONMENT

The environment plays a very important role in the corrosion resistance of a metal or alloy, a fact that has been demonstrated time and again by experience.

6.1.1. The nature of the environment

It is not easy to list a rigorous typology of environments with respect to corrosion phenomena observed on metals. Nonetheless, the following tendencies can be formulated:

- in *aqueous, ionic media*, the fundamental electrochemical reactions for corrosion of metals can take place;
- in *nonaqueous, nonionic, organic media*, no oxidation and reduction phenomena can take place, and therefore a different behaviour can be expected. However, certain totally dehydrated organic products such as alcohols and phenols can react very violently with aluminium at higher temperatures.

One should also distinguish between

- *gaseous media*, which are in general not very reactive at ambient temperature, except in the presence of moisture;
- *liquid media*, in which the contact with the metal is easy, especially when the liquid is wetting the metal. These are the media in which reactions are the most likely to take place; and
- *solid media* such as powders are generally not very reactive, except in the presence of moisture. Anhydrous inorganic salts theoretically should not react at all, if they are really anhydrous. Crystal water has no effect as long as it is trapped in the crystal, but it may be released at higher temperatures.

The action of organic products depends on the reactivity of their functional groups. For example, organic acids $RCOOH$ are more reactive than ketones R_1COR_2. Primary alcohols RCH_2OH are more reactive than secondary alcohols R_1CHOHR_2 and of course more reactive than tertiary alcohols $R_1COHR_2R_3$.

The effect of chlorinated derivatives depends on the position of the halogen atoms. As an example, benzene compounds are more reactive when substituted at a side chain than at the benzene ring.

6.1.2. Concentration

Concentration is an important factor in the corrosion of metals. In general, the reaction rate increases with the concentration of the corrosive agents. However, there is not necessarily a proportional relationship between their concentration in air, water, etc. and the corrosion rate.

In complex media where several anions and cations are present, synergies or antagonisms may occur between them. This happens especially in natural waters. Chlorides are aggressive, but their action can be slowed down, for example by precipitation of carbonates.

Figure B.6.1 shows that the effect of diluted solutions of phosphoric acid can be slowed down by the presence of sodium or ammonium ions.

6.1.3.   *Oxygen content*

The rate and form of the corrosion of certain metals such as iron depend to a large extent on the oxygen content of the water.

Oxygen is an oxidant and is corrosive in the sense that it depassivates the cathodes by starting up the cathodic reaction

$$2H_2O + O_2 + 4e^- \rightarrow 4OH^-$$

which favours the oxidation reaction at the anode, i.e. corrosion.

The situation is not as simple with aluminium, since corrosion is governed by the natural oxide layer, the formation of which requires oxygen.

The role of oxygen is, therefore, not a determining one. This is clearly shown by the example of desalination of seawater: the corrosion resistance is the same in aerated or

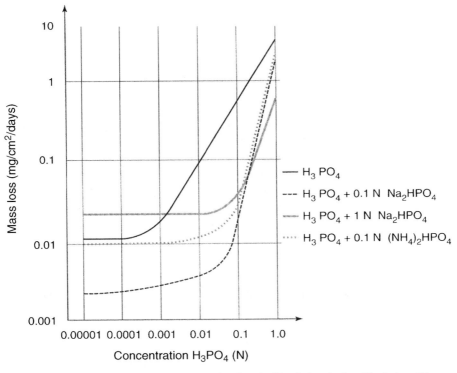

Figure B.6.1. Effect of sodium and ammonium ions in diluted phosphoric acid solutions [1].

deaerated seawater, if all the other parameters (temperature, salt concentration, flow rate, etc.) are kept constant.

6.1.4. *pH value*

According to the theory of S. Arrhenius, the electrical conductivity of pure water results from dissociation of part of its molecules (10^{-7} at 25 °C) according to the equilibrium:

$$H_2O \rightleftarrows H^+ + OH^-$$

By definition: $pH = -\log[H^+]$, and pH values thus range from 0 to 14 (Table B.6.1).

The pH is a characteristic value for aqueous media. In chemistry, electrochemistry and biology, this is a very important parameter because it helps understand reactions in aqueous media. The pH value is involved in the precipitation of hydroxides: for example, magnesia $Mg(OH)_2$ contained in water or seawater precipitates in alkaline media, generally on the metal's surface; this may modify its corrosion resistance (see Section B.5.5). The precipitation of carbonates is also related to the pH value.

The curve on Figure B.1.18, which shows the solubility of $Al(OH)_3$ as a function of pH, is a good example of the influence of pH on the solubility of a hydroxide such as aluminium hydroxide. The pH is a very important factor for the corrosion resistance of metals in aqueous solutions. It is well known that steel has a poor resistance to acidic media, and that aluminium has a poor resistance to highly basic media containing sodium or potassium hydroxide.

Pourbaix plotted electrochemical equilibrium diagrams of metals in water as a function of the potential E with respect to the hydrogen electrode, and as a function of pH (Figure B.1.10). Several domains can be identified in these diagrams: corrosion, passivation and immunity (see Section B.1.6).

However, the pH is not sufficient in order to predict the corrosion resistance of a metal or an alloy in aqueous solution. The nature of the acid (and thus of the anion associated with the proton H^+) and of the base (and thus of the cation associated with OH^-) also need to be taken into account. For example, mineral acids such as hydrochloric acid strongly attack aluminium. The rate of attack increases with concentration. On the other hand, concentrated nitric acid does not react with aluminium. Due to its oxidative action, it even

Table B.6.1. Range of pH values

Ratio $[H^+]/[OH^-]$	pH values	Medium
$[OH^-] = [H^+]$	$= 7$	Neutral
$[H^+] > [OH^-]$	< 7	Acidic
$[OH^-] > [H^+]$	> 7	Alkaline

contributes slightly to reinforce the natural oxide layer [2]. At a concentration higher than 50%, it is used for pickling treatments of aluminium.

The same applies to basic media: sodium hydroxide and potassium hydroxide, even at low concentrations, attack aluminium. On the other hand, at the same pH, an ammonia solution attacks aluminium only very moderately (see Section B.1.8).

At 20 °C, the dissolution rate of aluminium in sodium hydroxide solution (NaOH) at 0.1 g·l^{-1} (corresponding to a pH of 12.7) is 7 mm per year. The dissolution rate in ammonia at 500 g·l^{-1} (corresponding to a pH of 12.2) is 0.3 mm per year, which is 25 times less.

The pH of natural waters being close to neutral, there is no difference between waters whose pH differs only by a few tenths of pH units (see Section D.1.9).

6.1.5. Temperature

It is well known that increasing temperature leads to an increase in the rate of chemical reactions. In the case of the corrosion of aluminium, this applies to inorganic acids and bases (see Chapters E.4 and E.5), and also to certain organic media such as alcohols, phenols and chlorinated derivatives, especially when the temperature approaches their boiling point. However, in pure, distilled or poorly mineralised water, an increase in temperature will modify the form of corrosion because the natural oxide film can react with water to form a protective boehmite coating (see Section B.5.1). The corrosion resistance of aluminium in water depends on the temperature (see Section D.1.7).

As explained in Section B.6.4, the structure of certain aluminium alloys can change at prolonged heating at rather moderate temperatures, which may modify the mechanical properties and the resistance to corrosion. This is the case with alloys of the 5000 series containing more than 3.5% magnesium, when heated to 70–80 °C or more. The same applies to 7000 series alloys, if the service temperature is on the order of the artificial ageing temperature, i.e. more than 100–120 °C.

6.1.6. Pressure

Pressure has no influence on the corrosion of aluminium.

6.2. FACTORS RELATED TO THE METAL

Certain factors are related to the metal (or alloy) itself. Metallurgists try to adjust alloy compositions, transformation sequences and heat treatments in order to obtain the best possible corrosion resistance.

6.2.1. Alloy compositions

The influence of the main alloying elements is shown in Table A.3.1. Since the beginning of the 20th century, many studies have tried to quantify the influence of most of the metallic and metalloid elements on the properties of aluminium alloys, and especially on their corrosion resistance. An account of the results of these studies is beyond the scope of this book, especially because the composition of modern alloys takes advantage of all this metallurgical knowledge (the interested reader should refer to the bibliography compiled by M. Whitaker from BNFRMA in 1952 [3]).

Contrary to a common misconception, the purity of the base metal does not improve the corrosion resistance of aluminium. Metal with a very low iron and silicon content (1199) does not resist atmospheric corrosion better than 1070 or 1050. Only at much higher concentrations of iron and silicon (Fe > 0.50 and Si > 0.25), which was frequently found until the end of the 1940s, will the corrosion resistance be altered.

6.2.2. *Elaboration and transformation techniques*

Several modes of elaboration are used for aluminium alloys: casting, rolling, extrusion, etc. This is not an important factor.

Experience shows that aluminium casting alloys without copper generally have a better resistance to pitting corrosion than wrought alloys transformed by rolling or extrusion. This is probably due to the more resistant oxide layer of the as-cast surface compared to that of wrought semi-products. Often, the machined surface of a casting is more sensitive to pitting corrosion than the rest of the surface that has not been machined.

Roll casting does not significantly modify the corrosion resistance of alloys of the 1000 and 3000 series, and is said even to improve the resistance to pitting corrosion [4].

Cold working only has a minor effect on the corrosion resistance. At most, it changes the aspect of corrosion pits. In highly cold-worked metal, pits tend to propagate as so-called mole tunnels.

6.2.3. *Heat treatments*

Heat treatments of age-hardenable alloys—solution heat treatment, quenching and age hardening—lead to changes in the nature and the distribution of metallurgical constituents present in these alloys. They have an important influence on the susceptibility of these alloys to certain forms of corrosion, especially intercrystalline corrosion (see Section B.2.3) and stress corrosion (see Section B.2.5).

Alloys 2017 and 2024 are known for their high sensitivity to the quenching rate. They have to be quenched very quickly in cold water as soon as they leave the solution heat treatment furnace. Otherwise, if the quenching rate is low or if the water temperature is too high, a high susceptibility to intercrystalline corrosion will result. The corrosion rate increases greatly with the temperature of the quenching water, as shown by the results obtained on 2024 reported in Table B.6.2.

The temperature and duration of heat treatments also have a strong influence on the susceptibility to intercrystalline corrosion of these alloys, as shown in Figure B.2.10: in

Table B.6.2. Influence of quenching rate

Duration of the attack by NaCl (3%) + HCl (1%) (h)	Temperature of the quenching medium Mass loss (mg·dm^{-2})			
	20 °C	35 °C	50 °C	70 °C
4	15	15	16	21
8	20	22	27	37
12	31	59	69	128

2017A, a difference of 20 °C in the temperature of artificial ageing will be sufficient to sensitise or desensitise the metal against corrosion.

6.2.4. The surface state

The surface state has an influence on the corrosion resistance of aluminium. Experience has shown that scratched, scraped or ground surfaces are sites at which corrosion preferentially develops [5]. This can be observed very frequently on ground and machined surfaces of welded structures of tanks, for example after hydraulic testing.

Pickling (acidic or alkaline) often weakens the resistance to pitting corrosion of aluminium. The dissolution of the initial, rather thick oxide layer that covers cathodic intermetallics of the Al$_3$Fe type favours the development of pitting corrosion [6].

Unless necessary, acidic or alkaline pickling that modifies (or completely changes) the initial surface state should be avoided. Moreover, experience shows that it is often very difficult to properly rinse pickled surfaces, especially for intricate shapes or surfaces with recesses that are difficult to access. In these cases, pickling should be replaced by dry treatments: sand blasting, sanding, etc.

Chemical pickling treatments should be used only as a preparation for further surface treatments such as anodisation, lacquering, etc. Unless otherwise specified, they should never be used after forming operations or after welding.

6.3. FACTORS RELATED TO CONDITIONS OF USE

Factors related to conditions of use are often decisive. In many cases of corrosion under service conditions, it was not the alloy, which had often been chosen properly, but the details of construction and the service conditions that were responsible for corrosion.

6.3.1. Joining techniques

Except on alloys of the 7000 series without copper, MIG or TIG arc welding of weldable alloys, when performed under proper operating conditions with appropriate filler metal, does not decrease the corrosion resistance of the heat-affected zones (HAZ) and the welding bead [7]. These are not preferential corrosion sites, as has been shown in ship building, coastal equipment, etc.; nor is brazing a factor of corrosion, when done under appropriate metallurgical operating conditions. This is shown by the experience with automotive heat exchangers.

Other joining techniques such as riveting, bolting, clinching, etc., are not as such factors of corrosion, because they do not modify the metal structure like welding or brazing. However, their use is part of the principles of construction which will be discussed in Section 6.3.2.

6.3.2. Design principles

The term "design principles" here includes everything related to the final shape of a piece of equipment and to its fabrication. Inappropriate design may have a catastrophic effect on the corrosion resistance of the construction, even if the alloy was properly selected and the service conditions were appropriate [8, 9].

It is the responsibility of designers to steer clear of certain configurations which favour water retention and recesses where humid projections can accumulate, and where severe corrosion can develop [10]. Intermittent welds should not be used, or the gaps should be filled in order to prevent the penetration of moisture (Figure B.6.2).

6.3.3. Contacts with other metals and materials

Mechanical assemblies with other metals or graphite are possible, if the risk of galvanic corrosion is assessed properly and, if necessary, bypassed (see Section B.3.7).

The contact with nonmetallic materials does not lead to a risk of galvanic corrosion. Porous materials may favour the corrosion of aluminium (see Chapters G.3 and G.4).

6.3.4. Mechanical stresses

In certain cases, mechanical stresses may lead to stress corrosion (see Section B.2.5).

6.3.5. Maintenance

The periodic maintenance of equipment such as industrial equipment or in buildings increases its lifetime. Periodic elimination of dust and inorganic deposits significantly decreases the risk of pitting corrosion. Heat exchangers with aluminium are capable of a lifetime surpassing 10 years even in fertiliser plants with a high dust concentration, if maintained by water jet cleaning.

6.4.   THE CORROSION RESISTANCE OF ALUMINIUM ALLOYS

The corrosion resistance of aluminium mainly depends on the alloying elements, i.e. the series to which they belong. The types of corrosion they may undergo are listed in Table B.6.3.

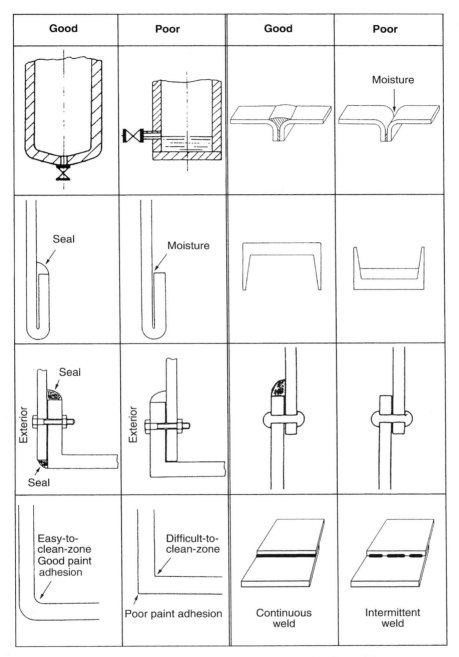

Good	Poor	Good	Poor

Figure B.6.2. Influence of design principles, examples of assemblies.

Table B.6.3. Types of corrosion of wrought alloy series

Series	Pitting	Uniform	Transcrystalline	Intercrystalline	Exfoliation	Stress corrosion
1000	×	×	×			
2000	×	×	×	×	×	×
3000	×	×	×			
5000	×	×	×	×	×	×
6000	×	×	×	×		
7000	×	×	×		×	×
8000	×	×	×			

6.4.1. Casting alloys

Alloys of the

–   40000 series that contain silicon but no copper: A-S3GT, A-S5G, A-S7G03 (42100), A-S7G06 (42200), A-S10G (43000) and A-S13 (44200),
–   50000 series containing magnesium: A-G3T (51100), A-G6 (51300),

show very good resistance to atmospheric corrosion.

The pitting depth on aluminium–silicon alloys does not exceed 200 μm after 10 years of exposure to marine or industrial atmospheres [11, 12]. The stress corrosion resistance of 42000 (A-S7G) alloys is excellent [13].

Alloys with magnesium have an excellent resistance to marine corrosion. They are used for upper works [14].

The use of alloys with copper of the 40000 family such as A-5UGT (21000 and 21100) should be avoided without protection, because their corrosion resistance is not satisfactory, especially in marine environment [15].

6.4.2. Wrought alloys

From a corrosion point of view, one must distinguish between:

–   strain-hardenable alloys of the 1000, 3000, 5000 and 8000 series,
–   age-hardenable alloys of the 6000 series, and
–   age-hardenable alloys of the 2000 and 7000 series.

■ Strain-hardenable alloys

These alloys have good corrosion resistance. Alloys of the 1000, 3000 and 5000 series are used for applications that require this level of resistance: construction, transport, various types of equipment, etc.

■ 1000 series

Within the composition limits of contemporary alloys, the iron and silicon content does not have a major influence on the corrosion resistance of semi-products of this family. Resistance to atmospheric corrosion is good. Pure alloys 1199 and 1190 are anodised when used outdoors.

■ 3000 series

Manganese leads to improved mechanical properties and corrosion resistance [16].

■ 5000 series

These are the strain-hardenable aluminium alloys with the best mechanical properties and corrosion resistance. Since 1960, they have been very widely used in marine applications such as shipbuilding, coastal equipment, and in road transport, for example as tanks for hydrocarbons and other liquid or pulverulent products.

Since the level of mechanical properties of these alloys increases with their magnesium content (Figure A.4.2), there was a natural tendency to push the magnesium level as high as possible, which is why rolled semi-products in A-G7 (7% manganese) and castings in A-G10 (11–12% manganese) called Aciéral were produced in the1950s. These products had such a high susceptibility to stress corrosion that they were quickly abandoned (Aciéral was fragile) [17].

In practice, in order to prevent these problems of extreme susceptibility to corrosion, the magnesium content of castings exposed to aggressive environments must be limited to 6%. According to the latest metallurgical knowledge and practical experience with this alloy series, the magnesium content of wrought alloys rarely exceeds 5%.

The solubility of magnesium in aluminium is very high at elevated temperatures, 15% at 450 °C, but at room temperature, it does not exceed 1% (Figure B.6.3). Therefore, magnesium will precipitate as soon as the temperature becomes less than the solidus temperature. As in all aluminium alloys, magnesium precipitation will start at the grain boundaries as Al_3Mg_2 (or Al_8Mg_5), commonly called β-phase. A micrographic observation can easily indicate whether the alloy is sensitised (Figure B.6.4).

For these alloys, precipitation tends to be continuous at grain boundaries. These intermetallics are anodic with respect to the grain: their potential is −1150 mV SCE, corresponding to a difference of approximately 300 mV with respect to the solid solution. That is quite high; therefore, in a corrosive environment, there is a real risk of intercrystalline corrosion and stress corrosion if precipitation is continuous at grain boundaries.

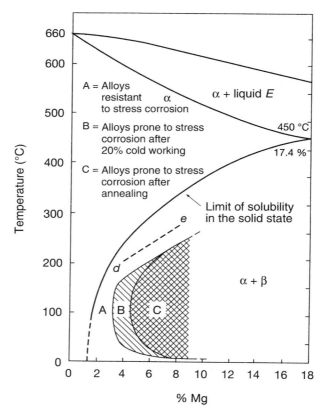

Figure B.6.3. Solubility of magnesium in aluminium [20].

Precipitation of the β-phase is quicker and more intense when

– the magnesium content is high,
– strain hardening is significant, and
– the service temperature is high.

As shown in Figure B.6.5, there is a temperature range, between 125 and 225 °C, in which the alloy is strongly sensitised. For alloys with a higher magnesium content, sensitisation may start at lower temperature, as low as 70 °C, as observed on rivets in A-G5 [18]. Temperature is not the only parameter which leads to precipitation of the β-phase; this is too often forgotten. Time, i.e. the duration of the exposure to elevated temperature, is also an important parameter, as shown in Figure B.6.6.

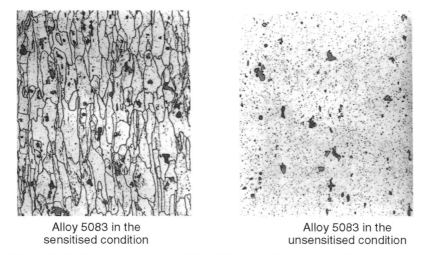

Alloy 5083 in the sensitised condition	Alloy 5083 in the unsensitised condition

Figure B.6.4. Micrographic structure of alloy 5083 in sensitised and unsensitised conditions.

This is not surprising, because the rate of migration of magnesium atoms to grain boundaries obeys the laws of diffusion in the solid state (aluminium matrix). Its temperature dependence follows the classic relationship

$$\tau = t \exp - (Q/T)$$

where τ is the diffusion distance, t time, T the absolute temperature, and Q a constant that depends on the element.

For a long time and rather arbitrarily, the limit operating temperature of alloys of the 5000 series containing more than 3.5% magnesium has been set at 65 °C. In fact, it is the product Time × Temperature that should be considered: at 65 °C, a continuous precipitation

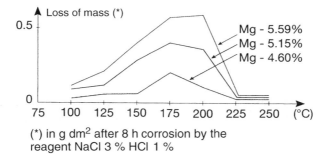

(*) in g dm² after 8 h corrosion by the reagent NaCl 3 % HCl 1 %

Figure B.6.5. Sensitisation of 5000 series alloys after maintaining them for 250 h at a given temperature. Influence of magnesium content and temperature [19].

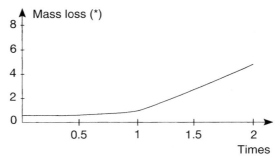

(*) in g dm² after 8 h etching in NaCl 3%, HCl 1%

Figure B.6.6. Sensitisation of 5000 series alloys at 65 °C. Influence of the duration of maintenance at that temperature [21].

at the grain boundaries of 5086 will form after 2 years (corresponding to 17 300 h), and several months are needed at 100 °C (see Figure B.6.7). The time effect is cumulative.

The document AD-Merkblatt W 6/1 "Aluminium and aluminium alloys malleable materials" from May 1982, published by the Vereinigung der technischen Überwachungsvereine e.V., D-4300 Essen 1, fixes the limit at 80 °C for the alloy AlMg4.5Mn, which is

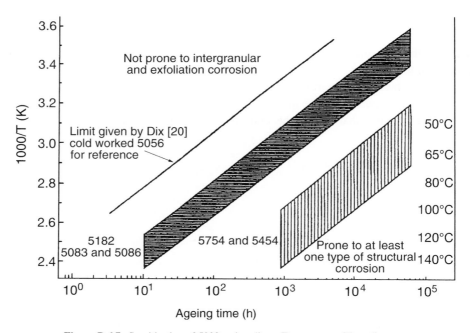

Figure B.6.7. Sensitisation of 5000 series alloys. Temperature–Time diagram.

equivalent to 5083. This document provides a tolerance up to 150 °C for periods not exceeding 8 h, under the condition that the operating pressure be reduced by one-half, and for periods not exceeding 24 h under the condition that the operating pressure is set at atmospheric pressure.

Sensitisation by precipitation at grain boundaries will not necessarily lead to corrosion. This depends on the environment, as shown by experience, even with heat exchangers operating in seawater at temperatures well above 65 °C. Certain diesel tank trucks have been used for over 20 years, filled each day at 65 °C and then operating for 8–10 h; this amounts to a cumulated holding time at 65–70 °C for more than 50 000 h.

Since the 1950s, all these parameters have been studied extensively [20]. Appropriate heat treatment during the manufacturing process may stabilise these alloys, leading to a discontinuous, pearl-chain-like precipitation of the β-phase at grain boundaries (Figure B.6.8) [21].

Temper H116 (and its equivalent H321) guarantees that alloys such as 5083, 5086, 5456 are prone neither to intercrystalline nor to exfoliation corrosion [22]. However, if the service conditions are such that the β-phase precipitates at the grain boundaries, these tempers will not prevent corrosion. This temper is widely used for ship building [23].

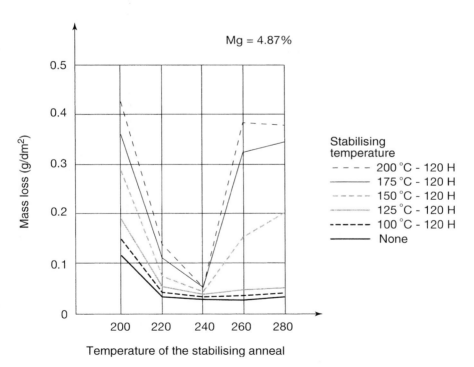

Figure B.6.8. Stabilising annealing of 5000 series alloys [19].

☐ **8000 series**

We will limit our discussion to alloys in which the level of iron and silicon has been purposely increased (8011, 8021) and will not deal with special or experimental alloys such as aluminium–lithium alloys.

These alloys are used in the manufacture of fins (thickness between 0.1 and 0.2 mm) for air–liquid heat exchangers. Adding iron results in an increase in the number density and in a better distribution of cathodic areas. In case of pitting corrosion, the number of pits will be increased, such that the pitting depth will remain very low, less than the thickness of the fins.

■ **Age-hardenable alloys of the 6000 series**

While slightly prone to intercrystalline corrosion, albeit generally only over a very small number of grain layers, these alloys are not prone to stress corrosion.

All of these rolled or extruded alloys have a good resistance to atmospheric corrosion.

The addition of copper, in excess of 0.50%, leads to a modification of the corrosion resistance of the alloys, such as 6056.

■ **Age-hardenable alloys of the 2000 and 7000 series**

Aeronautical construction is the main field of application of alloys of the 2000 series, as well as of copper-containing alloys of the 7000 series. The development of these metallurgically complex alloys as well as the requirements of the aeronautical industry in terms of resistance to intercrystalline corrosion and stress corrosion (and fatigue resistance) have triggered an impressive number of studies that have been conducted both by manufacturers of aluminium semi-products and by aeronautical constructors.[1]

These studies have resulted in metallurgical solutions (compositions, elaboration processes), in transformation schedules and heat treatments that are characteristic for the semi-products in these alloys, and which form a part of the know-how of manufacturers of semi-products.

Besides for aeronautical applications, which are beyond the scope of this volume, these alloys are mainly used in the mechanics industry. In most cases, they are used in the temper of delivery, because the users do not carry out further heat treatments. We will, therefore, limit our discussion of the corrosion resistance of these alloys to the application and assume that all aspects prior to corrosion (i.e. the influence of the composition, heat treatments, etc.) are taken into account by the manufacturer of the semi-product. This implies that if the user proceeds to any heat treatments during the further fabrication steps

---

[1] Since the first volume (1907) of the *Chemical Abstracts*, 300 abstracts discuss the corrosion of 2000 series alloys, and about 500 corrosion of 7000 series alloys.

of the product, he must strictly follow the supplier's recommendations; any deviation from these recommendations may result in a significant reduction in the corrosion resistance of the resulting products.

☐ 2000 series alloys

These were the first age-hardenable aluminium alloys; the development of their compositions, transformation schedules and heat treatments started in the 1920s.

As mentioned in Section B.6.2.3, the corrosion resistance of these alloys depends to a large extent on the heat treatment conditions: the quenching rate must be as high as possible and ageing conditions should tend towards overageing. Otherwise, these alloys will be sensitised to intercrystalline corrosion and stress corrosion.

It has long been known that even when delivered in appropriate tempers, their susceptibility to corrosion is, in general, too high to allow their use without protection in most environments: atmosphere, water, etc. [24]. Protections commonly used are paint, cladding with 1050A, and anodising. They are, however, not sufficient to compensate these alloys' intrinsically low corrosion resistance due to their high copper content.

The quenching rate depends on the thermal exchange between the hot metal and the cold water. For sheet a few millimetres thick, quenching is practically instantaneous. With increasing thickness, the quenching rate in the centre of the metal decreases and may reach values so low that sensitivity to intercrystalline corrosion will be perceptible.

In a thick bar 100 mm thick, the quenching rate decreases from the surface to the core (Figure B.6.9). Therefore, at a certain distance from the surface, sensitisation to intercrystal-

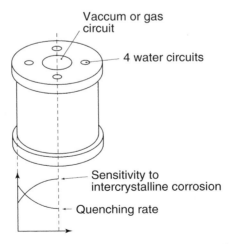

Figure B.6.9. Quenching rate and susceptibility to intercrystalline corrosion.

line corrosion will occur. For example, when machining a valve that is water-cooled by channels machined within this sensitised zone, exfoliation corrosion may occur in these channels. This example is not the result of a thinking process, but has actually been observed.

In spite of all kind of tricks used by the manufacturers of semi-products with high mechanical resistance, it is physically impossible to obtain a quenching rate in the core of the piece as high as that close to the surface. Designers should take this into account and select, if necessary, another aluminium alloy from the 6000 series (6082, 6005A, 6061, etc.). Experience shows that this is possible and preferable.

☐ Copper-containing alloys of the 7000 series

The best-known alloy among the copper-containing alloys is 7075. Owing to their very good mechanical properties, these alloys are used in more and more applications in the mechanics industry and for the manufacture of injection moulds for plastics.

They are prone to stress corrosion [25], especially in the short transverse direction, and to exfoliation corrosion. The remarks concerning the corrosion resistance of 2000 series alloys also apply to these alloys.

☐ Alloys of the 7000 series without copper

Unlike with copper-containing alloys, TIG and MIG welding is possible with alloys of the 7000 series without copper, such as 7020. During cooling, the HAZ is air quenched, and will exhibit mechanical properties close to a T4 temper. This is an advantage compared to alloys of the 5000 series in which the HAZ is in the annealed temper [26].

These alloys would have been very convenient for boiler making. However, it has been found that the HAZ is prone to exfoliation corrosion (Figure B.2.11). This is due to the precipitation of intermetallics of the Al_6Mn and $CrMnAl_{12}$ type, which will be aligned parallel to the rolling direction. Simultaneously, in the T4 temper, continuous zones enriched in Zn and Mg will form parallel to the grain boundaries; these zones are anodic [27, 28].

Welded 7020 is prone to intercrystalline corrosion [29] and in the T6 temper to stress corrosion [30].

In spite of a great deal of research, it has not yet been possible to control the exfoliation corrosion of 7000 series alloys without copper, which would otherwise have interesting properties for boiler making. Several metallurgical approaches are possible, consisting in modifying the composition, transformation schedules, etc. Another approach is the protection of the welded zones, which, however, is limited to applications that are subject of rigorous inspection and maintenance procedures. This explains the use of these alloys for military and space equipment. The protection of welded areas by thermal projection or plasma projection with aluminium–zinc alloys (containing less than 10% zinc) has also be envisioned [31, 32].

REFERENCES

[1] McKee A.B., Brown R.H., Resistance of aluminium to corrosion in solutions containing various anions and cathions, *Corrosion*, vol. 3, 1947, p. 595–612.

[2] Patrie J., Les phénomènes de passivation de l'aluminium immergé en milieu nitrique, *Revue de l'Aluminium*, vol. 194, 1952, p. 96, see also vol. 195, p. 5–11, vol. 196, p. 45–54 and vol. 197, p. 87–96.

[3] Whitaker M., A review of information on the effect of impurities on the corrosion resistance of aluminium, *Metal Industry*, vol. 80, 1952, p. 183–186, see also p. 207–212, 227–230, 247–251, 263–266, 288–289, 303–305, 331–332, 346–350 and 387–388.

[4] Nisancioglu K., Tusvik R., Corrosion of Stripo-cast aluminium alloys, *Aluminium*, vol. 64, 1988, p. 407–410.

[5] Principes à observer en vue d'améliorer les conditions de résistance à la corrosion de l'aluminium, *Revue de l'Aluminium*, vol. 77, 1936, p. 36–37.

[6] Lunder O., Nisancioglu K., The effect of alkaline-etch pretreatment on the pitting corrosion of wrought aluminium, *Corrosion*, vol. 44, 1988, p. 414–422.

[7] Blewett R.V., Skerrey E.W., The performance of aluminium welds in corrosive environments, *Metallurgia*, vol. 71, 1965, p. 73–81.

[8] Mears R.B., Brown R.H., Designing to prevent corrosion, *Corrosion*, vol. 3, 1948, p. 97–118.

[9] Elkington R.W., Design considerations for minimizing the corrosion of aluminium alloys, *Aluminium Industry*, vol. 5, 1986, p. 19–30.

[10] Rozenfeld I.L., *Crevice Corrosion of metals and alloys*, NACE, Houston, 1974, p. 373–398.

[11] Bowman J.J., Aluminum-base die-casting alloys, *ASTM Proceeding*, vol. 46, 1946, p. 225–232.

[12] Everhart J.L., Aluminium alloys casting, *Materials in Design Engineering*, vol. 47, 1958, p. 125–144.

[13] Speidel M.O., *Stress-corrosion cracking of cast aluminium alloys*, NATO, Advanced study Institute on stress-corrosion cracking, Denmark, 1975, p. 97–115.

[14] Les alliages de fonderie inoxydables, *Revue de l'Aluminium*, 1968, p. 882–893.

[15] Wood J., Harris D.A., Atmospheric corrosion tests on cast aluminium alloys, *British Foundryman*, vol. 74, 1981, p. 217–221.

[16] Develay R., Importance de l'addition de manganèse dans l'aluminium et les alliages de l'aluminium, *Revue de l'Aluminium*, 1978, p. 345–368.

[17] Perryman E.C., Haden S.E., Stress-corrosion of aluminium–7% magnesium alloy, *Journal of the Institute of Metals*, vol. 77, 1950, p. 207–235.

[18] Metcalfe G.J., Intercristalline corrosion of aluminium-magnesium rivets, *Journal of the Institute of Metals*, vol. 72, 1946, p. 487–500.

[19] Guilhaudis, A., *Influence de la teneur en magnésium sur la résistance à la corrosion sous tension des alliages Al–Mg*, rapport Pechiney SREPC, septembre 1959.

[20] Dix E.H., Anderson W.A., Shumaker M.B., *Development of wrought Aluminium–Magnesium alloy*, Alcoa, technical paper No. 14, 1958.

[21] Guilhaudis A., Traitements thermiques de stabilisation des alliages aluminium-magnésium à 5% contre les effets de chauffage à basse température, *Revue de l'Aluminium*, vol. 223/224, 1955.

[22] Czyrca E., Hack H.P., *Corrosion of aluminum alloys in exfoliation resistant tempers exposed to marine environments for two years*, Naval Ship Research and Development Center, report AD/A-002 234, November 1974.

[23] Brooks C.L., Aluminium–Magnesium alloys 5086 and 5456 H116, *Naval Engineers Journal*, 1970, p. 29–32.

[24] Rawdon H.S., *Corrosion embrittlement of duralumin. Practical aspects of problems*, US National Advisory Commision of Aeronautics, technical note No. 285, 1928, p. 1–11.

[25] Brown R.H., Sprowls D., Shumaker M.B., Influence of stress and environment on the stress-corrosion cracking of high strenght aluminium alloys, *Conference AGARD Proceedings*, vol. 53, 1969, p. 3.

[26] Chevigny R., Develay R., Guilhaudis A., Petrequin J., Les alliages aluminium–zinc–magnésium, *Revue de l'Aluminium*, vol. 334, 1965, p. 973–988.

[27] Reboul M., Bouvaist J., Étude du mécanisme de la corrosion feuilletante de l'alliage d'aluminium 7020, *Revue de l'Aluminium*, 1980, p. 41–53.

[28] Adenis D., Guilhaudis A., Relation entre l'état structural et la sensibilité à la corrosion feuilletante de l'A-Z5G, *Mémoires scientifiques de la Revue de Métallurgie*, vol. 64, 1967, p. 877–889.

[29] Baumgartner M., Kaesche H., Intercristalline corrosion and stress-corrosion cracking of AlZnMg alloys, *Corrosion*, vol. 44, 1988, p. 231–239.

[30] Reboul M., Dubost, B., Trentelivres, G., *Corrosion sous contrainte des soudures de réservoirs en 7020 du lanceur Ariane*, rapport Pechiney-CRV 1461, septembre 1980.

[31] Greenbank J.C., Andrews P., Birley S.S., Prevention of exfoliation corrosion in welded AlZnMg structures by thermal sprayed metal coatings, *Second Conference on Surface Engineering*, Stratford upon Avon, June 1987.

[32] Hepples W., Holroyd N.J., The corrosion protection of weldable 7XXX aluminium alloys by aluminium based arc sprayed coatings, *Aluminium Industry*, vol. 12, 1993, p. 14–19.

Part C

Atmospheric Corrosion of Aluminium

Chapter C.1
Atmospheric Corrosion

Chapter C.1
Atmospheric Corrosion

Due to its excellent resistance to atmospheric corrosion, the use of aluminium in construction, civil engineering, electrical power transmission lines [1] and transport has been increasing considerably since 1930. Nowadays, aluminium is the second most common metal, after steel, to be exposed to weathering, in all climate and geographic zones.

The resistance of aluminium to atmospheric corrosion has been a very important issue, and, since the early 1930s, has attracted a great deal of attention from corrosion experts working with the major aluminium producers in Europe and North America. The first tests of aluminium alloys in outdoor corrosion testing stations were performed in the United States in 1931 by ASTM [2].

Atmospheric corrosion has been the subject of many publications.[1] The first was by E. Wilson, who reported results of observations made on electrical cables exposed in London over 24 years, beginning in 1902 [3]. Since 1945, many national and international conferences in Europe and in the United States have been devoted to the atmospheric corrosion of aluminium (as well as that of other common metals and alloys such as steel, copper, etc.).

From the very beginning of the industrial history of aluminium, several outdoor applications were executed, some of which are still in use: the Eros statue, erected in 1893 on the top of the monument dedicated to the memory of Lord Shaftesbury at Piccadilly Circus in London [4], and the roof sheet of the San Gioacchino Church in Rome, build in 1898 (see Chapter C.5). Other applications, mainly in the building sector, were to follow during the 1920s and 1930s both in Europe and in the United States.

The widespread use of aluminium began in the 1950s in the building sector: claddings, curtain walls, roofing sheet, metallic fittings, as well as in civil engineering: street signs, road signs, street furniture, etc. These applications use both cast and wrought (flat rolled and extruded) products. These products can be used without any protection or can be protected by anodising, painting, lacquering, etc.

Our present knowledge of the resistance of aluminium to atmospheric corrosion has solid foundations, based on two complementary approaches:

- testing in outdoor corrosion testing stations in Europe and the United States (where most, and the oldest, of these stations can be found), often for a very long time: 10, 20

---

[1] Over 400 publications are referenced in *Chemical Abstracts*, from the first volume (1907) to volume 124 (1997).

years, and sometimes even longer, and under three classic atmospheric conditions—marine, industrial and rural;
—   experience with applications over several decades: some of the oldest and most prestigious have been reported by aluminium producers and transformers.

Our understanding of the atmospheric corrosion of metals is based on theoretical foundations that are a special case of corrosion of metals and alloys. Before discussing the specific aspects of aluminium, they need to be recalled here.

1.1.   THE NATURE OF ATMOSPHERIC CORROSION

Atmospheric corrosion is the attack of a metal (or an alloy) by the atmospheric environment to which it is exposed. This corrosion is caused by the simultaneous attack by rainwater or condensing water, oxygen contained in the air, and atmospheric pollutants. Atmospheric corrosion is a special type of corrosion because the electrolyte is represented by a thin film of moisture, whose thickness does not exceed a few hundred micrometres. It can be assumed that such a film is always saturated with oxygen, and that diffusion is not hindered. This type of corrosion may be intermittent, because it stops when the metal's surface is no longer humid. When immersed in water or in a salt solution, the metal is in permanent contact with the electrolyte, but the corrosion may be slowed down by the weak diffusion of oxygen to cathodic sites.

The first theoretical explanation of atmospheric corrosion of metals was given by Vernon [5] and Hudson [6] starting in 1923 and was completed later by Rozenfeld in the 1960s [7] and by Graedel for aluminium in the 1980s [8]. Vernon introduced the concept of the critical degree of moisture, the threshold below which practically no corrosion will occur. The value of this threshold depends on several factors such as the nature and concentration of atmospheric pollutants and the metal's surface condition.

REFERENCES

[1] Herenguel J., Examen de câbles conducteurs en Almelec et en aluminium-acier déposés après 15 à 25 ans de service, *Revue de l'Aluminium*, 1947, p. 357–360.
[2] Ailor W.H., ASTM Atmospheric corrosion testing: 1906 to 1976, *ASTM*, STP, 646, 1978, p. 129–151.
[3] Wilson E., The corrosion products and mechanical properties of certain light aluminum alloys as affected by atmospheric exposure, *Proceedings of the Physical Society of London*, vol. 30, 1926, p. 15–25.
[4] Sutton R.S., Selligmann R., An aluminium statue of 1893: Gilbert's Eros, *Journal of the Institute of Metals*, vol. 60, 1938, p. 67–74.

[5] Vernon W.H.J., First experimental report to the atmospheric corrosion research comite, *Transaction of the Faraday Society*, vol. 19, 1923–1924, p. 839–934.

[6] Hudson J.C., The effects of two years' atmospheric exposure on the breaking load of hard drawn non-ferrous wires, *Journal of the Institute of Metals*, vol. 44, 1930.

[7] Rozenfeld I.L., Atmospheric corrosion of metals, English language edition translated by Tytell B.H, *NACE*, Houston, 1972.

[8] Graedel T.E., Corrosion mechanisms for aluminium exposed to the atmosphere, *Journal of the Electrochemical Society*, vol. 136, 1989, p. 204c–212c.

Chapter C.2
The Parameters of Atmospheric Corrosion

Chapter C.2
The Parameters of Atmospheric Corrosion

Experience has shown that for a given metal or alloy, the resistance to atmospheric corrosion may differ significantly from one site to another. For example, the corrosion rate of galvanised steel may vary from 1 to 100 between a semi-arid zone and the atmosphere of a coastal industrial estate [1]. For a given condition, the resistance to atmospheric corrosion may also vary from one metal to another; this variation can be fairly substantial.

The resistance to atmospheric corrosion depends on several factors that are related to

- *climatic conditions of the site*: humidity, pluviometry, temperature, hours of sunshine;
- *pollution*: concentration of sulphur dioxide (SO_2), nitrogen oxides (NO_x), the quantity and nature of dust.

For the sake of convenience, these factors are usually studied separately. However, as we will see, interactions may exist between them. For example, pollution may lower the critical degree of relative humidity at which corrosion starts to develop. These factors may also have contradicting effects: rain increases humidity and thus the aggressiveness of the atmosphere, but washes away accumulated dust and may thus actually reduce the corrosion rate.

2.1. RELATIVE HUMIDITY

The rate of atmospheric corrosion is related to the relative level of humidity of air rather than the pluviometry of the site, which is one but not the only factor that determines the relative degree of humidity.

The level of relative humidity, also called the hygrometric degree of air, is the ratio between the vapour pressure e of the water and the maximum vapour pressure e_w (i.e. at saturation) that can occur at a given temperature

$$RH = 100 \times \frac{e}{e_w}$$

This equals the ratio of the quantity of water vapour actually contained in air and the maximum quantity at a given temperature, as calculated from the Clausius–Clapeyron equation.

At ambient temperature, air is considered to be

–   dry, if RH < 30%,
–   normal, if 50 < RH < 60%,
–   humid, if RH > 80%, and
–   saturated, if RH = 100%.

In deserts and arid zones, the relative level of humidity rarely exceeds 10–20%, while it ranges between 40 and 60% in temperate climates. During rainfall, it may reach 90–95%. In tropical zones it is close to 100% during the rainy season.

The "dew point" is defined as the temperature at which condensation will start. For a given level of relative humidity, this is the temperature at which air needs to be cooled in order to become saturated with moisture so that condensation will begin. The dew point depends on the level of relative humidity. The decrease in temperature varies from 0 to 6 °C in humid air (Table C.2.1).

Table C.2.1. Dew point and relative humidity

Ambient temperature (°C)	Relative humidity (%)	Vapour pressure of water	Dew point, τ (°C)	Temperature difference, $\Delta = T - \tau$ (°C)
10	100	9.2040	10	0
	90	8.2836	8	2
	80	7.3632	6	4
	70	6.4428	5	5
15	100	12.7840	15	0
	90	11.5047	13	2
	80	10.2264	11	4
	70	8.9481	9	6
20	100	17.5297	20	0
	90	15.7767	18	2
	80	14.0237	16	4
	70	12.2707	14	6
25	100	23.7530	25	0
	90	21.3770	23	2
	80	19.0020	21	4
	70	16.6270	19	6
30	100	31.8220	30	0
	90	28.6330	28	2
	80	25.4576	26	4
	70	22.2754	24	6
35	100	42.1770	35	0
	90	37.9593	33	2
	80	33.7416	31	4
	70	29.5239	24	6

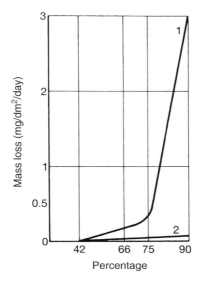

Figure C.2.1. Influence of moisture on atmospheric corrosion of aluminium (from Rozenfeld).
Curve 1: with 1% SO_2. Curve 2: without SO_2.

Vernon showed that there is a critical threshold of relative humidity below which common metals do not corrode because there is insufficient moisture to create an electrolyte film on the surface of the metal. On a freshly prepared surface and in unpolluted air, the critical threshold of humidity for aluminium is on the order of 66% (Figure C.2.1).

Atmospheric corrosion of metal develops in thin moisture films deposited on the metal's surface. The film thickness rarely exceeds a few hundred micrometres, except during the rainy season. It increases with the air's relative degree of humidity [2].

When dissolving pollutant gases, especially sulphur dioxide, and inorganic components that are present in dust, the moisture film will become a more or less conductive electrolyte, depending on the quantity and the nature of dissolved elements. Their concentration can exceed 100 $mg \cdot l^{-1}$. These thin films offer a very low resistance to oxygen diffusion. As a consequence, certain electrochemical reactions that involve oxygen may take place more readily than when the metal is immersed in a liquid.

Rozenfeld showed that the intensity of corrosion is very weak in thin films (Figure C.2.2). There are two reasons for this:

–   oxygen from the air contributes to the formation of the natural oxide layer, and thus the thinner the film, the easier its diffusion;
–   the low quantity of available water limits the dissolution of certain elements such as corrosion products, which will thus remain at the surface and protect it to a certain extent.

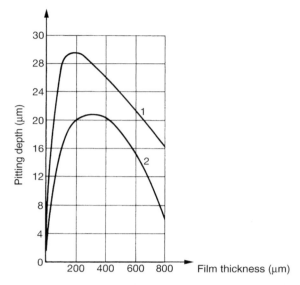

Figure C.2.2. Influence of moisture film thickness on pitting depth. Curve 1: condensed film. Curve 2: permanent moisture film (from Rozenfeld).

The critical threshold of relative humidity is not a constant value but depends on the metal's surface condition [3]. A rough surface with scratches or dust or with corrosion products or salts deposited on the surface will decrease the critical degree of relative humidity by capillarity [4]. These factors also favour condensation (Figure C.2.3) [5].

2.2.   RAIN

The chemical composition of rainwater depends on the atmosphere. It has been calculated that 1 l of rainwater collected at an altitude of 1000 m will wash out 326 m^3 of air and

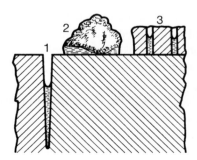

Figure C.2.3. Influence of deposit and surface condition: (1) microcracks, (2) dust, (3) porous corrosion product [5].

dissolve gaseous pollutants and soluble dust constituents present in the atmosphere. Rainwater thus contains inorganic components. Its normal pH is 5.6. This value results from equilibrium with carbon dioxide contained in the air. So-called acid rains have an even lower pH that can decrease to about 4. The acidity of these rains is due to the uptake of acidic compounds, mainly sulphates and nitrates [6].

The influence of rain on the atmospheric corrosion of metals in general, and aluminium in particular, is complex [7]. Rain maintains the level of humidity of air above the critical level at which corrosion starts to develop.

The first rains, in particular, may provide rather acidic inorganic compounds, but they also clean the surface and thus eliminate dust and acidic deposits originating from gaseous atmospheric pollutants that have accumulated prior to that rainfall. Furthermore, they dissolve or wash out corrosion products that could slow down corrosion.

Experience has shown that at a given location, surfaces that are exposed to rain generally have a better corrosion resistance than surfaces that are rarely or never cleaned by rain. This applies to the backside (directed towards the soil) of test pieces exposed in outdoor testing stations, but also to buildings, where parts that are rarely or never cleaned by rain always show more severe corrosion that those exposed to rain. From this point of view, the contemporary fashion in architecture consisting of orienting corrugated cladding sheet horizontally rather than vertically may eventually lead to degradation of those zones that are poorly cleaned by rainfall, and especially at their horizontal borders (Figure C.2.4).

Figure C.2.4. Cladding with horizontal corrugation.

Dust will accumulate on the upper part of the corrugation as everywhere, but this zone will not be rinsed by rain. When the level of humidity is high, and during condensation, the water film will dissolve inorganic components of the dust. This solution may not be washed away totally and may even remain on certain parts of the surface, which may thus corrode preferentially, because of the high concentration of more or less aggressive inorganic constituents. This is similar to the well-known "drop effect" at shard edges.

2.3. FOG

Fog absorbs gaseous air pollutants and contains moisture in excess of the critical threshold of relative humidity. Unlike rain, fog does not clean surfaces. It, therefore, is a much more aggressive environment for metals and other materials than rain. It is well known that humid rooms degrade more quickly than others do.

2.4. CONDENSATION

Condensation occurs when, at a given level of relative humidity, the temperature drops to the dew point τ. The lower the temperature difference $T - \tau$, the higher the relative degree of humidity (Table C.2.1). It amounts to $5-7\,°C$ at a relative humidity level of 80% and more, and to $20-25\,°C$ for a very low level of relative humidity (Figure C.2.5). Therefore,

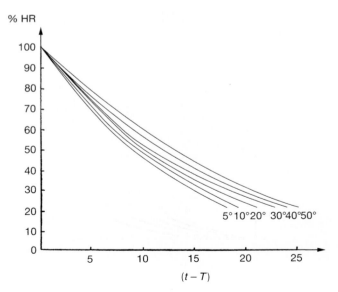

Figure C.2.5. Temperature difference $T - \tau$ as a function of RH (from Rozenfeld).

Table C.2.2. Quantity of water contained in saturated air at 1013 mbar

Temperature (°C)	Water content (g·m^{-3})									
	0	1	2	3	4	5	6	7	8	9
0	4.84	5.18	5.54	5.92	6.33	6.76	7.22	7.70	8.22	8.76
10	9.38	9.94	10.57	11.25	11.96	12.71	13.51	14.34	15.22	16.14
20	17.12	18.14	19.22	20.36	21.55	22.80	24.11	25.59	26.93	28.45
30	30.04	31.70	33.45	35.28	37.19	39.19	41.28	43.47	45.75	48.14

in very dry climates condensation is low, except when the temperature difference between day and night is very high.

The quantity of water contained in air (Table C.2.2) does not show a linear variation with temperature (Figure C.2.6). Therefore, for a given level of relative humidity, the quantity of condensed water will be higher, the greater will be the temperature difference $T - \tau$ and the temperature T.

In a given area, the frequency at which condensation occurs depends both on the level of relative humidity and the daily temperature difference; these two parameters are not related to each other. This frequency thus depends on local meteorological conditions. In many tropic and humid areas, the daily temperature differences may stay below $T - \tau$ for many days per year.

For a metallic surface, condensation does not depend solely on the difference $T - \tau$. The surface condition may also lead to condensation, by the simple effect of capillarity

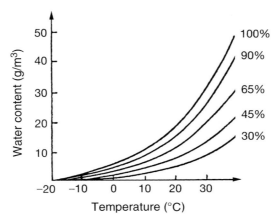

Figure C.2.6. Quantity of water vapour as a function of temperature and relative humidity level (from Rozenfeld).

resulting from surface irregularities (scratches) and superficial deposition (corrosion products, dust), even in the absence of a change in the ambient temperature. This effect is caused by a change in the surface tension of water at these surface irregularities. The saturated vapour pressure above a convex surface is lower than that above a plane surface. As a consequence, at a given level of relative humidity, saturation may occur above surface irregularities, resulting in condensation, which explains the damaging influence of scratches and surface deposits on metallic surfaces.

Experience shows that there are more condensation sites on common metals such as steel, aluminium, and zinc than on noble metals such as silver, gold, and platinum.

The effect of condensation on the corrosion resistance of metals depends on several factors such as the shape, speed, duration of condensation, and the level of relative humidity. Pitting corrosion is more intense under condensation than under a water film of the same thickness (Figure C.2.2). When condensation occurs quickly, or for longer periods of time, the small drops have a tendency to join and form a continuous film on the surface of the metal, under which corrosion is more dispersed and more superficial. A zone where condensation occurs will always corrode more severely than the others will, because the higher moisture level maintains corrosion; pitting, often superficial, will be denser. When condensation occurs at ceilings, pits in the centre of more or less grey aureoles can often be seen. In buildings, condensation contributes significantly to degrading the general appearance.

2.5. TEMPERATURE

The experience of many applications as well as tests in outdoor corrosion testing stations have shown that temperature does not have a major influence on atmospheric corrosion of metals; this applies, of course, to aluminium. It is generally admitted that between -26 and $+25\,°C$, temperature has no effect. Although tropical zones are more humid and hotter, no significant differences compared to tests in temperate zones have ever been reported. In fact, like most common metals, aluminium is more sensitive to the pollution level than to the temperature level. This explains why in temperate zones, the corrosion rate is only slightly influenced by seasonal variations in temperature. The effect of an increase in temperature during the summer is compensated in winter by higher emissions of sulphur dioxide due to combustion of fossil fuels (Figure C.2.7).

2.6. GASEOUS POLLUTANTS

Gaseous pollutants commonly present in polluted atmospheres are carbon dioxide (CO_2), sulphur dioxide (SO_2), and nitrogen oxides (NO_x). Locally, depending on the type of

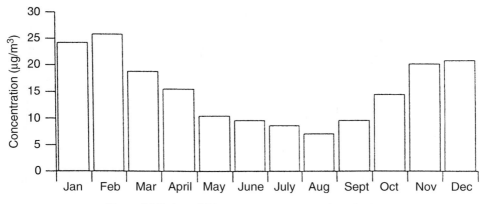

Figure C.2.7. Annual SO_2 concentration in the air in Paris [8].

industry and the efficiency of their filtering devices, other gaseous pollutants can be found such as

 - ammonia (NH_3), close to fertilising plants,
 - hydrogen sulphide (H_2S), close to sewage treatment plants or petrochemical plants,
 - chlorine (Cl_2) close to incinerators of domestic waste, and
 - vapour of organic solvents, close to manufacturing plants for paint, lacquers, etc.

2.6.1. Sulphur dioxide (SO_2)

Sulphur dioxide originates essentially from the combustion of gas oil and coal. Sources are, therefore, thermal power stations generating electricity, central or individual heating plants and industrial plants. Transport (cars, industrial vehicles) is not a significant source of sulphur dioxide, especially since in Europe the sulphur concentration in diesel fuel has been decreased from 0.20 to 0.05%.

The concentration of sulphur dioxide (SO_2) in the atmosphere varies within a very wide range, depending on the activity of the site, the type of collective or individual heating, the means of transport and the main wind direction. For typical atmospheres, the usual concentrations are

 - rural: $0-0.1$ mg$\cdot$m^{-3}
 - urban: $0.1-0.3$ mg$\cdot$m^{-3}
 - industrial: >0.3 mg$\cdot$m^{-3}

The sulphur dioxide concentration of the atmosphere of large cities varies within a wide range (Table C.2.3).

Table C.2.3. Sulphur dioxide concentration of urban atmosphere (1978) [9]

City	Average values (mg·m^{-3} of air)	Maximum values (mg·m^{-3} of air)
Paris	0.34–0.58	0.78
Marseille	0.18–0.33	0.66
Rotterdam	0.20–0.24	0.93
London	0.15–0.41	4.80
Basle	0.06	0.17
Munich	0.1–0.5	
Milan	0.59–0.70	

In non-industrial urban areas, sulphur dioxide emissions are much higher during winter than in summer. Over a long period, the sulphur dioxide concentration has decreased (Figure C.2.8) because of the anti-pollution policies, the evolution of the industrial structure resulting in the disappearance of heavy industries, and energy saving.

Sulphur dioxide is highly soluble in water (Table C.2.4) and dissolves easily in the moisture film present on the surface of the metal. With water, it forms sulphuric anhydride (H_2SO_3) and then sulphuric acid (H_2SO_4). This reaction is catalysed by soot and certain dusts. The medium becomes acidic: pH values of 3 or 4 are often found.

The moisture film will saturate quickly, at a level that depends on the level of relative humidity (Figure C.2.9). The sulphur dioxide content leads to a decrease in the critical threshold of relative humidity, as shown in Figure C.2.10.

Aluminium is much less sensitive to the attack by sulphur dioxide (SO_2) than steel, galvanised steel and zinc. It has been shown that during exposure to industrial atmosphere, aluminium absorbs 1 mg·dm^{-2} of SO_2, while steel absorbs between 22 and 55 mg·dm^{-2} of SO_2 under the same conditions [11].

The action of sulphur dioxide is moderate, even at high concentrations, and depends substantially on the level of relative humidity, as shown by results obtained on alloy 3003 (Table C.2.5). The concentrations of sulphur dioxide used during these tests were out of proportion with the highest pollution levels found in urban areas (Table C.2.3).

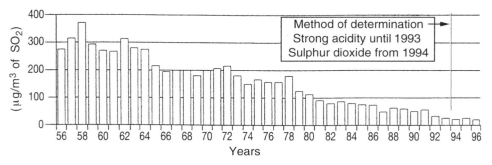

Figure C.2.8. Sulphur dioxide concentration in Paris since 1956 [8].

Table C.2.4. Solubility in water

Gas	0 °C	20 °C	25 °C
SO_2	228	113	94
CO_2	3.3	1.69	1.45

The reaction of aluminium with sulphur dioxide is not completely understood. In atmospheres that contain SO_2, the corrosion products of aluminium include aluminium sulphate $(Al_2(SO_4)_3 \cdot 18H_2O)$.

2.6.2. Carbon dioxide (CO_2)

Like sulphur anhydride, carbon dioxide originates from combustion of coal and oil. However, it is much less soluble in water, about 70 times less (Table C.2.4). It has no influence on the rate of atmospheric corrosion of aluminium.

2.6.3. Hydrogen sulphide (H_2S)

Only certain industries emit hydrogen sulphide: the petrochemistry and paper industries and possibly water purification plants. Except at certain particular sites, the concentration in air is usually low, on the order of one part per million (ppm).

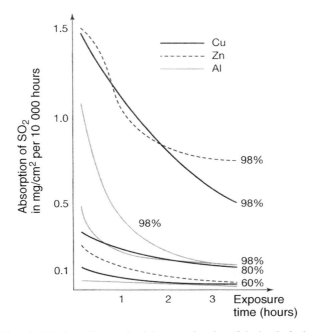

Figure C.2.9. Sulphur dioxide absorption on aluminium as a function of the level of relative humidity [10].

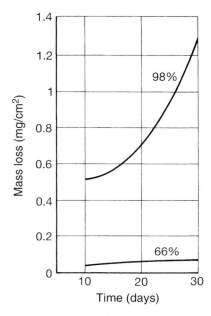

Figure C.2.10. Influence of the level of relative humidity on the atmospheric corrosion of 2024 in the presence of 1% SO$_2$ (from Rozenfeld).

Hydrogen sulphide has no effect on the atmospheric corrosion of aluminium, whatever its concentration and the level of relative humidity. Laboratory tests have shown that aluminium will not be attacked in air containing up to 5% of H$_2$S at 100% relative humidity [12]. Aluminium resists remarkably well in atmospheres charged with hydrogen sulphide such as the atmosphere close to petrol refineries and petrochemical plants in general.

2.6.4.   Ammonia (NH$_3$)

Ammonia originates from certain chemical productions (fertilisers, nitric acid, etc.) and from water purification plants.

Table C.2.5. Effect of sulphur dioxide

Concentration of SO$_2$ in air		Mass loss (mg·cm^{-2})	
(%)	(mg·m^{-3})	RH 66%	RH 98%
0.01	285	0.1	0.15
0.1	2850	0.1	0.60
1	28 500	0.1	1.80

Ammonia in air, even at high concentrations, has no effect on the atmospheric corrosion of aluminium. Aluminium can thus be used for roofing sheet and cladding panels in plants that produce ammonia, nitric acid and nitrogen containing fertilisers (urea, ammonium nitrate).

2.6.5. *Nitrogen oxides (NO$_x$)*

Combustion engines, mainly car engines, are the main source of nitrogen oxides. They emit nitrogen monoxide (NO), which is oxidised to nitrogen dioxide (NO$_2$). In the Paris area, their level is on the order of 50 mg·m^{-3}, a value that does not show significant seasonal variation.

Under the effect of solar radiation and in the presence of moisture and catalysts, nitrogen oxides are likely to form nitric acid that is diluted in the moisture film. Their effect on atmospheric corrosion of metals, including aluminium, is generally not well understood [13]. It can be inferred from experience with cladding panels and roofing sheet used in nitric acid and fertiliser plants that aluminium resists vapours containing nitrogen oxides well.

2.6.6. *Volatile organic compounds*

Volatile organic compounds include many products such as volatile hydrocarbons (alkanes, alkenes, aromatic compounds), carbonyl compounds (ketones, aldehydes), etc. In urban areas, they originate from motor vehicle exhaust gases, the evaporation of gasoline at filling stations, liquid fuels and industrial activities using solvents. Their concentration may reach 50 µg·m^{-3} in the atmosphere of large urban areas.

Given their organic nature and their low concentration in air, and given that most of them are insoluble in water, it is unlikely that these compounds have any effect on the atmospheric corrosion of aluminium. Experience has shown that a localised source of emission of chlorinated solvents, which used to exist in certain activities (dyeing, degreasing, etc.), could lead to severe corrosion of aluminium equipment exposed to this aggressive environment.

2.6.7. *Chlorine (Cl$_2$)*

In urban areas, chlorine emission originates mainly from the untreated smoke of household waste incineration plants burning plastic packaging, especially PVC-containing waste. In the atmosphere, chlorine in contact with moisture is transformed into hydrochloric acid.

Chlorine emissions accelerate the atmospheric corrosion of all common metals, including aluminium. The higher the chlorine concentration and especially the higher the level of relative humidity the more this will be accelerated. This has been shown by test results at high chlorine levels (which are out of proportion with those found in an atmosphere polluted by chlorine emissions) (Table C.2.6).

Table C.2.6. Effect of chlorine

Concentration of Cl_2 in air		Mass loss ($mg \cdot cm^{-2}$)	
(%)	($mg \cdot m^{-3}$)	RH 66%	RH 98%
0.01	317	0.05	0.05
0.1	3170	0.1	0.1
1	31 700	1.8	7.50

2.6.8. Chlorides (Cl^-)

Chlorides found in the atmosphere are of marine origin. The wind sweeping over the oceans carries them over tens or even hundreds of kilometres. Their effect is particularly significant in coastal areas, up to a distance of a few kilometres from the shore [14]. They play an important role in the corrosion of metals. The aggressiveness of an environment is closely related to the chloride content of the atmosphere.

Locally, chlorides may also stem from the pollution emitted by plants using chloride solutions for the production of chlorine or sodium hydroxide, or from rock salt mining.

2.7. DUST

Atmospheric dust has several origins:

- cosmic dust,
- dust from the erosion of soil and rock,
- soot and smoke from the combustion of coal or fuel (industry, power plants, etc.), or produced by motor engines of cars and industrial vehicles, and
- particles of plant origin.

The composition of dust deposits depends on the activity of the region under consideration:

- *In urban and industrial areas*, dust originates mainly from the combustion of coal in industrial furnaces and the combustion of fuel in car engines. They contain soot and inorganic compounds as minute particles with a diameter of $20-50 \ \mu m$. Dust deposits can be very significant in industrial areas: they have been estimated between 1.2 and 1.4 $kg \cdot m^{-3}$ per year in certain industrial areas.
- *In rural zones*, inorganic dust mainly originates from fertilisers: they contain calcium carbonate, calcium sulphate and ammonium sulphate.
- *In coastal areas*, dust contains salt particles carried by the winds over the oceans. One cubic meter of air can carry 5 mg of salt. The amount of salt deposited on the soil

depends on the distance from the shore. Measurements taken in Nigeria have shown that the deposit of salt, which amounts to about 1 $g \cdot m^{-2}$ per year at the shore, adds up to 10 mg at a distance of 1000 km from the coast. The influence of the marine environment is nearly imperceptible a few kilometres from the shore.

In the vicinity of certain plants such a concrete plants, fertiliser plants and steel plants, the nature of the dust depends on the type of industry. If not equipped with dust filters, these types of plants may give rise to considerable dust deposits.

Dust is noxious to the corrosion resistance of all metals, including aluminium:

– Dust favours condensation by decreasing the critical level of relative humidity [15].
– Soot catalyses certain reactions between the moisture film on the metal and gaseous pollutants, especially sulphur dioxide and nitrogen oxides.
– Inorganic constituents of dust will dissolve in the moisture film, depending on their solubility in water.
– Dust, whatever its composition, maintains a certain level of local humidity at which corrosion may develop.

Corrosion products can play a role similar to that of dust.

Experience has shown that at a given level for all other parameters (relative humidity, degree of pollution), dusty surfaces always show more severe corrosion than clean surfaces. Often, dust is a first-order parameter for atmospheric corrosion of aluminium and can thus be more important than the quality of the atmospheric environment.

This is why rain is so important. Rain cleans the surfaces and washes away dust and soluble corrosion products. From this point of view, rain has a beneficial effect on aluminium's resistance to atmospheric corrosion, and often it is easy to perceive the difference between surfaces cleaned by the rain and surfaces that are poorly (or not at all) cleaned by the rain. Periodic cleaning of aluminium surfaces of buildings that eliminates dust and deposits (including corrosion products) thus contributes to their resistance to atmospheric corrosion and maintains a clean surface aspect. Likewise, the difference between buildings and equipment that are properly maintained and those that are not at all or only poorly maintained can be easily seen.

The experience with air–liquid heat exchangers comprising tubes with fins in aluminium shows that dust is a decisive factor in the durability of this equipment, which is often exposed to industrial atmosphere [16]. Dust accumulation at the bottom of fins combined with humidity forms a kind of acidic cataplasm that attacks aluminium. The heat exchange performance as well as the lifetime of this type of equipment are increased by periodical cleaning that eliminates dust accumulated at the bottom of the fins.

2.8. ALTERNATING PERIODS OF DAMPNESS AND DRYNESS

It is well known that the corrosion rate of common metals and alloys globally increases when submitted to repeated cycles of immersion and emersion in seawater, compared to permanent immersion. This also applies to atmospheric corrosion when the metals are submitted to alternating cycles of rain and sunshine. However, in this case corrosion is not as severe as observed in the case of alternated immersion–emersion in salt solutions.

Whether the samples are placed on inclined stands in outdoor corrosion testing stations or in constructions (buildings, etc.), on sheet products, a significant difference in surface aspect is always observed between the faces oriented upside and downside.

As an example, an alloy 6060 exposed at a 45° angle to industrial atmosphere for 1 year shows a pitting depth of

–   40 μm on the face directed upside,
–   65 μm on the face directed downside.

The upside face is dried more quickly by the moving air and the sun, and the moisture films are thinner than on the other side, which is much less aerated and never cleaned. The pitting depth and the pitting density are lower under thin films than under thick films (Figure C.2.2). This is one of the factors that explain the different surface aspects of the two faces. The frequent reconstruction of the moisture film facilitates the access of oxygen to

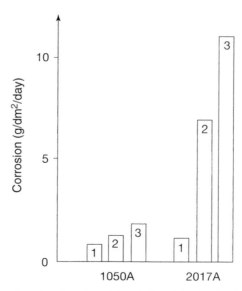

Figure C.2.11. Effect of periodic moistening with a 0.5 M NaCl solution for 30 days: (1) permanent immersion, (2) moistening once a day, (3) moistening six times a day (from Rozenfeld).

the metal surface, which promotes cathodic reactions and, as a consequence, anodic dissolution of the metal.

The influence of alternate periods of dampness and dryness depends heavily on the climate: when the relative moisture is high, the evaporation rate of the moisture films is slowed down, and in humid climates, the effect is therefore less pronounced than in a hot, dry climate: the higher the frequency of the dampness and dryness cycles, the greater the surface of the metal affected by pitting corrosion (Figure C.2.11). To put it differently: a high frequency of dampness and dryness will yield a more uniform surface appearance.

In general, aluminium and aluminium alloys, except alloys of the 2000 series and copper-containing 7000 series alloys, are much less sensitive to the effect of repeated dampness and dryness cycles than most other common metals.

REFERENCES

[1] Akimov G.V., Theory and methods of investigation of metals corrosion, *Izdatel'stvo. A. N. URSS*, 1945.
[2] Mikhailovski Y.N., Theoretical and engineering principles of atmospheric corrosion of metals, *Conference Atmospheric Corrosion*, Hollywood, October 1980.
[3] Brown P.W., Masters I.W., Factors affecting the corrosion of metals in the atmosphere, *Conference Atmospheric Corrosion*, Hollywood, October 1980.
[4] Tomashov N.D., Theory of corrosion and protection of metals, *Mc Millan*, New York, 1966.
[5] Boyd W.K., Fink F.W., *Corrosion of metals in the atmosphere*, Battelle, rapport AD-784 943, 1974.
[6] Lipfert F.W., Effects of acidic deposition on the atmospheric deterioration of materials, *Materials Performance*, vol. 26, 1987, p. 12–19.
[7] Aziz P.M., Godard H.P., Mechanism by which non-ferrous metals corrode in the atmosphere, *Corrosion*, vol. 15, 1959, p. 529t–533t.
[8] *Surveillance de la qualité de l'air en Île-de-France. Les résultats*, Airparif, 1996.
[9] Hasenberg L., Dechema, *Sulfur dioxyde Corrosion Handbook*, vol. 9, 1991.
[10] Sydberger T., Vannerberg N.G., The influence of the relative humidity and corrosion products on the adsorption of sulfur dioxide on metal surfaces, *Corrosion Science*, vol. 12, 1972, p. 775.
[11] Schikorr G., The sulfur dioxide repellent action of aluminum exposed to the influence of the atmospheres, *Aluminium*, vol. 43, 1967, p. 108–110.
[12] Shklovskii I.S., Investigations of metals corrosion, *Izdatel'stvo, A. N. URSS*, 1951, p. 241.
[13] Akimov A.G., Rozenfeld I.L. et al., Phase composition and structure of films formed on aluminium by interaction with an atmosphere containing water vapor and nitrogen peroxyde, *Zashchita Metallov*, vol. 17, 1981, p. 80–83.
[14] Duncan J.R., Ballance J.A., Marine salt's contribution to atmospheric corrosion, *ASTM*, STP 965, 1986, p. 316–326.
[15] Boyd W.K., Fink F.W., *Corrosion of metals in the atmosphere*, Battelle, rapport AD-784 943, August 1974.
[16] Wheeler K.R., Johnson A.B., May R.P., Aluminium alloy performance in industrial air-cooled applications, *ASTM*, STP 767.

Chapter C.3

Types of Atmospheres

Chapter C.3
Types of Atmospheres

In order to use, compare and extrapolate results obtained from outdoor corrosion testing stations,[1] corrosion specialists have defined a typology of atmospheres in which the test (or the use) takes place. This typology is based on the general climate of the zone and the presence of possible pollutants. Quantitative data have been collected, while the measurement methods and the equipment available have continued to progress.

3.1. THE CLASSIC TYPOLOGY

Corrosion experts have defined four typical atmospheres (Table C.3.1).
   This typology is based on two types of databases:

– *Meteorological data*: records of temperature, rainfall, hours of sunshine, major wind direction, etc.
– *Pollution data*: concentration of major identified pollutants.

   It also takes into account the economic activity of the site (agriculture, industry) and allows classifying the atmospheres as to their aggressiveness towards metals.
   Certain criteria are, however, difficult to quantify accurately, and the ranges are usually very wide.

3.2. STANDARDISED TYPOLOGY

In the 1970s, in order to facilitate the comparison of results, corrosion experts started to calibrate their outdoor testing stations on the basis of measurements taken on identical specimens of several metals and alloys such as steel, zinc, copper and aluminium. They were exposed in each participating testing station. These stations were thus classified based on their aggressiveness and the results compared.
   Another approach consists of establishing a link between the level of aggressiveness of a station, and more generally of a zone, and several meteorological criteria as well as

---

[1] The first tests were performed at the beginning of the 20th century, in 1906, by ASTM on coated steel products.

Table C.3.1. Typical atmospheres

Type of atmosphere	General characteristics	Rank of increasing aggressiveness
Rural	Low level of pollution. Inorganic dust: $CaSO_4$, $CaCO_3$, along with dust of plant origin	1
Tropical	Low level of pollution, or no pollution at all. Very high level of relative humidity, 80% or more	2
Urban industrial	Pollution with SO_2, CO_2. Dust from combustion of fuel and coal. Pollution due to motor vehicles	3
Marine, coastal	Humidity, salt-containing mist (NaCl)	4

the pollution levels. This makes it possible, at least in principle, to predict the aggressiveness of any location with respect to a metal (Figure C.3.1).

This approach was pursued from 1987 by the ISO's Technical Committee ISO/TC 156 within the ISOCORAG program, which followed up tests in about 50 corrosion stations in

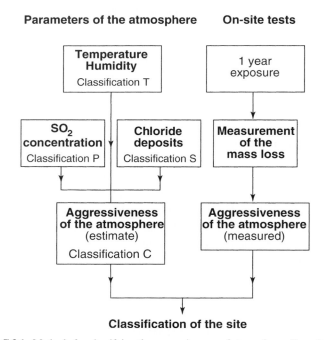

Figure C.3.1. Methods for classifying the aggressiveness of atmospheres (from Dean [3]).

Table C.3.2. ISO standards concerning atmospheric corrosion on metals

ISO standard	Title
9223	Corrosion of metals and alloys, corrosivity of atmospheres, classification
9224	Corrosion of metals and alloys, corrosivity of atmospheres, guiding values for the corrosivity categories
9225	Corrosion of metals and alloys, corrosivity of atmospheres, measurement of pollution
9226	Corrosion of metals and alloys, corrosivity of atmospheres, determination of corrosion rate of standard specimens for the evaluation of corrosivity

Europe, North America, South America, Japan and New Zealand [1]. This program led to four international standards (Table C.3.2) [2].

The aggressiveness of an atmosphere is determined from the following criteria:

- duration of exposure to humidity, T;
- pollution level by sulphur dioxide (SO_2), P;
- chloride deposits, S.

An index number is allotted to each criterion (Tables C.3.3–C.3.5).

From these criteria and from mass loss measurements taken on standard specimens exposed on site over 1 year, the aggressiveness may be classified, after statistical data analysis, into five classes. Table C.3.6 gives the corrosion rate of four metals: steel, zinc, copper and aluminium (1050) in each of these five classes.

This approach introduced a common, international reference system.

Three main criteria were used: the duration of exposure to humidity, the sulphur dioxide concentration and the amount of deposited chloride. Dust and corrosion products, which

Table C.3.3. Duration of exposure to humidity

Class	Duration of exposure to humidity (%)	Practical examples
T1	<0.1	Interior micro-climates after air-conditioning
T2	0.1–3	Interior micro-climates without air-conditioning, except non-air-conditioned interiors in humid climatic zones
T3	3–30	Exterior atmospheres of dry climatic zones, cold climatic zones and part of temperate climatic zones; non-ventilated shed in temperate climatic zones
T4	30–60	Exterior atmospheres in any climatic zone (except dry and cold): ventilated sheds in humid climates, non-ventilated shed in temperate climates
T5	>60	Parts of humid climatic zones: non-ventilated sheds in humid climates

Table C.3.4. Levels of sulphur dioxide pollution

Class	Deposition rate ($mg \cdot m^{-2} \cdot day^{-1}$)	Concentration ($mg \cdot m^{-3}$)
P0	<10	<12
P1	10–35	12–40
P2	36–80	41–90
P3	81–200	91–250

are also very important factors, were not taken into account. This certainly is one of the main limitations of this standardisation work [3].

Moreover, qualifying a site after an exposure period of only 1 year seems to be too short, at least in temperate climates that show substantial differences in meteorological conditions between summer and winter. The results may differ significantly after exposure for 1 year, depending on whether that period started in winter or summer, as shown by the results reported in Table C.3.7 [4]. At least 2 years would be necessary to neutralise the effect of the starting season.

For aluminium, the first years of exposure correspond to the transition period during which the deepening rate of pits has not yet stabilised (see Chapter C.5).

3.3. THE PREDICTIVE APPROACH

Several authors have proposed empirical equations for predicting the resistance of metals to atmospheric corrosion:

– depending on criteria related to meteorological conditions, for example [5]:

$$C = 0.76RH + 2.20T + 0.17L - 71.21$$

$$C = -0.036W + 4.58T + 0.51L - 69.50$$

where: C represents the annual corrosion in $g \cdot m^{-2}$ for aluminium, RH the relative humidity in percentage, T the average temperature in °C, L the number of rainy

Table C.3.5. Level of chloride deposits

Class	Deposition rate ($mg \cdot m^{-2} \cdot day^{-1}$)
S0	<3
S1	3–60
S2	61–300
S3	300–1500

Table C.3.6. Classes of atmospheric corrosivity

Class	Units	Carbon steel	Zinc	Copper	Aluminium
C1	$g \cdot m^{-2} \cdot year^{-1}$	≤ 10	≤ 0.7	≤ 0.9	Negligible
	$\mu m \cdot year^{-1}$	≤ 1.3	≤ 0.1	≤ 0.1	
C2	$g \cdot m^{-2} \cdot year^{-1}$	10–200	0.7–50	0.9–5	≤ 0.5
	$\mu m \cdot year^{-1}$	1.3–25	0.1–0.7	0.1–0.6	
C3	$g \cdot m^{-2} \cdot year^{-1}$	200–400	5–15	5–12	0.6–2
	$\mu m \cdot year^{-1}$	25–50	0.7–2.1	0.6–1.3	
C4	$g \cdot m^{-2} \cdot year^{-1}$	400–650	15–30	12–25	2–5
	$\mu m \cdot year^{-1}$	50–80	2.1–4.2	1.3–2.8	
C5	$g \cdot m^{-2} \cdot year^{-1}$	650–1500	30–50	25–50	5–10
	$\mu m \cdot year^{-1}$	80–200	4.2–8.4	2.8–5.6	

Notes: Corrosion rates expressed in $g \cdot m^{-2} \cdot year^{-1}$ have been converted to $\mu m \cdot year^{-1}$. The upper value is always lower or equal to the limit (example: $10 < x \leq 200$).

days per year, W the duration of exposure to humidity, with RH $\geq 80\%$ and $T > 0\,°C$;
– depending on pollution levels [6]:

The relative probability P (%) of pits of a given depth p equals

$$\log P = a + b \log t$$

where t represents the time (in months), a and b are coefficients that depend on the site's atmospheric parameters: concentration of chlorides (Cl^-) and sulphur dioxide (SO_2). The general equation is

$$\log P = -20.77 + 39.24 \log p - 18.01(\log p)^2 + 2.28[SO_2] + 8.12[Cl]$$
$$- 14.49[Cl]\log p + 6.27[Cl](\log p)^2 + \log t[4.06 - 8.92 \log p + 5.25(\log p)^2$$
$$- 2.78[Cl][SO_2] + 2.40[Cl][SO_2]\log p - 1.37[SO_2](\log p)^2]$$

Table C.3.7. Influence of the starting season of testing

Alloy	Atmosphere	Pitting depth after 1 year of exposure		
		Start in summer	Start in winter	Winter/summer ratio
1199 H14	Industrial	65	85	1.30
	Marine	42	68	1.61
5056 H14	Industrial	50	90	1.80
	Marine	25	60	2.40
6082 T6	Industrial	42	92	2.20
	Marine	30	70	2.33

where $[SO_2]$ represents the average concentration for the duration of exposure in $mg \cdot dm^{-2} \cdot day^{-1}$, $[Cl^-]$ represents the average concentration for the duration of exposure in $mg \cdot dm^{-2} \cdot day^{-1}$.

These equations are based on three or four criteria only, and do not take into account form factors (orientation of exposure: directed upwards or downwards) and the influence of dust accumulation. They somewhat simplify the complexity of atmospheric corrosion of metals.

REFERENCES

[1] Knotkova D., Boschek P., Kreislova K., *Results of ISO-CORRAG program: processing of one-year data in respect to corrosivity classification*, ASTM, STP 1239, 1995, p. 38–55.
[2] Vrobel L., Knotkova D., *Using the classification of corrosivity of atmospheres to extend the service life of materials, structures and products*, ASTM, STP 965, 1988, p. 248–263.
[3] Dean S.W., A new quantitative approach to classifying atmospheric corrosion: ISO breaks new ground, *ASTM Standardization News*, 1987, p. 36–39.
[4] Carter V.E., Campbell H.S., *The effect of initial weather conditions on the atmospheric corrosion of aluminium and its alloys*, ASTM, STP 435, 1968, p. 39–42.
[5] Feliu S., Morcillo M., Corrosion in rural atmosphere in Spain, *British Corrosion Journal*, vol. 22, 1987, p. 99–102.
[6] Otero T.F., Porro A., Elola A.S., Predicting of pitting probability on 1050 aluminium in environmental conditions, *Corrosion*, vol. 48, 1992, p. 785–791.

Chapter C.4

The Various Forms of Atmospheric Corrosion

Chapter C.4
The Various Forms of Atmospheric Corrosion

The most frequent forms of atmospheric corrosion of aluminium are

– pitting corrosion,
– galvanic corrosion, and
– filiform corrosion of coated products.

Uniform corrosion is not observed because it is so minute that it cannot be measured.

Other forms of corrosion of aluminium and its alloys such as exfoliation corrosion, stress corrosion, etc. are of course possible on age-hardenable alloys of the 2000 and 7000 series when exposed to weathering. However, the selection of alloys and their protection, if present, normally prevents these forms of corrosion from occurring in outdoor applications such as buildings or bridge railings.

4.1. PITTING CORROSION

Atmospheric corrosion develops in humid environments with, in general, a pH close to neutral. As a consequence, pitting corrosion is the most commonly observed form of corrosion on aluminium. As discussed below, experience over decades has shown that the rate of pit deepening decreases with time.

Experience has shown that atmospheric corrosion of aluminium does not always occur. There are many uses in which no pitting corrosion has been detected, even after many years of service. In other words, corrosion of aluminium exposed to weathering is not inevitable.

Anodising is a means to reduce or even prevent pitting corrosion, if the thickness is sufficient for the exposure environment (see Section C.5.5).

4.2. GALVANIC CORROSION

Except in rare cases, there is no homogeneity of materials, metals or alloys in buildings or in mechanically joined structures. For example, metallic fittings in aluminium are always joined with screws in stainless steel, while accessories such as hinges, filters, lift-off hinges are in stainless steel or chrome-plated steel, or even in brass. Under appropriate conditions, there is a risk of galvanic corrosion of aluminium.

Galvanic corrosion of aluminium in heterogeneous assemblies exposed to weathering obeys the rules given above (see Chapter B.3). It depends on several factors:

–   *The nature of the metals and alloys in contact with aluminium*: whatever the atmosphere, the most aggressive contacts for aluminium are assemblies with copper and copper alloys [1], lead, and steel (Figure C.4.1). The role of graphite-charged elastomers for the development of galvanic corrosion should be recalled here.

–   *The type of atmosphere*: marine atmosphere leads to the most severe galvanic corrosion [2] because of the presence of chlorides.

–   *The conductivity of the moisture film*: the higher the conductivity of the electrolytic medium, the better the cell set up between aluminium and the other metal works.

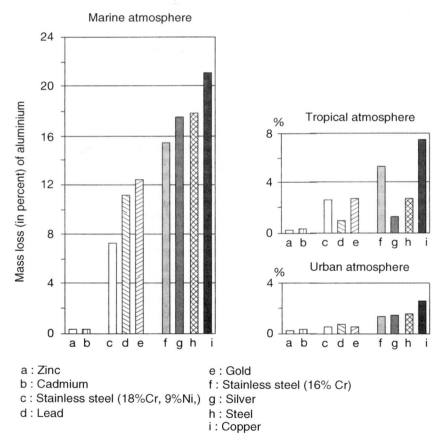

a : Zinc   e : Gold
b : Cadmium   f : Stainless steel (16% Cr)
c : Stainless steel (18%Cr, 9%Ni,)   g : Silver
d : Lead   h : Steel
                                                 i : Copper

Figure C.4.1. Galvanic corrosion of aluminium as a function of the type of atmosphere [3].

That is why a marine atmosphere, rich in chlorides, leads to the most severe galvanic corrosion of aluminium, under identical conditions for the all other parameters.

– *The frequency of moistening*: galvanic corrosion requires an electrolyte, which means that the contact area must be wet. Its intensity, therefore, depends on the local climatic conditions: rain, relative humidity, etc.

Atmospheric galvanic corrosion will always be limited to the contact area. Under appropriate conditions, it may lead to severe damage: roofing that is perforated around bolts or screws, electrical components that are corroded at contacts with components made of copper or copper alloys, etc.

In practice, contacts with stainless steel and zinc or cadmium-coated steel are the most common ones in constructions, and especially in metallic fittings. Experience throughout the world demonstrates that even without insulation between the two metals, galvanic corrosion does not lead to problems in these assemblies, if the design is such that any retention of moisture is avoided.

Experience in the construction industry and with various pieces of equipment exposed to bad weather in all climatic zones, such as road signs and landing stages of marinas, shows that galvanic corrosion of aluminium in contact with ordinary or stainless steel develops only in certain situations such as

– Areas where moisture is retained, and where rain water or condensed water can be trapped permanently or for long periods of time. This is often observed with embeddings that form a basin that can retain water. Galvanic corrosion is observed in contact with embedded steel pins.

– At the assembly points of roofing sheet and cladding panels, in substantially humid and aggressive environments. As an example, in coastal areas, strong galvanic corrosion around bolts can sometimes be found, because the felt used for insulation retains water, has disappeared or has been compressed, so that the aluminium comes into direct contact with the steel washers and bolts that are often rusty.

Particular micro-climates can favour the development of galvanic corrosion of aluminium in contact with steel. This can be observed in humid zones, close to factories that emit a great deal of dust: fertiliser plants, cement works, coal-fired power stations, etc. Experience shows that this situation, which is highly unfavourable for the resistance of materials, can be controlled to a large extent by suitable design and especially by frequent cleaning of accumulated dust.

The risk that is most often underestimated in assemblies with steel is the degradation of the appearance of aluminium constructions by rust runoff. While this has no impact on the corrosion resistance of aluminium, it degrades the overall appearance. Zinc or cadmium coatings on steel have a limited lifetime, depending on the aggressiveness of the environment. When they are exhausted, the steel will rust, and the situation is the same as in the case of direct contact between aluminium and unprotected steel.

Contacts between aluminium and other metals such as copper and cuprous alloys, lead, and tin, when exposed to very aggressive environments such as high level of humidity, frequent rain and dust, present a certain risk of galvanic corrosion of aluminium. They have to be protected or at least inspected in order to monitor the progression of possible galvanic corrosion and to intervene when necessary.

4.3. FILIFORM CORROSION

Filiform corrosion may occur on products coated with paint or lacquer, especially on lacquered products for buildings. It develops preferentially at coating defects (scratches, lacerations). Meteorological factors are of paramount importance because filiform corrosion occurs especially in marine and humid atmospheres and develops rather quickly, before the fifth year of exposure. While it does not affect the mechanical characteristics of aluminium semi-products, it degrades the overall appearance of a building (see Section B.2.6).

4.4. TARNISHING DUE TO WATER STAINING

Unprotected aluminium will progressively lose its brightness when exposed to weathering. After long outdoor exposure, the surface becomes more or less dull. The metal develops a certain patina, which is generally well accepted because it is uniform. The more polluted the atmosphere, the more intense and faster this alteration of surface aspect will be.

Like blackening (see Section D.1.5), this is an alteration of the visual properties of the natural oxide layer, due to its progression under the influence of atmospheric humidity. This is not corrosion. Water staining does not affect the resistance of aluminium to weathering. Anodising is a reliable means to avoid water staining of aluminium that is exposed to weathering.

4.5. WATER STAINING DUE TO STORAGE OF SEMI-PRODUCTS

Experience has shown that aluminium semi-products such as sheets and profiles can develop surface alterations during transportation or storage, when stacked one on the top of the other without insets. This alteration shows up as grey or white spots that often appear as aureoles. If two stacked sheets or profiles are stained, their contour is usually symmetrical and the same on the two opposing faces.

These spots are not actually corrosion, but only an alteration of the visual properties of the oxide layer caused by a complex reaction of the oxide layer with water trapped between the surfaces of the two sheets or profiles in contact. Magnesium alloys (5000 and 6000 series) are said to be more sensitive to this phenomenon, because of the presence of magnesium oxide (MgO) on the surface [4].

While this surface alteration has no effect on the mechanical characteristics of the metal and causes no subsequent corrosion in service, it nevertheless permanently alters the surface appearance. After surface treatment such as pickling, and also after anodising, more or less marked contrasts between the stained zones and the rest of the surface will remain. These semi-products are thus difficult to use in applications where surface aspect is paramount, i.e. where the surface appearance is the very reason for using aluminium.

In the case of cast (Properzi) drawing stock, these spots, which may be accompanied by superficial corrosion, can be the cause of rupture during subsequent cold drawing operations [5].

On semi-products to be welded, these spots need to be removed carefully in the welding zone by strong brushing with a stainless steel brush.

Aluminium and aluminium alloys have very good thermal conductivity, and they will react very quickly to any changes in temperature when in storage. Condensation will occur on the metal as soon as its surface temperature drops below the dew point. According to Table C.4.1, the dew point of an atmosphere at 18 °C and 50% relative humidity is 7 °C. So if the temperature of a metal being stored or truck-loaded in this atmosphere drops below 7 °C, condensation will take place.

Table C.4.1. Dew point (°C)

Temperature of the air (°C)	Relative humidity (%)									
	100	90	80	70	60	50	40	30	20	10
43	43	41	39	37	34	31	27	22	16	5
41	41	38	36	34	32	28	24	19	13	3
38	38	36	34	32	29	26	22	17	11	0
35	35	33	31	29	26	23	19	15	9	0
32	32	31	28	26	23	20	17	12	6	0
29	29	27	26	23	21	18	14	10	3	
27	27	25	23	21	18	15	12	7	2	
24	24	22	20	18	16	13	9	5	0	
21	21	19	17	15	13	10	7	3		
18	18	17	15	13	10	7	4	0		
16	16	14	12	10	7	5	2			
13	13	11	9	7	4	2	0			
10	10	8	7	4	2	0				
7	7	6	4	2	0					
4	4	3	1							
2	2	0								
0	0									

In practice, whenever the temperature of the metal drops below the dew point of the ambient atmosphere, condensation will occur. Similarly, in winter, condensation may occur during the transportation of semi-products that were packed in a workshop at a temperature between 20 and 25 °C. This means that sudden changes in the temperature, up or down, should be avoided, both during transportation and during storage.

The metal should be stored in an enclosed and aerated place and as far away as possible from openings, doors and windows to avoid excessive temperature changes. When packed before shipment, the box and packing paper should be at the same temperature. They must, therefore, be stored in the same place. Unpacked semi-products should not come into contact with floors, even concrete floors, because if poorly aerated, the floor will be colder. Sheet, plate and extrusions should be stored flat, supported on wooden wedges spaced close enough to prevent distortion.

Chemicals such as acids should not be stored in the same room, as their volatile vapours can cause more or less severe pitting corrosion to the metal when combined with atmospheric moisture.

Storage outdoors in all weathers may be inevitable for lack of space, etc., and experience shows that the metal may tarnish and suffer pitting corrosion, which is usually only very superficial (with a depth of a few micrometres), even at the seaside. However, outdoor storage close to sources of industrial or urban pollution such as dust, gas emissions, etc. is not recommended because in such an aggressive environment, the metal can suffer from very rapid and severe pitting corrosion.

REFERENCES

[1] Haynes G., Baboian R., *Atmospheric corrosion behaviour of clad metal*, ASTM, STP 965, 1988, p. 145–190.
[2] Godard H.P., Galvanic corrosion behavior of aluminium, *Materials Protection*, vol. 2, 1963, p. 38–47.
[3] Compton K.G., Mendizza A., Bradley W.W., Atmospheric galvanic couple corrosion, *Corrosion*, vol. 11, 1955, p. 383t–392t.
[4] Hinchiffle J.C.W., Storage corrosion of aluminium and its effects on finishing process, *Australian Corrosion Engineering*, 1972, p. 7–15.
[5] Langewerger J., Schwitzwasserkorrosion bei Lagerung und Transport von Properzi-Giesswalzdraht, *Revue suisse de l'aluminium*, vol. 31, 1981, p. 162–165.

Chapter C.5

The Resistance of Aluminium to Atmospheric Corrosion

Chapter C.5

The Resistance of Aluminium to Atmospheric Corrosion

The resistance of aluminium and aluminium alloys to atmospheric corrosion (weathering) is nowadays well known. This has been achieved through

- Monitoring over decades of several representative old applications, including the San Gioacchino church in Rome. Some of these are subject to regular inspections by aluminium producers [1].
- Results obtained in outdoor corrosion testing stations since 1935. These are mostly long-term tests (20 years and longer) in various types of atmospheres, especially marine [2, 3] and industrial.

5.1. THE DOME OF SAN GIOACCHINO CHURCH, A 100-YEAR-OLD EXPERIENCE

This is the oldest known application of aluminium in buildings that is still in service. The dome was covered in 1897 (Figure C.5.1). In order to obtain a satisfactory light level in the choir of the San Gioacchino church in Rome, the architects put numerous windows in the dome, which needed to be lightened. For this purpose, aluminium was chosen instead of lead, which had been commonly used as a roofing sheet for this type of building.

The 1.3-mm-thick aluminium sheets were manufactured in Neuhausen (Switzerland) in 1895. The composition of the metal corresponds to what was commonly achieved when the production of aluminium by molten salt electrolysis started (Table C.5.1). Very high in iron and silicon, the metal contained 98.3% aluminium. In the last 50 years, when unalloyed aluminium (1000 series) has been used for this type of application, it contains at least 99.5% aluminium.

The first inspection by Panseri in 1937 [4], as well as the second inspection in 1949 [5], on the occasion of the 50th anniversary of the church's completion, demonstrated the good corrosion resistance of this roofing. The micrographs included in the report show that the pitting depth did not exceed 100 μm. The areas of overlap between sheets did not exhibit preferential or more intense corrosion than the surfaces exposed to air.

I had the opportunity to inspect this dome in 1997. Visual inspection showed that the roof was in excellent condition and had taken on a yellowish grey patina. Superficial micropitting was present. Areas close to the (rusty) steel hinges did not exhibit galvanic

Figure C.5.1. View of the dome of the San Gioacchino church in Rome, built in 1897.

corrosion. The same applied to the numerous graffiti, some of which were drawn with a stencil or with the point of a knife (Figure C.5.2).

5.2. ATMOSPHERIC CORROSION OF ALUMINIUM OVER TIME

It has long been accepted that unprotected aluminium has a very good resistance to weathering, much better than steel (Figure C.5.3). Incidents of corrosion in service are infrequent. They are mainly the result of a poor choice of alloy or anomalous conditions of use.

Table C.5.1. Composition of the aluminium used for the dome (analysed in 1949 [5])

Element	%
Silicon	0.844
Iron	0.47
Copper	0.014
Titanium	0.009
Manganese	0.001
Aluminium (balance)	98.644

Figure C.5.2. Close-up view of the roof with steel hinges and graffiti.

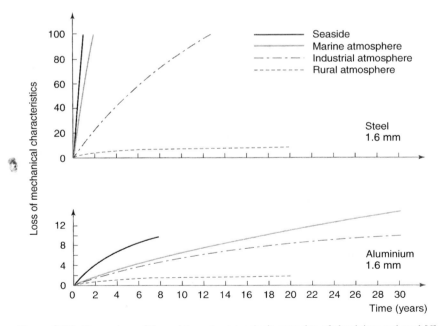

Figure C.5.3. Comparison of the resistance to atmospheric corrosion of aluminium and steel [6].

The excellent resistance to weathering of aluminium alloys of the 1000, 3000, 5000 and 6000 series without copper can be attributed to the properties of the natural oxide film (see Section B.1.8). The corrosion products are insoluble in slightly acidic or almost neutral media and tend to slow pitting corrosion once it has initiated. This explains the self-limiting effect of the atmospheric corrosion of aluminium, which has been observed for more than 50 years [7]. The curves in Figure C.5.4, representing the progression of pitting depth (maximum and mean values), were plotted in 1952 using the results of real applications and corrosion tests obtained in outdoor testing stations; the duration of service or testing ranges from 5 to 54 years, the last point corresponding to San Gioacchino church in Rome.

The general shape of these curves representing results over long periods of time is always the same, whatever the type of atmosphere: rural, industrial or marine. Three stages can be distinguished in atmospheric corrosion of aluminium (Figure C.5.5):

–   An *initiation stage*, during which pitting corrosion starts at fragile sites of the natural oxide film. Initiation of pitting is very fast, within a few days.
–   A *transition period*, which may last for several months or even years. During this period, the accumulation of corrosion products and deposits slow pitting corrosion. As shown by experience, in most cases the pitting depth no longer increases after 2 years.
–   A *stabilisation period*, during which the deepening rate of pits is very low. Corrosion will not necessarily stop. It may proceed on new sites, but according to the same

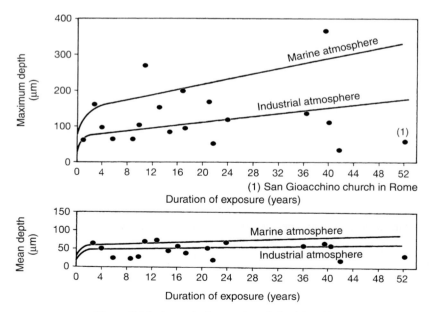

Figure C.5.4. Atmospheric corrosion of aluminium [8].

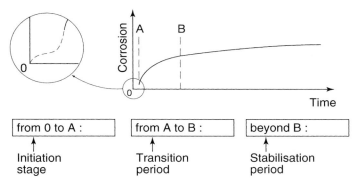

Figure C.5.5. Stages of atmospheric corrosion of aluminium [9].

mechanisms as the existing pits. As a consequence, for a given type of atmosphere, the maximum pitting depth will not be exceeded whenever a new pit is initiated. If corrosion is in progress, the density of pitting increases with exposure time.

5.3. WEATHERING RESISTANCE OF WROUGHT ALLOYS

There are three quantitative criteria for assessing the atmospheric corrosion of a metal:

- *Weight loss*. Atmospheric corrosion results in consumption of metal. The corresponding weight loss is an indicator of the intensity of atmospheric corrosion that can easily be determined by weighing the samples before and after exposure. Weight loss is a meaningful criterion only if the corrosion is uniform. However, on aluminium, uniform corrosion is extremely low; when corrosion occurs, it develops as pitting. Long-term tests in coastal corrosion testing stations over more than 10 years have shown that there is no correlation between the weight loss and the maximum depth of pitting [10]. Weight loss is, therefore, not a meaningful indicator for atmospheric corrosion of aluminium [11].
- *Changes in mechanical characteristics*: Ultimate tensile strength R_m, proof strength $R_{p0.2}$ and elongation $A\%$. This assessment method has been widely used for many years. Its disadvantage is that it is a relative method: on one hand, the pitting depth does not depend on the thickness of the sheet or profile; as a consequence, the relative variation in mechanical characteristics depends on the thickness of the exposed product. On the other hand, measurements on sheet with a thickness in the order of 1 mm have shown that the mechanical characteristics will decrease only if the pitting depth exceeds 5–6% of the thickness, i.e. 50–60 μm for a sheet with a thickness of 1 mm.

    As shown by the data listed in Table C.5.2, the mechanical characteristics do not change much for those alloys that show good resistance to atmospheric corrosion, namely

Table C.5.2. Evolution of the ultimate tensile strength (0.9 mm thick sheet)

Atmosphere	Reduction after 20 years of exposure (%)			
	1100 H14	3003 H14	6061 T4	2017 T3
Rural	0	0	0	0
Urban	7	8	12	7
Marine	8	7	20	20

strain-hardenable alloys. However, the drop is greater in the case of age-hardenable alloys.

– *The measurement of pitting depth* is the most representative assessment of atmospheric corrosion of aluminium. It is carried out under an optical microscope at low magnification (in the order of 10 × to 20 ×) by successive focussing on the rim and the bottom of the pit, the depth being given by the displacement of the lens. This method takes some time because at least five measurements have to be taken on the deepest pits that have been detected visually. These measurements can then be evaluated statistically (see Section B.4.4).

In many applications, and especially roofing sheet in buildings, perforation is the most unacceptable type of damage. Measuring the progression of pitting depth is the only method that gives an estimate of the lifetime and determines the minimum thickness required for a given application. In the following, results are always expressed as pitting depth (maximum and mean depth), and only refer to the face exposed to weathering.

Experience over more that 50 years acquired both in outdoor corrosion testing stations and by monitoring applications shows that the resistance to marine atmospheric corrosion of unprotected alloys of the 1000, 3000, 5000 and 6000 series is similar and this corrosion develops in the same manner over a period of 20 years or more (Figure C.5.6).

Tables C.5.3–C.5.7 list results of long-term exposure (20 years) of some of the most common alloys for outdoor use. These tests have been performed in French and American outdoor testing stations since 1932. A large number of publications have dealt with this issue, but choosing the progression of pitting depth as the only criterion for comparing industrial alloys over such a long period of time significantly restricts the number of results that can actually be exploited.

The analysis of these data shows that the maximum pitting depth never exceeds 0.5 mm after 20 years of exposure [12]. Within the same alloy series, there may be differences between alloys, in the order of 0.1–0.2 mm, at most, after 20 years of exposure.

The casting technology and the elaboration schedules of semi-products have undergone significant changes since 1935. On the basis of the available and usable results on several alloys, such as 1100, 3003, 6061 and 6082, it does not appear that the manufacturing conditions of semi-products, sheets or profiles have a significant

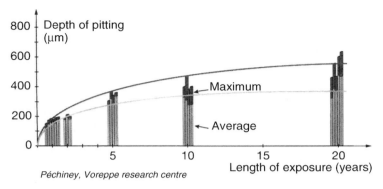

Figure C.5.6. Corrosion of alloys of the 1000, 3000, 5000 and 6000 series in marine atmosphere. Long-term exposure testing at the shore of the Mediterranean sea (Pechiney's outdoor corrosion testing station at Salin-de-Giraud).

influence on the resistance to atmospheric corrosion, in either direction. These comparisons based only on a few cases are too sketchy to be generalised. However, they confirm the experience of several decades that shows that the progress in the metallurgy of aluminium since 1950 has not significantly modified the resistance to atmospheric corrosion of aluminium alloys of the 1000, 3000, 5000 and 6000 series, in either direction. The experience with the dome of San Gioacchino church and other somewhat more recent applications confirms this a contrario.

Table C.5.3. Series 1000: marine atmosphere

Alloy	Temper	Year of exposure	Duration of exposure (years)	Site	Pitting depth (μm) Maximum	Minimum
1199	O	1980	20	Salin [1]	200	160
	H14	1980	20	Salin	200	160
	H18	1980	20	Salin	140	120
1070	O	1955	20	Salin	450	380
	H14	1955	20	Salin	300	280
1050	H14	1962	20	Salin	295	250
1100	H14	1932	20	Point Judith [2]	200	75
	H14	1935	20	Key West [2]	125	40
	H14	1935	20	La Jolla	350	200
	H16	1960	20	Durban [3]	130	
1200	O	1958	20	Salin	370	310
	H14	1958	20	Salin	370	300
	H18	1958	20	Salin	220	190

[1]Salin-de-Giraud, France.
[2]Testing stations in the United States.
[3]South Africa.

Table C.5.4. Alloy 3003: marine atmosphere

					Pitting depth (μm)	
			Duration of			
Alloy	Temper	Year of exposure	exposure (years)	Site	Maximum	Minimum
3003	O	1955	20	Salin	240	200
	O	1959	20	Salin	250	220
	O	1962	20	Salin	250	220
	H14	1932	20	Point Judith	200	75
	H14	1935	20	Key West	120	25
	H14	1935	20	La Jolla	255	105
	H18	1955	20	Salin	120	100
	H18	1959	20	Salin	225	280
	H18	1962	20	Salin	225	180

Table C.5.5. Series 5000: marine atmosphere

					Pitting depth (μm)	
			Duration of			
Alloy	Temper	Year of exposure	exposure (years)	Site	Maximum	Minimum
5005	H14	1955	20	Salin	130	120
	H14	1962	20	Salin	220	150
	H18	1955	20	Salin	130	100
	H18	1962	20	Salin	385	275
5754	O	1956	20	Salin	400	330
5086	H111	1981	20	Salin	400	300
5056A	O	1955	20	Salin	420	350
	H12	1955	20	Salin	350	300

Table C.5.6. Series 6000: marine atmosphere

					Pitting depth (μm)	
			Duration of			
Alloy	Temper	Year of exposure	exposure (years)	Site	Maximum	Minimum
6061	T4	1955	20	Salin	190	160
	T4	1962	20	Salin	225	180
	T4	1960	20	Salin	190	150
	T6	1955	20	Salin	100	80
	T6	1955	20	Salin	190	160
6082	T4	1955	20	Salin	240	200
		1960	20	Salin	260	220
		1980	20	Salin	260	180
	T6	1955	20	Salin	200	160
		1960	20	Salin	220	160
		1980	20	Salin	190	140

Table C.5.7. Industrial atmosphere

Alloy	Temper	Year of exposure	Duration of exposure (years)	Site	Pitting depth (μm)	
					Maximum	Minimum
1100	H14	1935	20	United States	275	100
	H14	1935	20		220	85
3003	H14	1935	20		175	70
	H14	1935	20		170	50
	H14	1951	20	Germany	140	90

For strain-hardenable alloys of the 1000 and 3000 series, the pitting depth is normally lower in soft tempers than in strain-hardened tempers; in the latter, the pitting depth increases with strain hardening from H12 to H19. The same applies to 5000 series alloys with a magnesium content not exceeding 3.5%. Above this concentration, a tendency to intercrystalline corrosion is observed, which increases with strain hardening. This tendency to intercrystalline corrosion can be limited by a stabilisation treatment (H3X) [13]. Alloys of the 6000 series show a resistance to atmospheric corrosion that is comparable to that of 3000 and 5000 series alloys.

It should be recalled here that copper-containing alloys of the 2000, 6000 and 7000 series are not intended for exposure to weathering and atmospheric corrosion without adequate protection such as specific anodising, cladding, etc. They are sensitive to intercrystalline corrosion, exfoliation corrosion and stress corrosion, and even more so in the T4 temper than in the T6 temper.

5.4. WEATHERING OF CASTING ALLOYS

It has been known for more than 50 years that silicon- and magnesium-containing casting alloys have an excellent resistance to atmospheric corrosion. The first testing results over 10 years in industrial atmosphere in the United States, published in 1946, showed that the pitting depth of samples of A-S5 (5% silicon) and A-S12 (alloy 44100, containing 1% iron) does not exceed 175 μm [14]. These results were confirmed by subsequent tests, also in industrial atmosphere over 13 years, that showed that the pitting depth does not exceed 200 μm on the alloys A-S13 (alloy 44100) and A-G5 (alloy 51300) [15].

In general, casting alloys without copper of the 40000 and 50000 series show an excellent resistance to atmospheric corrosion. They are widely used in construction and ship building (the upperworks of small crafts is made of anodised A-3GT (51100) and A-G6 (51300)), electrical engineering, etc.

The use of copper-containing alloys of the 20000 series and of alloys of the 40000 series A-S12U (47000), A-S7U3 (46200) and A-S9U3 (46400) should be avoided. From experience we know that severe corrosion can occur in urban and industrial atmospheres. Rupture of load-bearing pieces made of the A-U5GT alloy (43500) has been observed in aggressive atmospheres.

5.5. WEATHERING RESISTANCE OF ANODISED SEMI-PRODUCTS

The development of applications of aluminium in constructions such as metallic fittings, foldable walls, shop equipment, urban furniture stems from the excellent resistance to atmospheric corrosion of anodic layers. These layers aim at maintaining the surface appearance over a long period of time by preventing blackening and pitting corrosion.

On anodised products, alterations of the surface appearance can occur:

–   *In the short run*, within a few months up to a few years. This leads to iridescence [16], chalking, and powdering. These forms of ageing depend more on the anodising conditions (especially sealing) than on the climatic conditions [17]. Some can be eliminated by appropriate cleaning.
–   *In the long run*, pitting corrosion and fading of colours occur, especially in the case of coloration obtained by chemical impregnation. These alterations depend rather on the type of atmosphere (rural, marine, industrial) and on the thickness of anodisation, as far as pitting corrosion is concerned, and on the climatic conditions (exposure to sunlight, light intensity) as far the stability of colouring is concerned. Electrolytic colouring provides good stability [18].

Anodising delays the occurrence of pitting, but does not prevent it once it has initiated. As with other coatings, sharp angles are more sensitive to pitting corrosion than plane or bent surfaces. That is why sharp angles should be avoided as far as possible.

For architectural applications, the piece is anodised in a sulphuric bath, followed by sealing in boiling deionised water. Electrolytic colouring is obtained with nickel, cobalt and tin salts.

The long-term corrosion resistance of anodic coatings depends on a number of parameters that were the subject of many studies after 1955:

–   nature of the alloy,
–   anodising and sealing conditions,
–   film thickness, and
–   maintenance.

5.5.1. The nature of the alloy

Alloys of the 5000 and 6000 series form anodic coatings that resist atmospheric corrosion best. On the other hand, alloys of the 2000 series have the poorest performance (due to Al_2Cu intermetallics that locally weaken the anodisation layer).

5.5.2. Anodising conditions

If all other parameters are kept constant (such as alloy and thickness of the anodisation layer), the duration of the sealing treatment has an influence on the properties of the layer and its short-term ageing. Tests in outdoor corrosion testing stations have shown that the sealing treatment should be at least as long as the anodising treatment [19, 20]. Sealing in deionised boiling water, with or without added nickel salts, gives the best results over long periods of exposure time to marine or industrial atmospheres.

In Europe, the QUALANOD label was created for architectural products (directives concerning the quality label for anodic films on wrought aluminium intended to be used for architectural applications). Anodisers that adhere to this label must conform to a strict protocol of operating conditions [21]. Several methods are available for controlling the quality of anodised aluminium for architectural applications [22, 23].

5.5.3. Film thickness

This is the most important factor related to the resistance to atmospheric corrosion of anodised aluminium. The density of pitting decreases with increasing film thickness (Figure C.5.7). Numerous testing programs at several urban and marine testing facilities were carried out within the International Committee for the Development of Aluminium (CIDA). They have defined the minimum thickness of anodic coatings necessary to avoid pitting corrosion over a predefined period of service.

Comparative tests have shown that industrial atmosphere is more severe than marine atmosphere, because there are more or less abrasive dust particles and gaseous pollutants such as sulphur dioxide (SO_2) [24]. The minimal thickness must be 20 µm for industrial and marine atmospheres [25].

An annual decrease in thickness of 0.4–0.6 µm in marine atmosphere and of 0.7–0.9 µm in industrial atmosphere has been reported in outdoor testing stations testing samples with an anodic coating thickness of 15 µm [27]. This decrease in thickness probably results from a chemical attack or erosion by abrasive dust particles. Such a decrease in thickness is never observed in buildings, even after service times as long as 20 years or more [28]. It is generally estimated that the decrease in thickness of an oxide layer varies from 0 to 0.03 µm per year [29].

The properties of wrought products for buildings are the subject of several French and international standards [30]. The NF A 91-450 standard (AFNOR) defines the anodisation

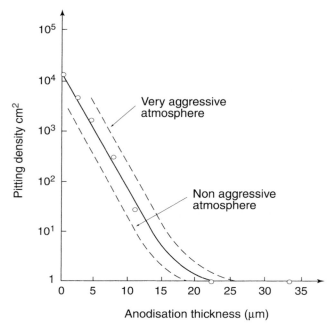

Figure C.5.7. Pitting density and thickness of anodisation on anodised 1100 after exposure to industrial atmosphere over 8 years [26].

classes as a function of thickness (Table C.5.8) [31]. The NF P 24-351 standard (AFNOR) defines the classes of anodisation for different types of interior ambient or atmospheres (Tables C.5.9 and C.5.10) [32].

The resistance of electrolytic coloration does not differ significantly from that of sulphuric colourless anodisation, if properly sealed. However, in the absence of regular maintenance, the dark bronze C34 and black C35 from the EURAS palette show a marked tendency to chalking and pitting [33].

Table C.5.8. Anodisation classes

Class	Average minimum thickness (μm)	Local minimum thickness (μm)
AA5	5	4
AA10	10	8
AA15	15	12
AA20	20	16
AA25	25	20

Table C.5.9. Anodisation classes and interior ambients

Type of ambient interior	Characteristics (hygrometry)	W/n [1] $(g{\cdot}m^{-3})$	Anodisation class
I_1	Weak	≤ 2.5	AA10
I_2	Average	$2.5-5$	AA10
I_3	Strong	$5-7.5$	AA15
I_4	Very strong	≥ 7.5	AA15
I_5	Aggressive		Specific testing

[1]W is the hourly amount of water vapour produced within the building, expressed in $g{\cdot}h^{-1}$. n is the hourly rate of air renewal, expressed in $m^3{\cdot}h^{-1}$.

5.5.4. *Maintenance*

Since the late 1960s, it has generally been admitted that anodised surfaces exposed to weathering must be cleaned in order to preserve their appearance and to avoid pitting corrosion of surfaces that are insufficiently cleaned by rain.

Like uncoated metal, an anodised surface will always corrode more on the reverse side and on faces that are poorly cleaned by rain. Dust and dirt enhance corrosion. For all these reasons, it is not possible to rely on the effect of rain or cleaning by plain water, although this was considered sufficient in the past.

Regular cleaning with appropriate neutral detergents that do not alter the anodic coating is necessary for the maintenance of anodised surfaces exposed to weathering and dust.

The frequency of cleaning depends on local conditions. The technical document DTU 33.1 "Travaux de bâtiments, menuiseries metalliques. Façades, rideaux, façades semi-rideaux, façades panneaux. Entretien Maintenance" recommends cleaning twice a year in urban, industrial or marine environments.

The absence of cleaning over several years can lead to severe damage of anodised surfaces that is difficult to recover.

Table C.5.10. Anodisation classes of exterior atmospheres

Atmosphere	Type of atmosphere	Anodisation class
E_{11} or E_{21}	Rural	AA15
E_{12} or E_{22}	Urban or industrial, normal	AA15
E_{13} or E_{23}	Urban or industrial, severe	AA15
E_{14} or E_{24}	Marine, 10–20 km from seaside	AA15
E_{15} or E_{25}	Marine, 3–10 km from seaside	AA15
E_{16} or E_{26}	Marine, less than 3 km from seaside	AA15
E_{17} or E_{27}	Mixed, normal	AA20
E_{18} or E_{28}	Mixed, severe	AA20
E_{19} or E_{29}	Aggressive [1]	Specific testing

[1]This is an environment in which the severity of the exposure described above is enhanced by certain effects such as very high corrosivity, abrasion, high level of moisture, high dust deposition, sea spray, etc.

REFERENCES

[1] Skerrey E.W., Long term atmospheric performance of aluminium and aluminium alloys, *Conference Atmospheric Corrosion*, Hollywood, October 1980, p. 329–352.

[2] Guilhaudis A., La tenue des alliages légers à la corrosion marine, *Revue de l'Aluminium*, 1952, p. 186–188.

[3] Guilhaudis A., Quelques aspects de la résistance des alliages d'aluminium en milieu marin, *Corrosion et Anticorrosion*, vol. 11, 1963, p. 404–419.

[4] Panseri C., Examination of a sheet of aluminium exposed for 40 years in an urban atmosphere, *Journal of the Institute of Metals*, vol. 63, 1938, p. 15–20.

[5] Panseri C., Gragnani A., Sul comportemento dell'alluminio esposto per alcuni decenni in atmosfera urbana, *Alluminio*, vol. 23, 1954, p. 627–637.

[6] Walton C.J., Sprowls D.O., Nock J.A., Resistance of aluminum alloys to weathering, *Corrosion*, vol. 9, 1953, p. 345–358.

[7] Sowinsky G., Sprowls D.O., Weathering of aluminum alloys, *Conference Atmospheric Corrosion*, Hollywood, October 1980.

[8] McGeary F.L., Englehart E.T., Ging P.J., Weathering of aluminium, *Materials Protection*, vol. 6, 1967, p. 33–38.

[9] Barton K., *Protection against atmospheric corrosion. Theories and methods*, Wiley, New York, 1976, p. 59.

[10] Reboul M., Forestier J., *Sensibilité à la corrosion atmosphérique d'alliages d'aluminium sans cuivre*, rapport Pechiney CRV 2581, 13 August 1992.

[11] Atteraas L., Haagenlud S., Atmospheric corrosion testing in Norway, *Conference Atmospheric Corrosion*, Hollywood, October 1980, p. 873–891.

[12] Lashermes M., Guilhaudis A., Reboul M., Trentelivres G., Thirty year atmospheric corrosion of aluminum alloys in France, *Conference Atmospheric Corrosion*, Hollywood, October 1980, p. 353–364.

[13] Lasehermes M., Tenue des alliages d'aluminium en atmosphère marine, *Revue de l'Aluminium*, 1982, p. 505–511.

[14] Bowman J.J., Aluminum-base die-casting alloys, *ASTM Proceeding*, vol. 46, 1946, p. 225–232.

[15] Wood J., Harris D.A., Atmospheric corrosion tests on cast aluminum alloys, *British Foundryman*, 1981, p. 217–221.

[16] Paulet J.F., Fuchs R., Gillich V., *Le phénomène d'iridescence sur les surfaces en aluminium anodisé et coloré*, Alusuisse Lonza, note FWEO, 1994.

[17] Furneaux R.C., Rigby W.R., Carter B.G., Mechanism of short-term superficial weathering of anodized aluminum, *Conference Interfinish*, Jerusalem, 1984.

[18] Patrie J., Outdoor corrosion performance of anodized and electrolytically coloured aluminium for architectural use, *Transactions of the Institute of Metals Finishing*, vol. 53, 1975, p. 28–32.

[19] Llowarch D., Croker B.V., Exposure tests on anodized and sealed aluminium specimens, *Corrosion Science*, vol. 3, 1963, p. 181–184.

[20] Guilhaudis A., Comportement de l'aluminium anodisé en atmosphère marine, *Conférence du Centre de recherches et d'études, océanographiques*, La Rochelle, 1965, p. 333–345.

[21] QUALANOD EURAS EWAA (European Association Anodisers Association) Zurich, BP 10 CH 8027.

[22] Patrie J., Les méthodes de contrôle applicables à l'aluminium anodisé en architecture, *Revue de l'Aluminium*, 1972, p. 869–877.

[23] Prati A., Définition d'une méthode d'essai de corrosion accélérée pour le contrôle des couches d'oxydation anodique, *3ᵉ Congrès de la Fédération européenne de la corrosion*, Bruxelles, juin 1963.

[24] Sheasby P.G., The properties of anodic oxidation coatings on aluminium, *Aluminium Industry*, vol. 2, 1983, p. 9–15.

[25] Campbell H., CIDA. *Anodized aluminum, atmospheric exposure tests. Euston 20 years*, Aluminium Federation Ltd, Birmingham B15 1 TW, report HC 8509/1, June 1987.

[26] Sowinsky G., Sprowls D.O., Weathering of aluminum alloys, *Conference Atmospheric Corrosion*, 21, Hollywood, October 1980, p. 297–328.

[27] Oelsner G., Results of natural, long time weathering of anodized aluminium, *Aluminium*, vol. 54, 1978, p. 530–531.

[28] Sheasby P.G., The weathering of anodized aluminium, *Conférence Aluminium* 2000, Modena, mars 1990.

[29] Faller F.E., Sautter W., Changes in film thickness of anodized aluminium parts in outdoor exposure, *Aluminium*, vol. 59, 1983, p. E8–E13.

[30] Patrie J., La normalisation des demi-produits en alliages légers anodisés destinés au bâtiment, *Courrier de la normalisation*, vol. 247, 1976, p. 53–57.

[31] *Anodisation de l'aluminium et de ses alliages. Couches anodiques sur aluminium, Spécifications générales*, norme AFNOR NF A 91-450, décembre 1981.

[32] *Menuiserie métallique. Fenêtres, façades rideaux, semi-rideaux, panneaux à ossature métallique. Protection contre la corrosion et préservation des états de surface*, norme AFNOR NF P 24-351, juillet 1997.

[33] Patrie J., Tenue à l'atmosphère industrielle et marine des demi-produits anodizes et colorés électrolytiquement destinés au bâtiment, *Revue de l'Aluminium*, 1975, p. 37–43.

Part D

Corrosion in Water

Unprotected aluminium resists corrosion very well in water. As early as at the end of the 19th century, aluminium replaced tin-coated copper in the manufacture of kitchen utensils. This was the first application of aluminium in consumer products. Other applications confirm this good resistance in water: roofing in buildings, street furniture, locks on rivers, irrigation tubes, heat exchangers, heaters for central heating, etc.

Aluminium has been used in contact with water for a century. Many studies[1] have investigated the corrosion resistance of aluminium in water, since Seligman and Williams' publication in 1920 [1]. In spite of all these references, aluminium has never been developed for tubes of water distribution grids or tanks used for the storage of fresh or brackish water. There are several reasons for this. The first is the unavailability of plumbing fittings (bends, T-fittings, taps) in aluminium. The second reason is that joining by brazing is not as easy as with copper. And finally, adhesive bonding is not as easy with aluminium tubes as with PVC or polypropylene tubes.

However, there is yet another reason: the corrosion resistance of aluminium in water is a topic that poses many problems for corrosion experts [2]. The great variety in the chemical composition of waters (there is no such thing as two identical waters) and the risk of pitting corrosion in contact with freshwater, especially stagnant water, makes it difficult to predict the corrosion resistance in a given water. Consequently, there has been no development of applications for aluminium in water distribution networks.

The corrosion resistance of aluminium depends on the type of water to which it is exposed. It differs significantly in freshwater, seawater, brackish water, etc. Many parameters come into play, most particularly

– the type of water,
– its composition,
– the temperature, and
– the flow of water.

This topic will be treated by type of water. This approach leads to a rather specific classification of the fields of application. The following types of water will be discussed:

---

[1] Slightly more than 400 summaries in the *Chemical Abstracts*, from volume 1 (1907) to volume 124 (1996), deal with this issue.

–   *Freshwaters*, with a total salinity generally not exceeding 1 g·l^{-1}. Rainwater, river waters and spring waters belong to this class.
–   *Distilled (high-purity) water*, also called deionised water. It is obtained by special processing of freshwater using ion exchange resins or distillation. Its salt content is very low, in general less than 1 mg·l^{-1}. High-purity water is devoted to specific uses: heat exchangers, chemical industry, surface treatment, etc.
–   *Brackish water*, with a much higher salinity, on the order of a few grams per litre, but much less than seawater. Most often, this water has been pumped from a great depth, but it can also result from processes in the chemical industry or mining.
–   *Waste waters* treated in water treatment plants. They have a low mineral content, but a very high content of organic matter in suspension, and their BOD (see Section D.1.2.4) is very high in the case of household wastewater. Industrial wastewater may contain a wide range of products that are dissolved or in suspension.
–   *Seawater*, with a salinity of approximately $30-35 \text{ g·l}^{-1}$.

REFERENCES

[1] Seligman R., Williams P., The action of aluminium of hard industrial waters, *Journal of the Institute of Metals*, vol. 23, 1920, p. 159–184.
[2] Jakson E.W., Aluminium vs corrosion by water, *Chemical & Process Engineering*, vol. 38, 1957, p. 391–393.

Chapter D.1

Freshwater

Chapter D.1
Freshwater

Freshwater has low mineral content: its total salinity does not exceed 1 $g \cdot l^{-1}$. River waters, spring waters and rainwaters are freshwaters.

The action of freshwater on aluminium depends on several factors inherent to the chemical composition of the water, its physicochemical properties and certain conditions such as temperature and flow speed. For classifying the influence of all these factors, we need to briefly recall certain aspects of the physical chemistry of water.

1.1. THE PHYSICAL CHEMISTRY OF WATER

Water is a powerful solvent, capable of dissolving many solid inorganic or organic compounds, liquids (if they are polar and contain a hydroxide group (OH^-)), and gases. Any water has a variable content of:

- inorganic salts,
- dissolved gases,
- matter in suspension, and
- organic matter.

The quantity of dissolved elements depends on the soils that have been crossed by the water and the products that may have been poured out there (pollution). Certain elements are in equilibrium with others, and their concentration may thus vary with physical conditions such as temperature.

Corrosion experts are not interested in all the elements dissolved in water, because some of them have no effect on the metals in contact with water. As an example, for aluminium, the presence of hydrocarbons is of no interest, and dissolved oxygen is hardly more important. On the other hand, the quantity of chlorides or the presence of ions of so-called heavy elements are important factors.

It should be emphasised that pollution and corrosion are not necessarily related, at least in the case of aluminium. A global evaluation of the pollution does not yield a corrosivity index of water. One must know the nature and the concentration of the pollutants to predict the corrosion resistance of aluminium in polluted water. A given water is, therefore, characterised by the quantity per unit volume of the contained products making up its composition. However, other physical and physicochemical parameters can have an important influence on the corrosion rate of a metal: temperature, electric conductivity, and pH.

1.2. THE ANALYSIS OF WATER

The typical water analysis certificate includes the concentration of inorganic salts, dissolved gases, matter in suspension, and organic matter that is in suspension or dissolved.

1.2.1. Inorganic salts
Inorganic salts are dissociated in cations (M^+) and anions (X^-). The results are generally expressed in:

– milligrams per litre (corresponding to ppm), unit: $mg \cdot l^{-1}$.
– milliequivalents per litre, unit: $meq \cdot l^{-1}$, which amounts to one thousandth of a gram equivalent divided by the valency of the element under consideration.

For example, the sulphate ion SO_4^{2-}

$$1 \ meq \cdot l^{-1} = \frac{96}{2 \times 1000} g \cdot l^{-1} = 48 \ mg \cdot l^{-1}$$

Table D.1.1 lists values in $meq \cdot l^{-1}$ for the four anions and the four cations that are usually analysed.

The ionic equilibrium of water requires the number of milliequivalents of anions to be equal to that of cations. Otherwise, the water analysis is incomplete or improperly done. There is no rule which defines the concentration ratios of the various anions and cations. Potassium and nitrates are not abundant in surface waters. The distribution of other elements varies greatly from one water to another. It depends on the geological layers that have been crossed by water, which explains why a very wide range of water compositions can be found.

Water can also contain ammonium (NH_4^+), nitrates (NO_3^-), phosphates (PO_4^{3-}), dissolved metals such as copper as (Cu^{2+}) and iron as (Fe^{3+}), generally as traces (a few milligrams per litre).

Table D.1.1. Millequivalent values

Cations	Value of meq ($mg \cdot l^{-1}$)	Anions	Value of meq ($mg \cdot l^{-1}$)
Calcium Ca^{2+}	20	Bicarbonates HCO_3^-	48
Magnesium Mg^{2+}	12	Chlorides Cl^-	35.5
Sodium Na^+	23	Sulphates SO_4^{2-}	48
Potassium K^+	39	Nitrates NO_3^-	62

Table D.1.2. Solubility of oxygen and carbon dioxide

Temperature (°C)	Solubility (mg·l^{-1})	
	Oxygen	Carbon dioxide
0	69.4	3350
20	43.4	1690
25	39.3	1450
40	30.8	973
60	22.7	576

1.2.2. Dissolved gases

In contact with the atmosphere, water dissolves a certain number of gases, mainly oxygen and carbon dioxide. Their solubility decreases with increasing temperature (Table D.1.2). Sulphur dioxide (SO_2), hydrogen sulphide (H_2S), and ammonia (NH_3) can also be found, in general as minute traces, except in wastewater where they are in high concentrations.

1.2.3. Matter in suspension

This is matter that is insoluble in water such as grains of very fine sand. The turbidity of water is an indicator for the quantity of matter in suspension.

1.2.4. Organic matter, soluble or in suspension

This is the organic matter that constitutes the bulk of the pollution in household wastewater. The pollution indices are the biochemical oxygen demand (BOD) and the chemical oxygen demand (COD).

■ BOD

BOD represents the amount of oxygen, expressed as milligrams per litre, that is consumed during a standard test (incubation at 20 °C in the dark) during a given period of time that ensures the biological oxidation of organic matter present in water. Complete oxidation is usually achieved in 21–28 days, but most often, BOD is determined after 5 days (BOD_5), which means the amount of oxygen consumed after 5 days of incubation. BOD represents only the biodegradable fraction of organic pollution.

■ COD

COD includes everything that can be oxidised, inorganic elements (sulphates, sulphites) as well as the major part of organic compounds, whether or not they are biodegradable.

It is expressed as milligrams per litre. The COD is no longer representative when the chloride content exceeds 2 g·l^{-1} [1].

In addition to the composition, two physicochemical characteristics of water need to be taken into consideration: pH value and electric conductivity.

1.2.5. pH
Most fresh waters have a pH close to neutral, between 6.5 and 7.5.

1.2.6. Electrical conductivity
The conductivity is determined as the conductance of a water column with a section of 1 cm^2 between two electrodes 1 cm apart. The unit of conductance is siemens, and conductivities are generally expressed as microsiemens per centimetre (μS·cm^{-1}). Resistivity is the inverse of conductivity; it is expressed as Ω·cm. The relationship between the two units is:

$$\text{resistivity } (\Omega \cdot \text{cm}) = \frac{10^6}{\text{conductivity } (\mu\text{S} \cdot \text{cm}^{-1})}$$

Electrical conductivity increases with the salinity of water (Figure D.1.1). For example, the resistivity of deionised water is several hundred thousand ohm centimetres, while that of municipal water is between 100 and 3000 Ω·cm, depending on the quantity and the nature of dissolved salts. Water with low resistivity is a good electrolyte in which electrochemical reactions take place more readily than in water with a very high resistivity. Seawater has a resistivity of a few tens of ohm centimetres.

1.3. TYPOLOGY OF WATERS

Inspection of many certificates of analysis shows that each water has a composition that differs from that of water originating from a nearby source. Nevertheless, a typology of waters can be set up as a function of composition-related criteria. Common properties exist for waters of common origin; these properties are a result of the nature of the soil crossed by water.

On this basis, the Department of Research and Development of EDF (French Electric Power Board) has defined six types of natural waters in France, according to composition-related criteria depending on their origin (Table D.1.3). Such a typology simplifies the study of the corrosion resistance of a metal, because only waters representative of those occurring in a given region (or even district) need to be considered. These waters can be prepared from inorganic salts and allow corrosion testing in the laboratory. The use of these standard waters also makes it possible to compare tests carried out in different laboratories.

Water type V is the most aggressive one, because it has the highest chloride concentration.

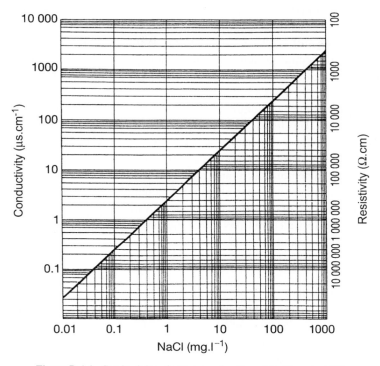

Figure D.1.1. Conductivity of water as a function of NaCl content [1].

Table D.1.3. EDF standard waters (in mg·l^{-1})

	Standard water					
	I	II	III	IV	V	VI
Origin	Seine valley	Massif Central	Loire valley	Magnesian	Coastal	Limestone
Mineralisation	Medium	Very low	Low	Medium	High	Very high
Ca^{2+}	86.0	8.4	42.0	99.0	125.0	228.0
Mg^{2+}	3.4	2.9	4.8	37.0	9.7	41.0
Na$^+$	5.5	3.7	14.0	10.0	69.0	31.0
K$^+$	4.7	1.6	2.3	4.7	4.0	8.0
HCO^{3-}	250.0	28.0	134.0	427.0	100.0	100.0
Cl$^-$	16.0	5.0	18.0	18.0	263.0	50.0
SO$_4^{2-}$	16.0	6.7	15.0	48.0	48.0	399.0
NO$_3^-$	–	7.4	8.7	3.7	6.0	12.0
Total salinity	381.6	63.7	238.8	647.4	624.7	1164.0

1.4. THE VARIOUS FORMS OF ALUMINIUM CORROSION IN NATURAL WATERS

Natural waters, both surface waters (river water, spring water) and seawater, generally have a pH close to neutral. Municipal waters mostly have a pH between 6.5 and 7.5. Distilled water is slightly more acidic, with a pH ranging from 6 to 6.5, and seawater is slightly alkaline, ranging from 8 to 8.2.

The corrosion resistance of aluminium depends on the stability of the natural oxide film that covers the metal. This film is most stable in the pH range of 6.5–7.5 (Figure B.1.18), which is why aluminium resists natural aquatic environments well, including weathering (see Chapter C.5). The dissolution rate of aluminium in natural waters at room temperature and up to elevated temperatures is minute and unmeasurable. On the other hand, in this pH range, at room temperature and up to about 80 °C, aluminium is prone to pitting corrosion. This is the common form of corrosion in contact with water. It should be noted that pitting corrosion can occur in water, but this is not inevitable: aluminium does not necessarily suffer from pitting corrosion in water.

1.4.1. Pitting corrosion

Pitting corrosion of aluminium in waters develops at preferential sites where the natural oxide film is less resistant because of heterogeneous features such as Al_3Fe intermetallics or defects related to very localised thinning or rupture of the natural oxide film. These sites are anodic with respect to their vicinity, and corrosion pits can develop according to the electrochemical mechanism described above (see Section B.1.2). Pitting is initiated by anions that penetrate into the defects of the natural oxide layer.

Since 1930, it has generally been accepted that among all anions, chlorides have the highest propensity to penetrate into the natural oxide layer, since they are small and very mobile [2]. They may replace oxygen atoms in the network of the natural oxide film. This leads to a decrease in the resistance of the film, which facilitates the release of aluminium atoms that diffuse into the water [3]. The aggressivity of a given water to metal in general, and to aluminium in particular, largely depends on its chloride content.

As in the case of atmospheric corrosion (see Section C.5.2), experience shows that in most cases, the penetration rate of pitting in contact with waters (freshwater, seawater) decreases with time. Trials conducted in the early 1950s with 25 different Canadian waters [4] showed that the deepening rate of pitting on alloy 1100 follows the equation:

$$d = kt^{1/3}$$

where d is the pitting depth, t the time and k is a constant that depends on the alloy and on the service conditions: nature of the alloy, temperature, flowing speed, etc.

This law has been checked by measuring the pitting depth at regular intervals over 13 years on an installation including over 100 km of pipes in alloy 5052 (unit length 15 m) for

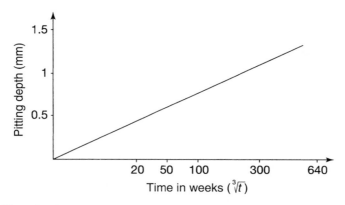

Figure D.1.2. Deepening rate of pitting of water pipes made of alloy 5052 [4].

water conveyance (Figure D.1.2) as well as on alloy 1050 with water at 90 °C over 90 days [5] (Figure D.1.3).

Experience shows that in certain cases, pitting corrosion in contact with water may develop during the first weeks of service. For example, in freshwater tanks, significant pitting corrosion can be observed after a few months of service as large pits, several millimetres in diameter and 1–2 mm in depth, covered by white pustules of alumina gel.

Users will usually and understandably anticipate perforation, based on the idea that the deepening rate of pitting is constant. Fortunately, this is never (or almost never) the case, because whatever the density of pitting corrosion, it obeys the $kt^{1/3}$ law. Several deep pits may develop during the first weeks of service, and then...nothing more for 25 years!

1.4.2. Galvanic corrosion

Aluminium and aluminium alloys are more electronegative than most common metals, with the exception of zinc, cadmium and magnesium (Table B.1.3). As a consequence,

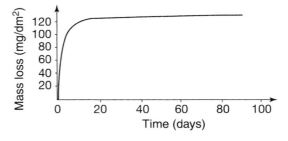

Figure D.1.3. Deepening rate of pitting of alloy 1050 in water at 90 °C [5].

aluminium may develop galvanic corrosion when in contact with most other metals and permanently immersed in water. The principles of galvanic corrosion have been described above (see Section B.3.7).

The higher the conductivity (or the salinity) of the water, the greater the intensity of galvanic corrosion of aluminium will be. In other words, galvanic corrosion is greater in seawater than in distilled water.

In practice, when assemblies of aluminium with other metals (including stainless steel) are permanently immersed, the contacts must be insulated in order to avoid galvanic corrosion. This precaution becomes more necessary as the conductivity of the water rises. In seawater, it is possible to neutralise galvanic coupling by using consumable anodes, if the geometry of the structure is suitable (see Section B.5.5). Consumable anodes are inefficient if the conductivity of the water is low or if the structure's geometry is complex, which is often the case inside the ballast tanks of ships.

Galvanic corrosion of aluminium in contact with galvanised or cadmium-coated steel is not observed, as long as the zinc or cadmium coating protects the steel.

For heterogeneous assemblies between aluminium and another metal, as well as for bolted homogeneous aluminium–aluminium assemblies, bolts made of anodised aluminium and sealed with potassium dichromate should be used whenever possible. This solution presents several advantages:

–   it eliminates any risk of galvanic corrosion,
–   it simplifies the assembly work, because there are no insulating gaskets, and
–   it saves weight compared to bolts made of ordinary or stainless steel. This can be an appreciable advantage for high-speed craft and any other use where weight is to be reduced.

1.4.3. Other forms of corrosion

All the other forms of corrosion of aluminium and aluminium alloys such as waterline corrosion, crevice corrosion, and stress corrosion can develop in water. They can be prevented by an appropriate selection of alloys, tempers, and possibly protections, as well as by an appropriate design that prevents hidden recesses and dead zones.

1.5. BLACKENING OF ALUMINIUM

Blackening is not a form of corrosion: it is only an alteration of the visual properties of the outmost oxide layer [6]. It does not alter the corrosion resistance of aluminium in water.

This unavoidable phenomenon has been observed since kitchen utensils made of aluminium began to be used [7]. To prevent blackening, it used to be recommended to boil

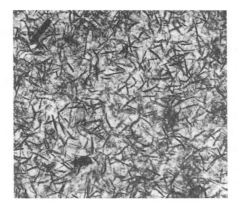

Figure D.1.4. Clear oxide layer formed on 3003 by immersion in boiling distilled water (50 000 ×) [9].

milk in the utensils prior to their first use [8]. This old recipe is only somewhat effective. Blackening is not irreversible and can disappear in acidic media. For example, a blackened saucepan may recover the brilliant surface aspect of aluminium after preparing sauerkraut in it.

Blackening of aluminium is due to the structure of the uppermost layer of the natural oxide film. Upon immersion of a sample of 3003 in boiling distilled water, a layer of well-crystallised boehmite fibres will form (Figure D.1.4). When doing the same experiment with natural tap water, the layer will have a totally different aspect: it will be amorphous, black, and built from porous cells [12] (Figure D.1.5).

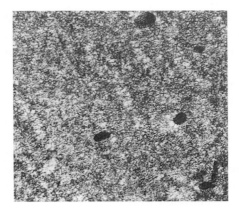

Figure D.1.5. Grey-black oxide layer formed on 3003 by immersion in boiling tap water (50 000 ×) [9].

The alloy composition is not a first-order factor. Magnesium, which is anodic with respect to aluminium, is known to reduce the sensitivity to blackening. For example, 1070A (Si < 0.20 and Fe < 0.25) is more prone to blackening than an alloy containing 5% magnesium, elaborated with 1070A and having the same amounts of silicon and iron. Alloying elements or additives that are cathodic with respect to aluminium favour blackening. No aluminium alloy that resists blackening is known yet.

On the other hand, water composition is a factor of paramount importance: blackening occurs if the water contains bicarbonates HCO_3^- and if its pH is between 8 and 9. Blackening is said to be caused by the adsorption of bicarbonate ions at the porous natural oxide film. Blackening may also appear when aluminium is coupled with a less electronegative metal in water: silver, stainless steel, or copper. On the other hand, aluminium that is cathodically protected by a sacrificial magnesium anode does not blacken in water [10].

Blackening can be eliminated with acidic solutions containing 10% phosphoric acid or 1% tartaric acid and 1% sodium fluoride at 60 °C [11]. Cleaning agents for dishwashers usually contain additives that help prevent blackening of aluminium utensils. Good results are obtained by using slightly acidified rinsing water for kitchen utensils.

1.6.   THE ACTION OF WATER ON ALUMINIUM

The action of water on aluminium depends on several factors related to the composition and the physical conditions (temperature, agitation, etc.), the influence of which will be discussed below.

1.6.1.   *The influence of chloride concentration*

Since 1930, it has generally been accepted that among all anions, chloride ions have the highest power of penetration into natural oxide film, because they are small and very mobile. Their dimension is close to that of oxygen [12]. Chlorides (Cl^-), as well as fluorides (F^-), bromides (Br^-) and iodides (I^-) belong to the anions that activate corrosion of aluminium in water, while sulphates (SO_4^{2-}), nitrates (NO_3^-) and phosphates (PO_4^{3-}) hardly activate it or do not activate it at all [13].

Chlorides may substitute oxygen atoms in the alumina network. This leads to a decrease in the film's resistivity, which facilitates the release of aluminium atoms that diffuse into the water [13]. The aggressivity of a medium with respect to metals in general and aluminium in particular depends to a large extent on its chloride and sulphate concentration.

Taking into account the penetration power of a natural oxide film by anions, in particular chlorides, the density and depth of pitting generally increases with increasing chloride concentration (Figure D.1.6). The action of chlorides is independent of the action of the associated cation (or cations, if several are present).

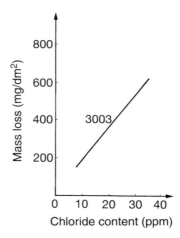

Figure D.1.6. Influence of chloride concentration on the corrosion resistance of 3003 in water.

At a given chloride concentration (Cl⁻), there is no perceptible difference between the action of a solution of potassium chloride (KCl), sodium chloride (NaCl) or ammonium chloride (NH₄Cl).

1.6.2. The influence of sulphate concentration
In water, sulphates are associated with calcium, magnesium and sodium. Being bulkier and less mobile than chlorides, they penetrate less easily into the oxide film [13]. Their action on the corrosion resistance of aluminium in water is weaker than that of chlorides (Figure D.1.7). While they have no influence on the pitting density, they are said to have a tendency to lead to a significant increase in pitting depth and to inhibit the formation of boehmite layers [14]. As in the case of chlorides, the action of sulphates does not depend on the nature of the associated alkali or earth alkali anion.

1.6.3. Influence of calcium concentration
Calcium is associated in water with carbonates, and more rarely with sulphates. Calcium, as carbonate or bicarbonate, has no influence on the corrosion resistance of aluminium in water, even at concentrations as high as 500 mg·l⁻¹ of calcium carbonate (CaCO₃), or even higher [15].

1.6.4. Influence of carbonate concentration
The so-called hard waters are not more aggressive towards aluminium and aluminium alloys than waters having a low concentration of carbonates or bicarbonates. The latter is

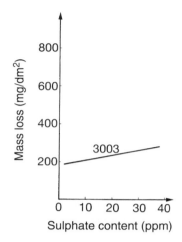

Figure D.1.7. Influence of sulphate concentration on the corrosion resistance of 3003 in water.

even said to have a slightly inhibiting effect, because a continuous carbonate film forms on the metal as soon as the temperature increases.

1.6.5. Influence of freshwater treatment

Chlorination is the most common method for biological purification of water. Water is chlorinated by bubbling gaseous chlorine though water, or by adding sodium hypochlorite (household bleach), lime chloride, etc. Municipal water and swimming pool water is disinfected using this treatment. Cooling water in condensers of electric power stations and industrial plants is chlorinated in order to avoid the proliferation of algae, shells and biological contamination.

These treatments are at a level of a few parts per million (ppm) chlorine (which in water transforms into chloride) or sodium hypochlorite.

This addition of chloride is generally very low compared to chlorides that are naturally contained in water. These treatments are thus unlikely to have any influence on the corrosion resistance of aluminium in freshwater.

Tests have shown that the corrosion resistance of aluminium in water is not influenced by the injection of 10 ppm chlorine, which is well above the usual level in biological water treatment plants (Figure D.1.8) [16]. Aluminium resists water and the chlorine environment of swimming pools very well. When unprotected, it will develop the usual water staining, and when anodised at 20 μm, neither pitting corrosion nor water staining will develop for several years [17].

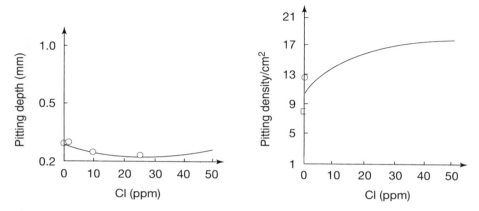

Figure D.1.8. Influence of adding chlorine to water on the corrosion resistance of 1200 [16].

1.6.6. *Concentration of metals dissolved in water*

No open or closed water circuit is homogeneous in terms of materials, i.e. contains only components in aluminium or in stainless steel or cuprous alloys. Materials for the components of a circuit are selected according to feasibility, cost and availability criteria. For radiators of a central heating installation in aluminium alloy, the boiler may be in stainless steel, the pumps in stainless steel or cuprous alloys, and the piping in ordinary steel or copper. The same applies to water distribution circuits.

When one of the components of an open or closed circuit happens to corrode, one (or more, in the case of an alloy) of the elements will be dissolved. When thermodynamically possible (potentials of the electrochemical reactions), the ions of this metal may be reduced in contact with other metals present in the circuit, which is why liquids used in closed circuits contain inhibitors to avoid corrosion of all metals and alloys present in the circuit, especially copper and copper alloys, since copper ions Cu^{2+} can cause severe corrosion of aluminium. This applies especially to the cooling circuits of car engines.

Certain metal cations dissolved in water become reduced by aluminium. This oxidation of aluminium leads to pitting corrosion. With copper, the following reactions can take place:

$$3Cu^{2+} + 6e^- \rightarrow 3Cu$$

$$2Al \rightarrow 2Al^{3+} + 6e^-$$

$$\overline{3Cu^{2+} + 2Al \rightarrow 2Al^{3+} + 3Cu}$$

Table D.1.4. Action of copper present in the water of London [18]

Cu^{2+} content, $mg \cdot l^{-1}$ (ppm)	Initiation period of pitting (h)	Mass loss $(mg \cdot dm^{-2})$	Pitting density $(au \cdot dm^2)$	Mean depth (mm)
0	31–46	27.6	4	0.39
0.01	31–46	18.0	4	0.42
0.02	31–46	17.6	8	0.39
0.05	31–46	13.6	8	0.17
0.2	7–22	13.0	4	0.12
0.5	0.75–1	32.8	13	0.25
1	0.50–0.75	44.4	16	0.37
2	0.50–0.75	57.6	36	0.26
5	<0.50	64.4	60	0.23
10	<0.50	68.4	16	0.13
20	<0.50	76.0	64	0.19
30	<0.50	90.8	100	0.10

Cathodic sites of aluminium at which these reactions take place are rich in iron, probably especially the intermetallic compound $FeAl_3$.

Copper particles will deposit on aluminium and result in pitting corrosion. They can also be reoxidised, be dissolved as Cu^{2+} and then be reduced once again by aluminium. All in all, in the presence of copper salts dissolved in water, severe pitting corrosion may occur on aluminium.

Several studies have been undertaken in order to determine the threshold concentration in water above which there is a risk of aluminium corrosion. Studies carried out by BNFRMA (British Non-Ferrous Metals Research Association Laboratory) during the 1950s [18] with London water have shown that the concentration threshold above which copper has an effect is on the order of 0.2–0.5 ppm ($mg \cdot l^{-1}$).

According to these results (Table D.1.4), copper acts more on the density of pitting than on pitting depth, and it reduces the initiation period. Copper is fixed rapidly on aluminium, within less than 24 h [19]. In fact, the threshold is probably higher, on the order of 1 ppm. At very low concentrations, from 0.01 to 0.2 ppm, copper seems to have a tendency to slightly reduce pitting corrosion. Recent trials [20] have confirmed these results (Figure D.1.9) for temperatures between 20 and 50 °C.

The influence of copper dissolved in water depends on other factors such as water composition and pH. It can be harmful to the corrosion resistance of aluminium, even when associated with an organic anion. The action of copper sulphate solutions is discussed in Section E.6.6.

Other metals can be dissolved in water; they can be of natural origin, such as in the case of iron-bearing water, or originate from the corrosion of other metals present in the circuit or from discharge of industrial wastewater.

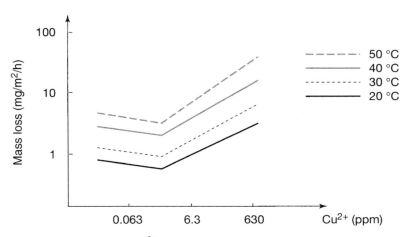

Figure D.1.9. Influence of Cu^{2+} ions on the corrosion resistance of 6063 in water [20].

Among common metals that can be found in this way, a distinction can be made between:

– *those that attack aluminium* by a reduction according to the process described above:
   • copper,
   • mercury,
   • lead;
– those that have no effect:
   • manganese,
   • cobalt;
– *those that form a film* on aluminium without attack:
   • iron,
   • chromium,
   • zinc.

National or European directives regulate the presence of metallic cations in drinking water. The maximum allowed concentration of chromium and mercury within the European Union is

– chromium content < 0.005 ppm,
– mercury content < 0.001 ppm.

Industrial wastewater or waters operating in closed circuits can contain ions of heavy metals capable of corroding aluminium.

1.7. THE INFLUENCE OF TEMPERATURE

Temperature has a very important influence on the corrosion resistance of aluminium in freshwater, whether natural or distilled. The same applies to seawater.

Several temperature domains corresponding to different forms of corrosion need to be distinguished. Their limits are not defined strictly, but may depend on the composition and the nature of the water.

1.7.1.   *Reactions of natural oxide films with water*

The corrosion resistance of aluminium in water depends on the reaction of the natural oxide film with water.

The natural oxide film is composed of two layers (Figure B.1.14):

– an *internal layer*, normally amorphous, in contact with the metal;
– an *external layer*, whose structure changes in contact with water as a function of temperature. Up to 70 °C, the reaction of the oxide film with water leads to the formation of bayerite ($Al_2O_3 \cdot 3H_2O$). Above 70 °C, boehmite ($Al_2O_3 \cdot H_2O$) is formed (see Section B.5.1); its thickness may reach a micrometre [21]. This layer is protective.

Water reacts with the barrier layer, which transforms into bayerite or boehmite depending on the temperature, and thus increases the thickness of the external oxide layer. The barrier layer will rebuild instantaneously from underlying aluminium, and therefore this reaction does not limit the overall corrosion rate. The corrosion rate in water, which follows a parabolic law, is controlled by the external film and depends on its thickness (Figure D.1.10).

The corrosion rate of aluminium in water, therefore, depends on the respective kinetics of three reactions:

– the dissolution rate of the external film,
– the conversion of the barrier layer into bayerite or boehmite, and
– the formation of the barrier layer from the metal.

The corrosion rate can be zero or very low in aggressive aqueous media because a steady state has built up, which corresponds to the equilibrium between these three reactions [22].

1.7.2.   *Temperature and forms of corrosion*

During the 1950s, when aluminium was considered as a cladding material for nuclear fuel, a large number of tests in pure water up to 350 °C were carried out in France,

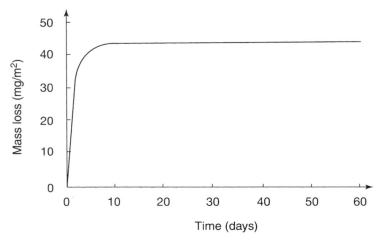

Figure D.1.10. Corrosion of 1200 in water at 50 °C [23].

the United States and Great Britain. These tests showed that the forms of corrosion are temperature-dependent (Table D.1.5).

It should be noted that the limits of the temperature ranges may vary significantly, by several tens of degrees, depending on the composition of aluminium and water.

In freshwater up to 60–70 °C, the dominating corrosion tendency is pitting. At higher temperatures, the pitting depth sharply decreases, while the density may increase in certain cases (Figure D.1.11). Above 70 °C and up to 150 °C, the propensity of pitting corrosion progressively disappears.

Whether in distilled water, freshwater or seawater, above 60–70 °C, the formation of more or less coloured layers with a structure similar to that of boehmite is observed. The thickness of these oxide layers increases with temperature and with time. They can

Table D.1.5. Temperature and forms of corrosion

Temperature domain	Forms of corrosion
< 100 °C	Pitting corrosion (in freshwater, the tendency to pitting corrosion decreases above 60–70 °C)
100–150 °C	General corrosion
150–250 °C	General corrosion and intercrystalline corrosion
> 250 °C	Intercrystalline corrosion with destruction of the metal

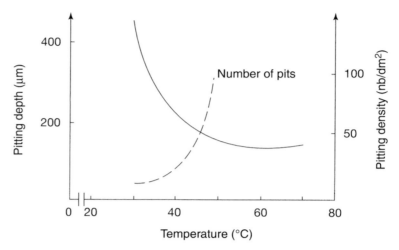

Figure D.1.11. Influence of the water temperature on corrosion [4].

reach several micrometres at 100 °C in distilled water [24]. These layers protect aluminium against corrosion in water, including seawater (see Section B.5.1).

The increase in temperature has several effects:

– the solubility of dissolved gases decreases,
– the solubility of carbonates and sulphates (with calcium and magnesium) decreases. They precipitate and form a thin layer on the metal's surface. This film can provide further protection, in addition to the boehmite layer. Carbonates and sulphates can also be found in the natural oxide layer.

Globally, aluminium's corrosion resistance in freshwater is better at a temperature between 70 and 100 °C than at room temperature.

Above 150 °C, whatever the nature of the water, intercrystalline corrosion develops rapidly at the surface and in depth (Figure D.1.12): it may reach several hundred micrometres within 50 h. Intercrystalline corrosion of aluminium in overheated water vapour between 300 and 350 °C was already observed in 1929, and it was found that the more pronounced it was, the purer the aluminium was [25].

Intercrystalline corrosion is accompanied by a modification of the structure of the barrier layer that tends to crystallise. This form of corrosion leads to exposure of grains and swelling of the metal. Pitting corrosion is also observed. Above 250 °C, a sharp increase in corrosion is found: the kinetics changes to a linear law (Figure D.1.13). A rapid destruction of the metal is observed.

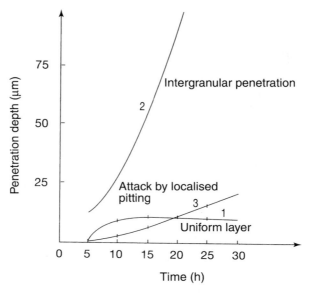

Figure D.1.12. Corrosion rate in water at 165 °C [26].

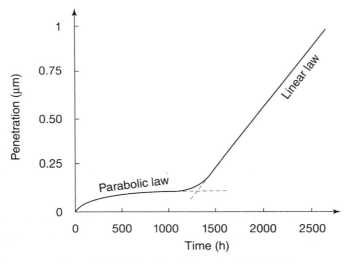

Figure D.1.13. Corrosion rate in water at 363 °C [30].

1.7.3.   Influence of aluminium composition

Alloying elements in aluminium may have an influence on the resistance of the barrier layer and the underlying metal, but not on that of the external layer. The purer the metal, the more pronounced the sensitivity to intercrystalline corrosion [27].

Metallurgists have proposed AlFeNi alloys containing roughly 1% iron and nickel, such as 8001 (Fe 0.5, Ni 0.9). The addition of iron and nickel forms uniformly distributed, very small intermetallic precipitates of Al_3Fe and Al_9NiFe that reduce the sensitivity to intercrystalline corrosion in hot water [28, 29].

1.7.4.   Temperature limits

The upper limit of the temperature for use of aluminium in freshwater depends on several parameters:

–   the *mechanical characteristics*, which begin to weaken at 150 °C for strain-harden-able alloys,
–   the *effects of prolonged heating* above 100 °C on the mechanical characteristics of age-hardenable alloys, and above 70–80 °C on the resistance to intercrystalline co-rrosion of 5000 series alloys with more than 3.5% magnesium (see Section B.6.4),
–   the *sensitisation to intercrystalline corrosion* of aluminium and all aluminium alloys at high temperatures.

In summary, it is impossible to define a precise temperature limit for the use of aluminium and aluminium alloys in contact with water, because it also depends on the application. However, 150 °C seems, in many cases, to be a limit that is difficult to exceed with aluminium.

It should be remembered that design rules define the conditions that must be fulfilled by certain equipments subject to regulations such as pressure vessels. These design rules apply to aluminium alloys and specify their temperature of use.

1.8.   THE INFLUENCE OF OXYGEN DISSOLVED IN WATER

Oxygen dissolved in water has less influence on aluminium corrosion than on steel corrosion. It maintains corrosion by depolarisation of the cathodes, but in the case of aluminium, it also contributes to the formation of the natural oxide layer when corrosion occurs [31].

Whatever the fundamental aspect of this question, experience shows that the resistance of aluminium is not directly related to the amount of oxygen dissolved in water, and that resistance does not differ significantly in aerated and deaerated water. The presence of oxygen leads at most to a more localised corrosion, but has no influence on pitting depth. A well-known fact demonstrates the favourable role of oxygen: the resistance of aluminium roofing sheet to rain water is excellent. For example, a sheet in 5754 has excellent resistance to weathering.

In a freshwater tank made of the same alloy in the same temper, there will be a fair probability of finding pitting corrosion with stagnant water; within 6 months, these pits often have a large diameter exceeding 1 mm and are rather dense. The difference in resistance between a sheet in 5754 exposed as a roofing sheet or used in a tank seems to be mainly due to aeration, and therefore to the regular uptake of oxygen. Experience shows that a tank containing water that is replaced at regular intervals has a better resistance than a tank with stagnant water.

1.9. THE INFLUENCE OF pH

The stability of natural oxide film, which governs the corrosion resistance of aluminium, depends on pH (Figure B.1.18). Most natural, untreated and unpolluted surface waters have a pH between 6.5 and 7.5. Since the solubility of alumina is minute and is practically constant in this pH range (Figure D.1.14), pH is not an important factor for the corrosivity of natural waters.

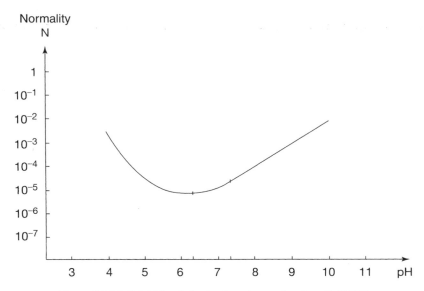

Figure D.1.14. Solubility of alumina in water as a function of pH [32].

1.10.   THE INFLUENCE OF WATER MOVEMENT AND FLOW SPEED

Based on experience, it is known that the corrosion resistance of aluminium in moving (or frequently replaced) water is always better than in stagnant water, if all the other parameters are kept constant. Tests over 1 week in freshwater at 20 °C have shown that the density and depth of pitting decreases with increasing flow speed (Table D.1.6) [33]. Water movement regularly eliminates corrosion products and uniformises the cathodic and anodic zones by removing a possible local excess of H^+ and OH^- ions. In an open circuit, moving water is aerated, and oxygen uptake contributes to repairing the oxide layer [34]. In a closed circuit, the movement of the liquid prevents the formation of deposits under which corrosion can easily develop. Aluminium can withstand a water flow speed up to $2.5-3$ m·s^{-1} without any risk of erosion.

Pitting corrosion in stagnant water has a different aspect than in moving water. Pits are often disseminated, of large diameter ($1-5$ mm), covered with voluminous white pustules of alumina gel, and sometimes with a deposit of hard, light yellow scale made up of carbonates. When scraping this deposit, a pit can be found, the depth of which may exceed 1 or 2 mm (Figure D.1.15). Analysis has shown that an acidic medium is retained under these scale nodules [35].

This type of corrosion is often found in water tanks of small craft, and in tanks whose water is changed only rarely. In such a situation, an attempt should be made to eliminate the corrosion products using appropriate mechanical means such as brushing or high-pressure water spray.

In any case, pickling should be avoided, because efficient rinsing of the interior of a tank is always very difficult: the retention of acidic or alkaline pickling chemicals may lead to severe corrosion in recesses. Furthermore, the water should be replaced regularly, for instance every month. In most cases, this will be sufficient for stopping (or slowing down) pitting corrosion.

Table D.1.6. Influence of the flow speed of water on pitting

Speed (m·min^{-1})	Mean density	Depth (mm)		
		Maximum	Minimum	Average
0.3	244	220	100	148
0.6	145	150	80	107
0.9	26	100	50	79
1.2	58	140	60	90
1.5	25	80	40	50
1.8	15	60	20	35
2.1	50	60	10	29
2.4	0	0	0	0
3	0	0	0	0

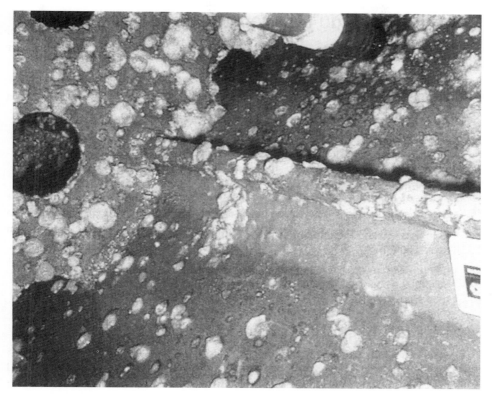

Figure D.1.15. Aspect of pitting in a tank with stagnant water.

1.11. PREDICTING THE RESISTANCE OF ALUMINIUM IN FRESHWATER

Several equations have been proposed for predicting the corrosion rate of aluminium as a function of certain parameters related to water composition:

– The *Pathak–Godard equation* concerning the deepening rate of pitting [36]:

$$\log p = 2.5 - 0.28 \log(SO_4^{2-}) + 0.18 \log(Cl^-) - 0.20 \log((pH)^2 \times 100)$$
$$- 0.42 \log(30\,000/R) - 0.06 \log(Cu^{2+} \times 10^3)$$

where p is the number of weeks necessary to obtain a penetration of 1 mm, R the resistivity in $\Omega \cdot$cm, and (SO_4^{2-}), (Cl^-), (Cu^{2+}) are concentrations expressed in mg·l^{-1} (ppm).

This equation is based on numerous tests on samples of 3003 in 67 waters of increasing aggressivity. Calculations based on this equation are in good accordance with experimental results in very aggressive water, but for waters of medium aggressivity, they yield a higher pit deepening rate than what is observed in reality.

–   The equation of T. Sakaida, H. Ikeda and Z. Tanabe [37] concerning the mass loss and the deepening rate of pitting:

$$W = \{0.254(SiO_2) - 0.062(CaCO_3)\}(Cl^-) - 1.617(SiO_2) + 0.997(CaCO_3) - 5.251$$

$$K = 0.205(Cl^-)^2 + \{-0.430(SO_4^{2-}) - 0.174(CaCO_3) + 16.341(Cl^-)\} + 0.238(SO_4^{2-})^2$$
$$-6.850(SO_4^{2-}) - 0.214(SiO_2)^2 + 11.885(SiO_2) + 0.021(CaCO_3)^2 - 107.261$$

where W is the mass loss in $mg \cdot m^{-2} \cdot day^{-1}$, K the deepening rate of pits in $mm \cdot day^{-1/3}$ and (Cl^-), (SO_4^{2-}), (SiO_2) and $(CaCO_3)$ are concentrations expressed in $mg \cdot l^{-1}$ (ppm).

1.12. EXAMPLES FOR THE USE OF ALUMINIUM IN CONTACT WITH FRESHWATER

In general, strain-hardenable wrought alloys of the 1000, 3000 and 5000 series and age-hardenable alloys of the 60000 series, as well as silicon-containing casting alloys (series 40000 without copper) and magnesium-containing casting alloys 51000 (A-G3T) and 51300 (A-G6) have good resistance to corrosion in freshwater and can be used in most cases without protection.

1.12.1. *Irrigation and conveyance of water*
Irrigation is the most common application in open circuits: seam-welded tubes in 3003, 3004, 5052, extruded tubes in 6060, with fittings such as valves in 44100 (A-S13) and 42100 (A-S7G) are used.

Several experiments with aluminium pipelines several dozen kilometres long, some embedded and some not, were conducted during the 1950s and 1960s in Canada [38, 39]. Besides these experiments, there are few examples of aluminium equipment for the conveyance of freshwater.

This can be explained by several factors:

–   the well-established use of copper pipes for water distribution networks in residential blocks is in itself a reference which no other material can match;
–   the lack of experience, and therefore the difficulty of easily predicting the penetration rate of pitting in contact with water of a given composition, as well as the absence of an economic motivation;

– the more and more widespread use of polymer pipes for the distribution of drinking water at room temperature, or of hot water in bathrooms.

1.12.2. Central heating

Radiators made of aluminium were manufactured in Switzerland and Sweden during the 1940s and in France at the beginning of the 1950s [40]. However, it was not before the early 1970s that radiators made of casting alloy A-S13 (44100) with up to 2% copper based on secondary aluminium, and those made of extrusions in alloy 6060 became widespread. This application has been widely developed in Europe, and no corrosion problems have been detected. Laboratory tests in closed circuits with standard waters have confirmed the excellent resistance of aluminium radiators.

After a few weeks of service at a temperature between 60 and 70 °C, a reddish-brown deposit becomes incrusted; it is made up of a mixture of calcium carbonate, iron oxide (Fe^{3+}) and aluminium hydroxide ($Al(OH)_3$) [41]. The iron oxide stems from the superficial corrosion of steel pipes. This deposit protects aluminium against possible attack of copper ions [42]. When pitting occurs, its depth is approximately 0.5 mm and does not increase after a few months of operation [43].

No galvanic corrosion is observed at the contact with fittings made of cast iron, or with connecting nuts made of steel. The only precaution recommended by the manufacturers is to avoid direct contact with copper pipes by inserting a steel nut between the aluminium radiator and the adduction pipe.

1.12.3. Condensing boilers

The recovery of the latent heat of water vaporisation and the latent heat of gases by condensation above their dew point (52–56 °C) leads to energy savings on the order of 15%. Many feasibility studies and corrosion studies were carried out in the early 1980s in order to develop gas boilers for domestic heating as well as unit heaters for domestic or industrial heating.

Condensates resulting from the combustion of natural gas are an aggressive medium. Their pH is acidic (around 4). They contain:

– chlorides (Cl^-): 12–20 ppm,
– sulphates (SO_4^{2-}): 15–50 ppm, and
– nitrogen oxides (NO_x): 20–100 ppm.

The condenser is a gas–liquid heat exchanger for boilers (Figure D.1.16) and a gas–gas heat exchanger for unit heaters. Aluminium offers several advantages for this:

– the possibility to manufacture cast condensers, condensers with tubes and fins (offering the specific advantage of a large surface within a limited volume), or plate condensers for unit heaters,

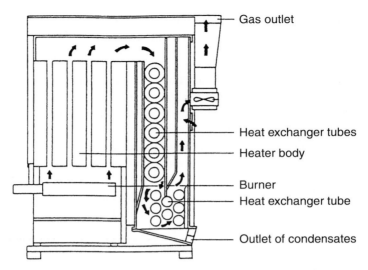

Figure D.1.16. Principle of a Remeha condensing gas boiler [44].

–   much better heat-exchange performance compared to stainless steel, and
–   lightness of the condenser.

   The condenser's surface is exposed to several possible forms of corrosion: uniform corrosion (due to the very acidic pH of the condensates), pitting corrosion, and crevice corrosion (Figure D.1.17). Tests on alloys 1060, 1100, 5052, 5754, 6061, 6082, A-S13 (44100) and A-S10G (43100) have shown that with condensates containing 100 ppm chlorides, the uniform corrosion is about 100 μm for each heating season, and that pitting corrosion reaches a maximum depth of roughly 400 μm [45, 46].

   In condensers of unit heaters, made of sheet in 3003 and 5754, a maximum pitting depth of 60 μm with 3003 and 80 μm with 5754 was found after an equivalent service period of 12 weeks with condensates containing 35 ppm chlorides and having a pH of 4 [47]. The plates further exhibited generalised water staining.

   All tests performed in several European and American laboratories have shown that condensers can be manufactured using unprotected, wrought products in 3003 or 5754 (rolled products), or in 6061 or 6082 (extrusions), as well as castings in A-S7G (42100), A-S10G (43100) or A-S13 (44100) in boilers and unit heaters fuelled with natural gas.

1.12.4.   Various types of equipment

Aluminium is used for the construction of locks [48], landing stages and various types of equipment on rivers, lakes etc.

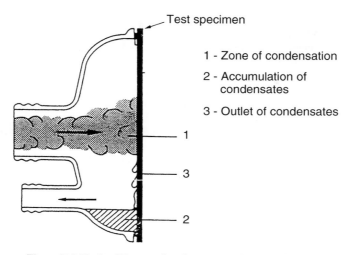

Test specimen

1 - Zone of condensation

2 - Accumulation of
condensates

3 - Outlet of condensates

1

3

2

Figure D.1.17. Possible corrosion sites on a condenser surface [45].

1.13. DISTILLED WATER

Aluminium alloys of the 1000, 3000, 5000 and 6000 series as well as casting alloys A-S7G (42100) resist very well in deionised or demineralised water, without contaminating it. Many installations such as storage tanks and distribution grids have been manufactured in aluminium [49]. Refined aluminium 1199 is prone to intercrystalline corrosion in deionised water, even at low temperatures, i.e. around 50 °C [50].

In experimental nuclear reactors, tanks for heavy water, light water or reception chambers for neutron fluxes have been made in aluminium alloy.

REFERENCES

[1] *Mémento technique de l'eau*, 9th edition, Degrémont, Paris, 1989.

[2] Britton S.C., Evans U.R., *Journal of the Chemical Society*, 1930, p. 1773.

[3] Heine M.A., Keir D.S., Pryor M.J., The specific effects of chloride and sulfate ions on oxide covered aluminium, *Journal of the Electrochemical Society*, vol. 112, 1965, p. 24–32.

[4] Godard H.P., The corrosion behavior of aluminium in natural waters, *The Canadian Journal of Chemical Engineering*, vol. 38, 1960, p. 167–173.

[5] Groot C., Troutner V.H., *Corrosion of aluminium in tap water*, USAEC, report HW 43 085, June 1956.

[6] Sawyer D.W., Brown R.H., Resistance of aluminium to fresh water, *Corrosion*, vol. 3, 1947, p. 443–457.

[7] Czochralski J., Blackening of aluminium cooking dishes, *Z. Metallkunde*, vol. 12, 1920, p. 430-443.

[8] Altenpohl D., Einiges zur sog. Brunnenwasserschwärzung des Aluminiums und ihrer Verhütung, *Metalloberfläche*, vol. 9, 1955, p. 118–121.

[9] Bourbon R., Adenis D., Moriceau J., *Relation entre le noircissement à l'eau des alliages d'aluminium et la structure des couches d'alumine*, rapport Pechiney CRV, avril 1966.

[10] Reboul M., Évolution de l'aspect de surface de l'aluminium brut de transformation en milieux naturels, *Revue de l'Aluminium*, 1976, p. 485–491.

[11] Godard H.P., Jepson W.B., Rothwell M.R., Kan R.L., *The corrosion of light metals*, Wiley, New York, 1967, p. 208.

[12] Britton S.C., Evans U.R., *Journal of the Chemical Society*, 1930, p. 1773.

[13] Dacres C.M., *An investigation of the influence of various environmental factors upon the aqueous corrosion of aluminum alloys*, American University, Washington DC, PhD, 1977.

[14] Godard H.P., Torrible E.G., The effects of chloride and sulfate ions on oxide films growth on aluminium immersed in aqueous solutions at 25 °C, *Corrosion Science*, vol. 10, 1970, p. 135–142.

[15] Bell W.A., Effect of calcium carbonate on corrosion of aluminium in waters containing chloride and copper, *Journal of Applied Chemistry*, vol. 12, 1962, p. 53–55.

[16] Doyle D.P., Godard H.P., Influence of additions to water on the corrosion behavior of aluminium, *Oil in Canada*, vol. 14, 1962, p. 42–44.

[17] Arnold G.G., The effect of swimming pool atmospheres on aluminium, *Anti-Corrosion Materials & Methods*, vol. 19, 1972, p. 5–9.

[18] Porter F.C., Hadden S.E., *Effect of water temperature and copper content on the pitting of aluminium*, BNFRMA, rapport RRA, February 1952.

[19] Davies D., Pitting of aluminium in synthetic waters, *Journal of Applied Chemistry*, vol. 9, 1959, p. 651–659.

[20] Bazzi L., Kertit S., Hamdani M., Effet de l'addition d'ions Cu^{2+} sur le comportement à la corrosion de l'alliage de l'aluminium 6063 en milieu neutre, *Revue de métallur-gie-CIT/ Science et génie des matériaux*, décembre, 1994, p. 1835–1843.

[21] Bryan J.M., Action of boiling distilled water on aluminium, *Journal of Society Chemical Industry*, vol. 69, 1950, p. 169–173.

[22] Troutner V.H., The mechanism and kinetics of aqueous aluminium corrosion. I. Role of corrosion product film in the uniform aqueous corrosion of aluminium, *Corrosion*, vol. 15, 1959, p. 9t–12t.

[23] Troutner V.H., Observations on the mechanism and kinetics of aqueous aluminium corrosion. II. Kinetics of aqueous aluminium corrosion, *Corrosion*, vol. 15, 1959, p. 13t–16t.

[24] Coriou H., Grall L., Huré J., Lelong P., Herenguel J., Corrosion de l'aluminium et de certains alliages dans l'eau à haute température, *Revue de Métallurgie*, vol. 53, 1956, p. 775–790.

[25] Guillet L., Ballay M., *La corrosion des alliages d'aluminium dans la vapeur d'eau surchauffée*, Comptes rendus, Académie des sciences, vol. 189, 1929, p. 551–552.

[26] Herenguel J., Lelong P., Les mécanismes d'attaque de l'aluminium de haute pureté par l'eau à température élevée, *Revue de l'Aluminium*, vol. 35, 1958, p. 991–998.

[27] Karlsen K.M., Mechanism of aqueous corrosion of aluminium at 100 °C, *Journal of the Electrochemical Society*, vol. 104, 1957, p. 147–154.

[28] Coriou H., Grall L., Hauptman A., Huré J., Influence du silicium sur la résistance d'alliages aluminium-fer à la corrosion par l'eau à haute temperature, *Revue de Métallurgie*, vol. 55, 1958, p. 968–975.

[29] Videm K., Corrosion of aluminium alloys in high-temperature water. A survey, *Journal of Nuclear Materials*, vol. 1, 1959, p. 145–153.

[30] Dillon R.L., Observations on the mechanisms and kinetics of aqueous aluminium corrosion. Part II. Kinetics of aqueous aluminium corrosion, *Corrosion*, vol. 13, 1957, p. 13t–16t.

[31] Hackerman N., Effect of temperature on corrosion of metal by water, *Industrial and Engineering Chemistry*, vol. 44, 1952, p. 1752–1755.

[32] Bryan J.M., The mechanism of the corrosion of aluminium, *Chemistry & Industry*, 1948, p. 135–136.

[33] Wright T.E., Godard H.P., Laboratory studies on the pitting of aluminium in aggressive waters, *Corrosion*, vol. 10, 1954, p. 195–198.

[34] Jakson E.W., Aluminium vs corrosion by water, *Chemical & Process Engineering*, vol. 38, 1957, p. 391–393.

[35] Porter F.C., Hadden S.E., Corrosion of aluminium alloys in supply waters, *Journal of Applied Chemicals*, vol. 3, 1953, p. 385–409.

[36] Pathak B.R., Godard H.P., Equation for predicting the corrosivity of natural fresh waters to aluminium, *Nature*, vol. 218, 1968, p. 893–894.

[37] Sakaida T., Ikeda H., Tanabe Z.I., *Estimation of corrosion rate of aluminium in tap water*, Sumitomo Light Metal, technical report 26, 1985, p. 221–229.

[38] Godard H.P., Aluminium pipeline case history data, *Materials Protection*, vol. 2, 1963, p. 101–104.

[39] Ailor W.H., William J.R., A review of aluminium corrosion in tap water, *Journal of Hydronautics*, vol. 3, 1969, p. 105–114.

[40] Grimal M., Le radiateur en aluminium, *Revue de l'Aluminium*, vol. 180, 1951, p. 341–344.

[41] Vargel, C., *Tenue à la corrosion des radiateurs Altrois P*, rapport Pechiney CRV, janvier 1979.

[42] Isenberg C., Korrosionverhalten stranggeprebter Radiatorprofile aus AlMgSi$_{0.5}$ in geschlossenen Warmwasserheizkreisen, *Aluminium*, vol. 54, 1978, p. 270–273.

[43] Lashermes, M., *Essais de radiateurs en profilés*, rapport Pechiney CRV, juin 1983.

[44] Jannemaan T.B., Condensing boilers in the Federal Republic of Germany. Corrosion behavior and performance data, *Proceedings of the 1987 International Symposium on Condensing Heat Exchangers, Ohio*, 1987, p. 17–33.

[45] Kobussen A.G., Oonk A., Hermkens R.J., Corrosion of condensing heat exchangers and influence of the environment, *Proceedings of the 1987 International Symposium on Condensing Heat Exchangers, Ohio*, 1987, p. 1–16.

[46] Searle M., The design and installation of domestic condensing boilers in the UK, *Proceedings of the 1987 International Symposium on Condensing Heat Exchangers, Ohio*, 1987, p. 75–93.

[47] Vargel C., Solar P., *Radiateurs de chauffage central*, rapport Pechiney CRV 1048, octobre 1975.

[48] Les portes des écluses de Gand, *Revue de l'Aluminium*, vol. 248, 1957, p. 1124–1125.

[49] Binger W.W., Marstiller C.M., Aluminium alloys for handling high purity water, *Corrosion*, vol. 13, 1957, p. 591t–596t.

[50] Hensler J.H., Corrosion of wrought aluminium in demineralized water, *Australasian Corrosion Engineering*, vol. 7, 1963, p. 23–28.

Chapter D.2
Brackish Waters and Wastewater

Chapter D.2
Brackish Waters and Wastewater

2.1. BRACKISH WATERS

The salinity of brackish waters depends on their origin. It is always on the order of a few grams per litre, somewhere between that of freshwater and seawater. These waters most often contain sodium chloride, sodium sulphate, and sometimes magnesium chloride. When originating from deep boring, they come out at high pressure, sometimes at an elevated temperature (50–70 °C), and are loaded with hydrogen sulphide and iron sulphide. Brackish water may also result from the dilution of seawater with river water or ground water. In all cases, experience has shown that they are very aggressive to metals in general. The use of aluminium equipment cannot be envisioned without prior testing of the corrosion resistance of aluminium in contact with these waters.

Waters of geothermal origin are brackish waters. They can lead to severe corrosion of aluminium alloys [1].

2.2. WASTEWATER

Aluminium has a high resistance to the atmospheres of wastewater treatment plants containing the following gaseous effluents:

- hydrogen sulphide (H_2S),
- ammonia (NH_3),
- carbon dioxide (CO_2), and
- methane (CH_4).

Much of the equipment in these plants is made of aluminium alloys: cladding panels, roofing, guardrails, footbridges, ladders, stairs, ventilation shafts, etc.

Urban household wastewaters have a slightly alkaline pH (7.5–8.5) and carry a heavy load of organic matter (Table D.2.1).

Tests over 1 year have shown that in these waters, the resistance of alloys 3003, 5052, 5456, 5083 and 5086 is excellent: the annual decrease in thickness is less than 1 μm, and pitting corrosion is virtually nonexistent. In the waters that are rejected after treatment, the same alloys have a tendency to pitting corrosion, with a depth of 0.5 mm [2].

Many urban household water treatment plants are equipped with scraping bridges, valves, cofferdams, baffle plates and overflow blades made of aluminium alloys: 3003,

Table D.2.1. Typical composition of urban wastewater

Parameters	Concentration ($mg·l^{-1}$)	Decantable fraction (%)
Dry extract	1000–2000	10
Matter in suspension	150–500	50–60
BOD_5	100–400	20–30
COD	300–1000	20–30
Total organic carbon (TOC)	100–300	
Kjeldall nitrogen	30–100	10
Ammonia nitrogen	20–80	0
Detergents	6–13	0
Phosphorous	10–25	10

5052, 5754, 6061 and 6063 [3, 4]. These materials are very resistant. After several years of service, they generally exhibit many pits that are rather evenly distributed over the surface of the piece. Surfaces close to the waterline (air–water limit) are often covered by concretions that are more or less hard and by viscous deposits under which pitting corrosion is slightly more developed than on totally immersed surfaces.

On scraping bridges and immersed equipment, the effect of galvanic coupling between immersed aluminium parts and accessories and fittings (bolts, etc.) that are often made of stainless steel must be limited. Such a conductive medium favours galvanic corrosion. As many heterogeneous contacts as possible must be insulated.

Overflow blades made of aluminium have a somewhat shorter service lifetime, between 8 and 10 years, because they are subject to galvanic corrosion with steel pins embedded in concrete. When the immersed aluminium surface is small compared to the surface of other metals, there is a risk that aluminium is consumed rapidly, which is what happens to overflow blades when they are the only aluminium equipment in such a plant.

While the resistance to urban wastewaters is well known nowadays, the resistance in contact with industrial wastewater cannot easily be predicted, because it depends on the nature of the discharge. Each piece of equipment is an individual case.

REFERENCES

[1] Larsen-Basse J., Corrosion tests in Hawaiian geothermal fluids, *Proceeding International Congress Metal Corrosion*, 1984, p. 641–648.
[2] Ailor W.H., Metals in wastewater treatment, *Journal of Environmental Engineering Division*, 1974, p. 295–309.
[3] Siegrist M., L'aluminium dans les stations d'épuration des eaux usées, *Revue suisse de l'aluminium*, 1972, p. 227–233.
[4] *Aluminum, the modern material for sewage treatment equipment and plants*, Aluminium Association 1970.

Chapter D.3

Seawater

Chapter D.3
Seawater

The first marine applications of aluminium date back to the early 1890s. Naval architects had perceived the advantages of this metal for shipbuilding [1]. Among the first vessels were the *Mignon*, a yacht 12 m long and purchased by Alfred Nobel in 1902 [2], and several military vessels, including two 18-m-long torpedoboats, *Le Foudre* and *Le Lansquenet*, ordered in 1895 by the French Admiralty from the British shipyard Yarrow. The metal came from the Froges plant in the Dauphiné country [3]. This example was followed in the same year by the Russian and American Navy. At the end of 1895, the *Defender*, equipped with aluminium plating [4], won the America Cup.

All these experiences were rather short-lived, because these vessels suffered very soon (after a few months) from severe corrosion: the alloy of the plates contained 6% copper! In order to obtain acceptable mechanical strength, copper had been added to the primary aluminium in a concentration between 6 and 10% (some producers proposed adding 2–3% nickel, and even tungsten).

The planking sheets had been riveted with copper alloy rivets on steel frames. As a result, intense galvanic corrosion led to severe damage to the sheets in aluminium–copper alloy, which is already by itself highly prone to corrosion.

It was not until the 1930s that naval architects could envision the use of aluminium in shipbuilding. Weight savings were required for the superstructures of the Navy's vessels [5]. Manufacturers had created aluminium–magnesium alloys, the corrosion resistance of which was a decisive advantage for their success. In 1936, a model of a section of the vessel *Alumette* (length, 4 m; width, 3.3 m; height, 1.5 m) was put into seawater by Alcoa. The sheets were made of 5052, and profiles and rivets in 6053 [6].

Since the beginning of the 1950s, marine applications of aluminium have been widely developed.

Experience shows that the corrosion resistance in natural seawater is rather different and better than that in brackish water with the same total salinity. This is because seawater is an aqueous medium that represents an equilibrium of

-   dissolved inorganic salts, approximately $30-35$ g·l^{-1},
-   matter in suspension,
-   dissolved gases, including 5–6 ppm of oxygen,
-   living organic matter, and
-   decomposing organic matter.

This forms a very complex medium, in which the influence of each factor, whether chemical (such as composition), physical (such as temperature) or biological, on the corrosion of metals cannot be readily separated, unlike in salt solutions [7].

Likewise, corrosion results obtained in artificial seawater, prepared in the laboratory from distilled water and the aimed concentration of dissolved inorganic salts, are known to be generally different from those obtained in natural seawater, even if its mineralisation (quantity and nature of dissolved salts) is close to that of natural seawater. For aluminium, experience shows that laboratory results obtained with artificial seawater are generally more severe than those obtained with natural seawater. All the biological components that are involved in the corrosion behaviour of metals in contact with seawater are missing in artificial seawater.

3.1. CHARACTERISTICS OF SEAWATER

3.1.1. Salinity

The great oceans (Atlantic, Indian, Pacific), linked together in the Southern hemisphere, all have nearly the same salinity: between 32 and 37.5 g of dissolved salts per litre. The main inorganic species for seawater with a salinity of 35 $g \cdot l^{-1}$ are listed in Table D.3.1 [8].

The salinity of closed or isolated seas can differ from that of the oceans (Table D.3.2). Salinity can also vary over the year, depending on the season.

The aggressivity of seawater and of marine environments, in general, is due to the abundance of chlorides (Cl^-). It contains about 19 $g \cdot l^{-1}$ such as potassium chloride (NaCl) and magnesium chloride ($MgCl_2$).

Results of corrosion testing as well as experience show that the corrosion resistance of immersed aluminium is the same whatever the ocean or sea. Comparative tests using the same alloys immersed in the North Sea ($16-17$ $g \cdot l^{-1}$ chlorides, annual temperature range,

Table D.3.1. Inorganic species in seawater

Species	Concentration ($g \cdot kg^{-1}$)	$meq \cdot l^{-1}$
Chlorides (Cl^-)	19.353	545.15
Sulphates (SO_4^{2-})	2.712	28
Bicarbonates (HCO_3^-)	0.142	2
Bromides (Br^-)	0.067	
Sodium (Na^+)	10.760	468
Magnesium (Mg^{2+})	1.294	11.3
Calcium (Ca^{2+})	0.413	21
Potassium (K^+)	0.387	

Table D.3.2. Salinity of isolated seas

Sea	Total salinity (g·l⁻¹)
Baltic Sea	8
Caspian Sea	13
Black Sea	22
Irish Sea	32
Mediterranean Sea	41
Persian Gulf	57

0–18 °C) and in the Persian Gulf (26–34 g·l⁻¹ chlorides, annual temperature range, 17–30 °C) show that there is no significant difference between these two locations, although the salinity differs by a factor of two [9].

If urban or industrial effluents pollute the water of harbours, the corrosion resistance of aluminium can be modified, provided that the pollutants have an influence on the corrosivity of seawater. As in the case of freshwater, the nature of the pollutants must be known in order to predict the corrosion resistance of aluminium in a polluted harbour.

Diluted estuary seawater is generally more aggressive to materials than that of the open sea. This paradox can be explained in several ways: precipitation of calcium and magnesium carbonates does not occur in diluted seawater, biological activity is slowed down, and domestic or industrial effluents can modify the physicochemical equilibria and possibly themselves cause corrosive action.

3.1.2. Dissolved oxygen

Seawater contains 6–8 mg·l⁻¹ (ppm) dissolved oxygen. The oxygen concentration of seawater varies significantly with the depth of the ocean (Figure D.3.1), but also with the temperature and the biological activity of the medium. For example, at the bottom, where decomposition of organisms takes place, the oxygen concentration is depleted because of the strong biological oxygen demand (BOD) necessary for maintaining this decomposition.

The role of oxygen in the corrosion resistance of aluminium in seawater is the same as in freshwater. It accelerates corrosion by depolarising the cathodes, but also contributes to the repair of the natural oxide layer that protects the metal [10]. The experience with desalination of seawater shows that the resistance of aluminium is the same in aerated and deaerated seawater, from room temperature up to 120 °C [11].

3.1.3. Temperature

The aggressivity of seawater towards metals could be expected to increase with a rise in temperature; in other words, the corrosion resistance in tropical oceans should be less than in arctic seas. This is far from being true, because an increase in temperature has

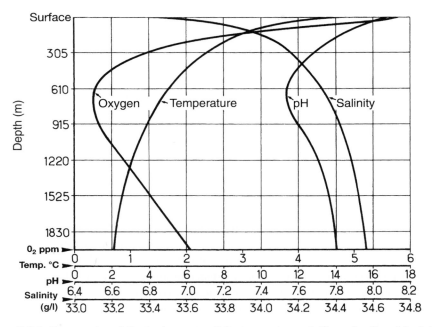

Figure D.3.1. Concentration of dissolved oxygen, salinity, temperature and pH as a function of depth [12].

contradicting effects on corrosion: the solubility of oxygen decreases, the biological activity increases, and magnesium and calcium carbonates precipitate at increasing temperature, thereby creating a protective film.

No significant differences in the corrosion resistance of aluminium have been observed between warm and cold oceans.

3.1.4.   pH

The pH of seawater close to the surface is very stable and is approximately 8.2. It depends very slightly on the photosynthetic activity of plankton and marine algae. It also depends on depth (Figure D.3.1).

The pH of seawater remains within the domain of stability of the natural oxide film. This explains the good corrosion resistance of aluminium in seawater.

3.1.5.   Flow speed

As in freshwater, stagnation of seawater has an unfavourable effect on the corrosion resistance of aluminium. Agitation of seawater tends to improve the corrosion resistance and leads to a decrease in pitting depth. The experience with heat exchangers of seawater

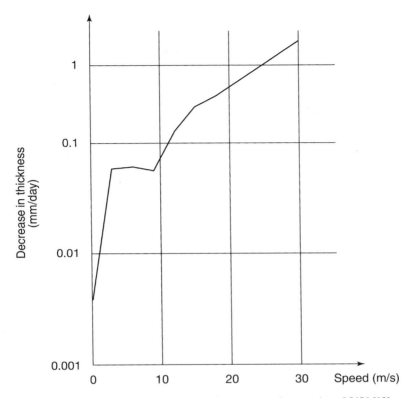

Figure D.3.2. Influence of the flow speed of seawater on the corrosion of 5456 [13].

desalination plants shows that aluminium very well withstands flow speeds up to 2.5 m·s^{-1} at temperatures below 100–120 °C, provided that the geometry of the circuitry does not lead to cavitation.

At high flow speeds, starting at 9 m·s^{-1} (17.5 knots), tests on 5456 H117 at room temperature have shown that the oxide film becomes unstable and that corrosion starts at intermetallics, which are laid bare and swept away by the liquid flow. Corrosion becomes uniform and regular (Figure D.3.2). The annual decrease in thickness amounts to 60 μm per year at a flow speed of 9 m·s^{-1}, and to 350 μm per year at 20 m·s^{-1} (39 knots) [13].

3.1.6. *Biological activity*
Seawater is a living medium. It has a very high level of biological activity that also contributes to the corrosion resistance of metals (for better or worse).

Whenever a piece of metal is immersed in seawater, within a few seconds it will be covered by a viscous, biological humour, a so-called zooglea, on which then all the marine matter built up from plants, marine animals, especially molluscs with or without a shell such as barnacles, corals, algae, sponges, etc., will develop.

It should be recalled that this plant or animal matter would not get fixed on a metal whose salts are toxic for these organisms. This is the case of copper, mercury and tin, which is why the most efficient antifouling paints, nowadays prohibited, were organic salts of copper or mercury.

Aluminium has no antifouling effect because its salts, including alumina ($Al(OH)_3$), are not toxic for marine organisms. As a consequence, marine biological matter (algae and molluscs) will cover aluminium immersed in seawater very quickly if local conditions allow their growth.

Marine flora and fauna that develop on the surface constitute so-called marine fouling. When a barnacle (or any other mollusc) attaches itself to a metal, it acidifies the confined medium that it creates locally on the metal. On aluminium, this results in a rather brilliant and highly visible mark at the site occupied by the mollusc (Figure D.3.3). Since aluminium is insensitive to crevice corrosion, these traces do not undergo more than a very superficial pickling. In addition, pitting corrosion may be observed. However, the pitting depth is not greater than the depth of pitting that develops outside the areas occupied by the molluscs. The fixation of algae or other plant varieties has no special effect on the corrosion resistance of aluminium.

Figure D.3.3. Appearance of sheets after elimination of marine fouling.

3.2. MARINE ENVIRONMENTS

Corrosion resistance depends on the position of the metal with respect to the sea. It can be immersed, semi-immersed or placed at a certain level above the water. In each case, the environment is different, depending on the exposure, and consequently, the corrosion resistance of the metal may be different.

3.2.1. Exposure to marine atmosphere

This is the case of ships and equipment at the seaside. The metal is exposed to marine winds that are more or less loaded with salt, to high marine humidity, and possibly to mould that may develop on the metal. The corrosion resistance under these conditions has been discussed in Part C.

3.2.2. Immersion in surface seawater

This is the case of hulls, buoys and supports of landing stages. The structure is immersed in a few meters of very aerated water that is constantly being replaced because of the action of waves and tides. When the structure is not floating, part of it is subject to immersion–emersion cycles that correspond to the tides. This zone always corrodes more than the zones that are immersed permanently.

3.2.3. Immersion in deep seawater

Certain physicochemical characteristics of seawater such as salinity, temperature and oxygen concentration vary as a function of the immersion depth. This has been shown by measurements taken in the Pacific Ocean off California between the surface and a depth of 2000 m (Figure D.3.1).

Certain equilibria are modified: for example, the precipitation of calcium carbonate does not take place at a great depth (the metal is not covered by a scale film), which explains the lesser corrosion resistance of certain metals and alloys in deep seawater.

The development of several applications at the sea bed level such as bathyscaphs, small experimental submarines and submarine detection beacons has triggered studies of the corrosion of metals in deep waters. Their results are reported below.

3.2.4. Immersion in deep waters

Several studies have been undertaken since 1965 to investigate the corrosion resistance of aluminium up to a depth of 2000 m in the Pacific Ocean [14]. These tests have shown that the corrosion resistance in deep waters is comparable to that observed at the surface. The forms of corrosion are the same. Pitting depth on 5000 series alloys is of the same order of magnitude whatever the depth [15]. On other alloys (3000 and 6000 series), it can be higher. The role of oxygen has been put forward as an explanation for these differences.

Table D.3.3. Influence of immersion depth and sediments on the corrosion resistance of 5000 series alloys

		Pitting depth (μm)	
Duration of test (d)	Depth (m)	Water	Sediments
181	1.5	125	
197	713	435	570
402	722	810	337
1064	1615	1155	1195
123	1719	300	900
751	1719	1215	740
403	2066	1250	790

More recent tests up to a depth of 5000 m in the Atlantic Ocean [16] and up to 3000 m in the Indian Ocean [17] have shown that the corrosion resistance of aluminium does not depend on the immersion depth. Several exploration devices have been made of aluminium such as the *Aluminaut* made in 7079 T6 [18].

The presence of abundant deposits of organic or inorganic sediments could be expected to have an influence on the corrosion resistance of certain metals and alloys, because of the release of sulphur dioxide, carbon dioxide and ammonia as a consequence of the bacterial decomposition of these sediments. Experience with supports of landing stages embedded in the mud and in the sediments at the coast shows that the corrosion resistance of 5000 series alloys is not significantly affected by this medium. This also applies to great depths (Table D.3.3) [19].

3.3. FORMS OF CORROSION IN SEAWATER

Almost all forms of corrosion described in Chapter B.2 can occur in seawater, because of the high aggressiveness of this medium, particularly rich in chlorides, and the very low electrical resistivity of seawater that facilitates most electrochemical reactions (Figure D.3.4).

The resistivity of freshwater is between 1000 and 3000 $\Omega \cdot cm^{-1}$, depending on its salinity. The resistivity of seawater is only 10 $\Omega \cdot cm^{-1}$. Seawater, therefore, facilitates ionic conductivity. However, for alloys commonly used for marine applications, the two most common forms of corrosion are pitting corrosion and galvanic corrosion.

3.3.1. *Pitting corrosion*

The most frequent form of corrosion is pitting corrosion, because the pH of seawater (8–8.5) is within the range in which aluminium and its alloys are prone to this form of

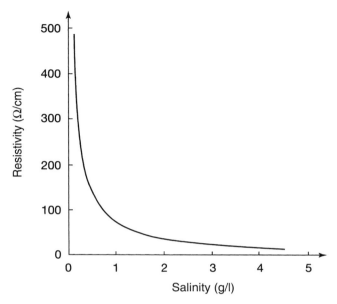

Figure D.3.4. Resistivity of seawater.

corrosion. As in freshwater, the pitting corrosion rate of aluminium follows the $kt^{1/3}$ law. Pitting, therefore, tends to slow down over time, as shown in Figure D.3.5.

Over the past 50 years, many results of immersion tests in seawater have been published [20, 21]. Table D.3.4 summarises the maximum pitting depths after 5 and 10 years

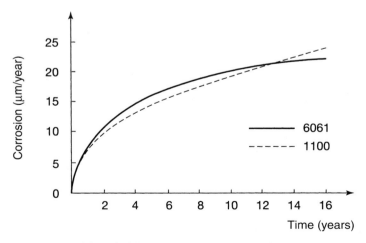

Figure D.3.5. Corrosion rate in seawater [22].

Table D.3.4. Resistance during immersion in seawater

Alloy	Maximum pitting depth (μm)	
	5 years	10 years
1199	90	
1100 H14	<50	1000
3003 H14	<50	250
	130	1170
	330	530
	530	300
5052 H34	<50	120
	150	300
5056	250	1250
	250	625
	300	375
	400	1000
5083	<50	150
	120	1300
5086 H112	500	
	800	
5154	375	
5454 H34	330	
5456 H321	200	
	280	
	550	
6051 T4	270	400
6051 T6	260	500
	450	600
6053 T6	80	90
	160	250
	400	260
6061 T4	50	350
	80	100
	200	325
	350	325
	500	825

of immersion of several alloys of the 1000, 3000, 5000 and 6000 series. It should be noted that

- the maximum pitting depth rarely exceeds 1.5 mm after 10 years of immersion in seawater,
- there is no relation between the maximum pitting depth after 5 and 10 years, and
- the performance of 5000 and 6000 series alloys is comparable.

Table D.3.5. Influence of immersion–emersion under the influence of tides

		Pitting depth (μm)	
Alloy	Duration (ans)	Total immersion	Immersion–emersion
1199	5	90	230
5083 O	1	750	700
5083 H323	2	300	50
5083 H323	4	30	20
5154 H34	2	20	20
5154 H38	5	250	175
5456 H321	5	950	820

The effect of alternating immersion and emersion according to a tidal rhythm has no influence on the pitting depth, as shown by the results summarised in Table D.3.5.

3.3.2. Galvanic corrosion

The very high electrical conductivity of seawater favours the development of galvanic corrosion, which is why galvanic coupling in heterogeneous assemblies between aluminium and other metals must be neutralised.

In order to achieve this, three solutions can be envisioned:

–   *insulate the two metals in presence* according to the principle given in Figure B.3.3;
–   *paint both metals* in order to cover the surfaces, and especially the cathodic surface. The proper adherence of the paint should be checked, and surfaces should be repainted regularly;
–   *neutralise the galvanic coupling* between aluminium and the other metals by using consumable anodes. This solution is generally adopted on ships. The anodes aim at neutralising coupling between the aluminium hull and the propulsion system, such as shafts made of stainless steel and propellers made of stainless steel or bronze (see Section B.3.11).

Anodising aluminium does not prevent galvanic coupling. The use of fittings made of aluminium alloys with high mechanical characteristics (6060, 7075), anodised and sealed with dichromate, considerably simplifies bolted assemblies of aluminium components.

Other forms of corrosion such as crevice corrosion and stress corrosion may also occur. They can be prevented by selecting alloys that are suitable for this type of environment, by specific protections and by appropriate design avoiding recesses and dead areas.

3.4. MARINE APPLICATIONS OF ALUMINIUM

Immersion tests in seawater at many places, some of which have exceeded 10 years, as well as more than 50 years of experience with marine applications of aluminium demonstrate that casting alloys without copper of the 40000 series: A-S7G (42100), A-S10G (43100); the 50000 series: A-G3T (51100), A-G6 (51300), as well as wrought alloys of the 5000 and 6000 series have an excellent resistance to corrosion in the marine environment and seawater.

These alloys can be used *with no protection*, except when fouling must be prevented. The use of antifouling paints containing copper or mercury must be absolutely prohibited, because they induce pitting attacks (by reduction of copper and mercury cations on aluminium) that can jeopardise the use of the aluminium equipment. Antifouling paints based on organo-tin compounds may be used, unless prohibited by regulations.

Copper-containing alloys of the 2000, 6000 and 7000 series, as well as copper-containing casting alloys of the 20000 series (A-U5GT) and the 40000 series (A-S9U3) cannot be used without protection: paint, anodisation, etc. They are in fact much too prone to pitting corrosion as well as to other forms of corrosion (exfoliation, stress corrosion).

Aluminium has a very large number of marine applications [23].

3.4.1. Shipbuilding

Although a few passenger ferries were built in aluminium before 1939, one in Great Britain, another in Canada [24], the first modern applications in shipbuilding were superstructures of civil or military ships, in order to save weight in the upper part of the ship: funnels [25], wheelhouses, upper decks [26], etc. Bolting made the junction between steel structures and aluminium structures. In order to avoid galvanic corrosion, great care was taken to insulate these two metals. Since steel–aluminium transition joints have come onto the market, assembling is done by welding, which simplifies the joining (see Figure B.3.4).

Ships with aluminium hulls were developed from the end of the 1950s. The length of civil or military ships rarely exceeded 15 m. This field of application has grown ever since: Servicing boats for offshore platforms, patrol boats, fishing boats, pleasure boats and ferries. The size of aluminium ships has increased steadily, up to 120 m in 1995 [27]. The need of weight saving for fast ferries, the good corrosion resistance of alloys of the 5000 series (5754, 5083, 5086) and 6000 series (6005A, 6082) as well as the expertise of welding aluminium semi-products explain the development of aluminium in ship building [28].

3.4.2. Coastal equipment

Since 1970, all marinas have landing stages made of aluminium. There are over 250 km in France, made of extrusions in 6005A or 6060, and rolled products in 5754, 5083 or 5086. This equipment requires no maintenance.

The use of aluminium has been extended to many other applications: locks [29, 30], road signs, various buoys, fish farming (pontoons for fish cages), etc.

3.4.3. *Desalinisation of seawater*

Aluminium has many decisive advantages for desalination of seawater applications: good thermal conductivity, good resistance to corrosion in seawater, availability of seam-welded and extruded tubes in corrosion-resistant alloys in contact with seawater, easy manufacture of tubular heat exchangers in aluminium, etc. Finally, aluminium is less expensive than cuprous alloys that have a good corrosion resistance to seawater up to 100–120 °C.

Tests on pilot plants have shown that the alloys of the 3000, 5000 and 6000 series satisfy the requirements of desalinisation plants:

– seawater flow speed up to 2.5 m·s^{-1} in seam-welded tubes. Unmachined welding beads resist corrosion very well, and erosion is not observed at these flow speeds;
– brine concentration up to twice as high as in seawater, under the temperature and flow speed conditions of a desalinisation plant;
– brine temperature up to 120 °C;
– heterogeneous assemblies between vessels in steel and tubing in aluminium. It is necessary and sufficient to provide proper and controlled insulation between steel and aluminium during assembly, and to compensate the difference in thermal expansion between the two metals in order to avoid galvanic corrosion. Aluminium resists rusting very well under the conditions of seawater desalination.

Long-term tests in circuits have identified certain important aspects:

– the oxygen concentration of seawater has no influence: the corrosion resistance is the same in aerated seawater (containing about 8 ppm oxygen) and in seawater that has been degassed in a vacuum (oxygen level < 0.1 ppm). This applies whatever the temperature of the stage (20–120 °C). This has been checked on exchanger tubes operating with aerated and deaerated seawater [31];
– oxide layers of variable colour, but mostly in a bronze-like colour, develop from 60 °C in aerated or deaerated seawater on alloys of the 3000, 5000 and 6000 series. The higher the temperature, the faster their growth. They can reach 5 μm at 60 °C and 20 μm after 10 000 h of operation at 100 °C. They are composed of colloidal boehmite together with carbonates, magnesium oxides, etc. [32]. This scale contributes to the good corrosion resistance of aluminium in warm seawater;
– treatments of seawater so as to control its pH (in order to keep it at 6–6.5 by elimination of carbonates), or chlorinating (in order to avoid fouling) have no influence on the corrosion resistance of aluminium.

All these results obtained in laboratories or pilot plants have shown that the use of exchangers in aluminium for the desalination by distillation is possible. The highest performance alloys in terms of corrosion resistance are those of the 3000 series (3003) and 5000 series (5454, 5754, 5083, and 5086).

On alloys with a higher magnesium content, such as 5083 and 5086, grain boundary precipitation of the phase Al_3Mg_2 is observed, starting at 75 °C. This precipitation is unavoidable and does not necessarily lead to corrosion, if the medium does not lead to corrosion, and especially when the conditions are such that protective boehmite coatings are formed. This has been shown on the Salin-de-Giraud circuit. It is advisable to select alloys containing slightly less magnesium.

Aluminium exchangers are used especially in plants with a capacity of 4000–5000 $m^3 \cdot day^{-1}$ (and less), operating with compressed vapour.

3.4.4.   Ocean thermal energy conversion

In tropical zones, the temperature difference of seawater at the surface (27–30 °C) and at a depth of 1000 m (4 °C) is sufficient for use as the hot and cold reservoir of a thermal engine. The idea of Ocean Thermal Energy Conversion (OTEC) was worked out in 1881 by d'Arsonval and taken up by Georges Claude, who in 1934 attempted, unsuccessfully, to set up such a machine off Cuba.

During the 1970s, the idea of producing electrical energy in tropical zones by means of OTEC was reconsidered in the United States, Japan and Europe. Two approaches can be envisioned:

–   *open circuit*: surface water is evaporated and drives a turbine coupled with an alternator and is then condensed in contact with cold seawater pumped from a depth of 800–1000 m;
–   *closed circuit*: ammonia (NH_3) is used as a fluid. In a first exchanger, it evaporates in contact with seawater taken from the surface, and after having driven the turbine, it condenses in a second exchanger that is cooled by deep seawater. Such a closed circuit limits the volume of vapour.

The second approach implies two seawater–ammonia exchangers. Two materials are in competition for the manufacture of these exchangers: titanium and aluminium. Copper is excluded because of the presence of ammonia.

Many studies were carried out between 1975 and 1980 in order to qualify the materials for the manufacture of these exchangers. Several questions had to be addressed:

–   the corrosion resistance in contact with seawater from the surface and from the ocean depths;

- influence of ammonia leaks on the corrosion resistance in hot or cold seawater contaminated by ammonia. Ammonia leads to the precipitation of magnesium hydroxide $(Mg(OH)_2)$ and to the alkalisation of seawater up to pH 10;
- influence of mechanical or physicochemical cleaning methods of fouling that reduces thermal exchange.

The tested aluminium alloys have very good corrosion resistance in contact with surface seawater. For example, the extrapolated decrease in thickness after 30 years would be on the order of 200 μm for alloys 3004 and 5052. In contact with deep seawater, the same alloys exhibit pitting corrosion with a depth up to 200 μm after 3 months of operation [33].

This different behaviour in contact with seawater from the surface and from a depth of 800–1000 m is probably due to the lower pH and the lower temperature of deep seawater [34]. It should, however, be noted that rather few results are available with deep seawater.

Ammonia leaks have only a limited effect on the resistance of aluminium: above a concentration of 30%, uniform corrosion on the order of 100 μm per year is observed; between 30 and 70% pickling occurs, and at even higher concentrations, superficial micropitting appears (Figure D.3.6).

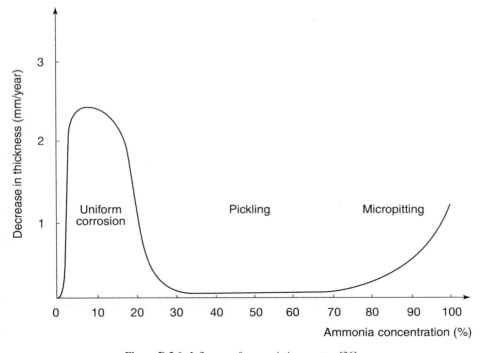

Figure D.3.6. Influence of ammonia in seawater [36].

The mechanical cleaning by circulation of foam rubber balls inside the piping has very little effect on the corrosion resistance. The same applies to chlorination [35].

REFERENCES

[1] McGuire J.C., Aluminum: its alloys and their use in ship construction, *The Aluminum World*, vol. 2, 1895, p. 49–56.
[2] Renié C., Charles D., Le métal de l'Ange, *Voiles et Voiliers*, vol. 229, 1990, p. 70–75.
[3] Guérin R., Les possibilités d'application de l'aluminium et de ses alliages dans la marine, *Revue de l'Aluminium*, vol. 12, 1926, p. 204–209.
[4] Hobson R.P., Notes on the yacht Defender and the use of aluminium in marine construction, *US Naval Institute Proceedings*, vol. 23, 1897, p. 523–562.
[5] Sielski R.A., The history of aluminium as a deckhouse material, *Naval Engineers Journal*, vol. May, 1987, p. 165–172.
[6] Mears R.B., Brown R.H., Resistance of aluminium-base alloys to marine exposures, *Transactions of the Society of Naval Architects Engineerws*, vol. 52, 1944, p. 91–113.
[7] Boyd, W.K., Fink, F.W., *Corrosion of metals in marine environments*, Battelle, report MCIC-78–37, March 1978.
[8] Lyman J., Abel R.B., Chemicals aspects of physical oceanography, *Journal of Chemical Education*, vol. 35, 1958, p. 113–115.
[9] Huppatz W., Meissner H., Effect of the temperature and salt content of sea water on the corrosion behavior of aluminium, *Werkstoffe und Korrosion*, vol. 38, 1987, p. 709–710.
[10] Fink, F.W., *Corrosion of metals in sea water*, OSW, report No. 46, December 1960.
[11] Vargel, C., *Utilisation de l'aluminium dans les installations de dessalement par flash distillation. Examen après 16 400 heures d'essai*, rapport Pechiney CRV 678, 13 juillet 1972.
[12] Reinhart, F.M., *Corrosion of materials in hydrospace*, US Naval Civil Engineering Laboratory, report R504, December 1966.
[13] Gehring G.A., Peterson M.H., Corrosion of 5456 H117 aluminium in high velocity sea water, *Corrosion*, vol. 37, 1981, p. 232–242.
[14] Reinhart, F.M., *Corrosion of materials in hydrospace V. Aluminium alloys*, US Naval Civil Engineering Laboratory, report AD-683334, January 1969.
[15] Ailor W.H., Evaluation of aluminium panels exposed in the Pacific Ocean at 2 340 feet for six months, *Metallurgia*, vol. 75, 1967, p. 99–108.
[16] Ulanovskii I.B., Corrosion of metals in the Atlantic Ocean, *Zashchita Metallov*, vol. 15, 1979, p. 697–700.
[17] Sawant S.S., Wagh A.B., Corrosion behavior of metals and alloys in the waters of the Arabian Sea, *Corrosion Prevention & Control*, vol. 37, 1990, p. 154–157.
[18] Brooks C.L., *Aluminium alloys for pressure hulls*, Metals Engineering Quarterly, ASM, 1965, p. 19–22.
[19] Reinhart, F.M., Jenkins, J.F., *Corrosion of alloys in hydrospace. 189 days at 5900 feet*, US Naval Civil Engineering Laboratory, technical note N-1224, April 1972.
[20] Wagner R.H., Bonewitz R.A., *Catalog information on the performance of aluminium in sea water*, Alcoa, report NXS-3, April 1978.
[21] Schumacher M., *Seawater corrosion Handbook*, Noyes Data Corporation, New Jersey, 1979.

[22] Southwell C.R., Alexander A.L., Humer C.W., Corrosion of metals in tropical environments. Aluminium and Magnesium, *Materials Protection*, vol. 4, 1965, p. 30.

[23] Vargel C., L'aluminium et la mer, Matériaux et Techniques, vol. juin, 1986, p. 233–244.

[24] *The application of aluminium alloys to marine uses*, Aluminium Development Association London, 1948.

[25] La marine, *Revue de l'Aluminium*, vol. 211, 1954, p. 220–226.

[26] Les nouveaux paquebots italiens, *Revue de l'Aluminium*, vol. 212, 1954, p. 235–238.

[27] First Stena HSS 1500 delivered by Finnyards, *Fast Ferry International*, April 1996, p. 15–24.

[28] Vargel C., Les navires à grande vitesse. Le choix des alliages d'aluminium, *Colloque SFM*, Sollac, Fos-sur-Mer, 18–19 mars 1996.

[29] Corrosion test of aluminium alloys for water gate construction, *Journal JLMA*, 1980, p. 1–31.

[30] The corrosion resistance of aluminum alloys in sea water, *Alluminio*, vol. 41, 1972, p. 107–108.

[31] Vargel C., Mirabel M., Emploi de l'aluminium dans les installations de dessalement par distillation, *4th International Symposium on Fresh Water from the Sea*, Heidelberg, septembre 1973, p. 295–306.

[32] Wanklyn J.N., Wilkins N.J., The corrosion of aluminium alloys in hot brine, *3rd International Symposium on Fresh Water from the sea*, Dubrovnik, septembre 1970, p. 617–629.

[33] Larsen-Basse J., Corrosion of aluminium alloys in ocean thermal energy conversion seawaters, *Materials Performance*, vol. 23, 1984, p. 16–21.

[34] Dexter S.C., Localized corrosion of aluminium alloys for OTEC heat exchangers, *Oceano Science and Engineering*, vol. 6, 1981, p. 109–148.

[35] Lewis R.O., Influence of biofouling countermeasures on corrosion of heat exchangers materials in sea water, *Materials Performance*, vol. 21, 1982, p. 31–38.

[36] Bonewitz, R.A., The performance of aluminium alloys in ammonia-sea water solutions, *Conference OTEC 5*, Washington, 1978.

Part E

The Action of Inorganic Products

Part E

The Action of Inorganic Products

Chapter E.1

Oxides and Peroxides

Chapter E.1
Oxides and Peroxides

Oxides of metals and metalloids are chemically very stable.

Metal oxides are normally white or coloured powders, depending on the nature of the metallic element. They constitute most of the ores from which metals such as iron, aluminium, and titanium are extracted. They form the basis of refractory and ceramic industries and are widely used as pigments in paints such as white zinc oxide or titanium oxide pigments.

Peroxides are also widely used, for example, hydrogen peroxide, a powerful oxidant.

Oxides of light metalloids are gaseous under standard temperature and pressure conditions, such as the oxides of carbon and nitrogen. Oxides of the heavier metalloids are normally solids; this is the case of silica, SiO_2, and phosphoric anhydride P_2O_5, but with the exception of sulphur dioxide SO_2, which is gaseous.

1.1. OXYGEN

■ ADR numbers

–   Oxygen compressed [1072]
–   Oxygen, refrigerated liquid [1073]

■ Action on aluminium

The role of oxygen vis-à-vis aluminium has several aspects:

–   electrochemical aspects in the fundamental reactions that are responsible for the corrosion of aluminium (see Section B.1.2);
–   chemical aspects of the formation of the oxide film on the metal surface in the liquid or solid state (see Section B.1.8).

Fire resistance is dealt with in Chapter G.7.

The enthalpy of formation of the oxide Al_2O_3:

$$2Al + \frac{3}{2}O_2 \rightarrow Al_2O_3, \qquad \Delta H_0 = -1675 \text{ kJ·mol}^{-1}$$

is one of the highest of any metal oxide, the reason why aluminium is such a powerful reducing agent.

However, the massive metal does not react with oxygen as long as it is covered by its natural oxide film, which forms spontaneously (see Section B.1.8). Only in the finely divided state, i.e. in the form of powder or granules, does aluminium burn spontaneously in contact with air or oxygen. The solid rocket boosters of certain rockets contain blocks of aluminium powder.

In the presence of gaseous oxygen, pure or diluted with nitrogen or other inert gases, aluminium is very stable at a wide range of temperatures. As a consequence, alloys are fused and elaborated under the natural atmosphere of the casthouse.[1] Heat treatments are normally carried out in furnaces under air up to temperatures in the order of 600 °C for durations of up to 24 hours in the case of homogenisation treatments. Heat treatment practices using controlled atmosphere are reserved for cases where very specific quality requirements must be met.

Liquid oxygen liquid (-217 to -182 °C) has no action on aluminium [1]. Owing to their good thermal conductivity and their high mechanical characteristics at cryogenic temperatures, alloys of the 5000 and 6000 series are widely used for the construction of heat exchangers and for storage tanks for liquid oxygen and for gaseous oxygen at room temperature.

1.2. OZONE

Dry ozone O_3 does not react with aluminium used for the manufacture of electrical ozonisers. In the presence of moisture, ozone can give rise to slight pickling.

1.3. METAL OXIDES

■ ADR numbers

–   Arsenic acid, liquid H_3AsO_4 [1553]
–   Arsenic acid, solid H_3AsO_4 [1554]
–   Barium oxide, BaO [1884]

■ Action on aluminium

In general, metal oxides, with the exception of those mentioned below, do not react with aluminium. This is the case of the oxides of titanium, zirconium (zircon), chromium, molybdenum, tungsten, iron, nickel, zinc, cadmium, aluminium (alumina, corindon), and silicon (silica).

---

[1] With the exception of alloys supplemented with a significant amount of an uncombined element that is very reactive towards oxygen, such as lithium.

Refractory materials formed from mixtures of sintered oxides have no corrosive action on aluminium. Like all more or less finely divided or granulated oxides, they can have an abrasive action on the metal. For decades, alumina has been transported in trucks made of alloy 5086. In mining plants (such as iron or chromium mining), equipment made in aluminium such as electrical cables, transport equipment, roofs and cladding panels for buildings are used. Mineral pigments such as zinc oxide, titanium oxide and chromium oxide are used in the formulation of paints for aluminium.

The oxides of lithium Li_2O, sodium Na_2O, potassium K_2O, which are readily decomposed in water to the corresponding base lithium hydroxide LiOH, sodium hydroxide NaOH, potassium hydroxide KOH are very aggressive towards aluminium, even in the presence of no more than traces of moisture.

Calcium oxide (lime) CaO has very little effect on aluminium. In the presence of moisture, a superficial attack of the metal can be observed.

Magnesium oxide (magnesia) MgO and barium oxide BaO have no effect, even at high temperatures. These products are dried on aluminium trays and stored and transported in aluminium containers.

Arsenic trioxide As_2O_3 does not react with aluminium even in presence of moisture. Arsenic pentoxide As_2O_5 does attack aluminium because arsenic acid H_3AsO_4 is readily formed.

The same is observed with phosphoric anhydride P_2O_5, which forms phosphoric acid H_3PO_4 in the presence of moisture. On the other hand, phosphates, or more precisely phosphate minerals, which are normally a mixture of phosphoric anhydride P_2O_5 and lime CaO have only a very moderate action on aluminium.

Tests on 1100 and 5754 with a mineral composed of

- P_2O_5: 34.5%,
- CaO: 51.5%,
- SiO_2: 1.3%

have shown that the dry or humid mineral attacks these alloys only very superficially. This experience has demonstrated that it is possible to use equipment made of aluminium alloys such as wagons in phosphate mining.

1.4. OXIDES OF COPPER CuO, Cu_2O, SILVER Ag_2O, MERCURY HgO AND LEAD PbO, Pb_2O

■ **ADR numbers**

- Mercury oxide HgO [1641]

■ **Action on aluminium**

Like copper salts, the oxides of copper, lead, silver and mercury are readily reduced in the presence of water, according to an electrochemical reaction of the type:

$$Pb_3O_4 + H_2O + 2e^- \rightarrow 3PbO + 2OH^-, \qquad E_0 = +0.25 \text{ V}$$

$$PbO + H_2O + 2e^- \rightarrow Pb + 2OH^-, \qquad E_0 = -0.58 \text{ V}$$

In the presence of humidity, particles of the reduced metal on the surface of aluminium will form so-called micro-batteries, leading to pitting corrosion. Therefore, in the presence of humidity, the contact with products containing these oxides should be avoided. The use of antifouling paints based on copper, lead or mercury salts[2] should be strictly prohibited on structures made of aluminium alloys. Experience has shown that the effect of these paints on the hull of ships is disastrous.

1.5. CARBON MONOXIDE CO AND CARBON DIOXIDE CO₂

■ **ADR numbers**

- Carbon monoxide, compressed CO [1016]
- Carbon dioxide, liquid CO_2 [1013]
- Carbon dioxide, refrigerated liquid CO_2 [2187]

■ **Action on aluminium**

Carbon monoxide CO has no action on aluminium, even at very high temperatures of 500–600 °C.

Carbon dioxide CO_2, also called carbonic gas, dry or saturated with humidity, has no action on aluminium [2]. On the other hand, water saturated with carbon dioxide can act on aluminium, possibly leading to water staining (see Chapter D.3). Even at a high temperature (400 °C), carbon dioxide, anhydrous or saturated with humidity, does not attack aluminium at a higher pressure (40 bar).

Gas cylinders made of 6061 have been used for several decades for the storage of carbon dioxide under pressure, especially for beer.

The action of carbon dioxide as a gaseous atmospheric pollutant is treated in Section C.2.6.

---

[2] Due to the noxious effect on the environment, especially on oysters, mussels, etc., their use is generally prohibited or strictly limited.

1.6. SULPHUR DIOXIDE SO$_2$

■ **ADR number**

– Sulphur dioxide [1079]

1.6.1. Liquid and gaseous

Gaseous or liquefied sulphur dioxide has no action on aluminium in a very wide range of temperatures, from -50 to $+400\,°C$. At $400\,°C$, the annual decrease in thickness is estimated at 0.05 mm per year [3]. The presence of humidity does not significantly modify the resistance of aluminium. In liquid sulphur dioxide containing 1% water, the decrease in thickness is on the order of 0.01 mm per year on 1199 and 0.02 mm per year on 5754 [4].

Aluminium is used for refrigerators operating with sulphur dioxide, and for the transportation and storage of sulphur dioxide.

1.6.2. In solution

■ **ADR number**

– Sulphurous acid H$_2$SO$_3$ [1833]

■ **General information**

Sulphur dioxide is highly soluble in water:

– 113 g·l^{-1} at 20 °C
– 94 g·l^{-1} at 25 °C
– 54 g·l^{-1} at 40 °C

Its solutions are acidic, with a pH of about 2; they contain sulphurous acid of the theoretical formula H$_2$SO$_3$.

■ **Action on aluminium**

Solutions of sulphurous acid attack aluminium and the higher the concentration of sulphur dioxide the more severe the attack. At 40 °C, in a solution at 10 g·l^{-1}, the corrosion rate is 0.19 mm per year, and at a concentration of 20 g·l^{-1}, it amounts to 0.32 mm per year. In practice, the use of aluminium in contact with solutions having more than 50 g·l^{-1} is not recommended.

The action of sulphur dioxide as a gaseous atmospheric pollutant is discussed in Section C.2.6.

1.7. SULPHURIC ANHYDRIDE SO₃

In the absence of any trace of humidity, sulphuric anhydride has no action on aluminium. In the presence of water, sulphuric anhydride will be converted into sulphuric acid and will, therefore, become very aggressive to aluminium (see Chapter E.5).

1.8. NITROGEN OXIDES NO₂, N₂O₄

■ ADR numbers

–   Dinitrogen tetroxide N_2O_4 [1067]
–   Nitrogen dioxide NO_2 [1067]

■ General information

Several nitrogen oxides, NO_2 or N_2O_4, are known that are involved in the production of nitric acid. Dinitrogen tetroxide N_2O_4 is used as a propellant for missiles (i.e. as a fuel), where it is stored in tanks made of weldable high-resistance aluminium alloys: 2219, 6061, 5086 and 7020.

■ Action on aluminium

Tests and experience have shown that nitrogen dioxide NO_2 and dinitrogen tetroxide N_2O_4, whether liquid or gaseous, have no action on aluminium from low temperatures ($-100\ °C$) up to relatively high temperatures and pressures (200 °C and 20 bar) [5].

At a humidity level of less than 0.2% in dinitrogen tetroxide, a film may form on the metal; it provides protection against possible pitting corrosion.

Tests have shown that in the absence of humidity, the dissolution rate of 5086 is below 10 μm per year in dinitrogen tetroxide. At a humidity level of 0.2%, it amounts to 1.25 mm per year [6].

1.9. PEROXIDES

1.9.1. Hydrogen peroxide H₂O₂

■ ADR numbers

- Hydrogen peroxide, aqueous solution, stabilised [2015]
- with more than 60% hydrogen peroxide [2015]
- with not less than 20% but not more than 60% hydrogen peroxide [2014]
- with not less than 8% but less than 20% hydrogen peroxide [2984]

■ General information

Hydrogen peroxide is one of the most powerful oxidants, together with the peroxides of alkali metals (sodium, potassium). Hydrogen peroxide is used for whitening textiles and as fuel for rockets and missiles.

The most common solution contains 30 wt% (110 vol%). It is slightly acidic, with a pH of 4 at that concentration.

Hydrogen peroxide is delivered with stabilisers, corrosion inhibitors, nitrates and a phosphate buffer. These additives do not modify the behaviour of aluminium.

■ Action on aluminium

Aluminium, except copper-containing alloys of the 2000 and 7000 series, is one of the rare common metals that do not catalyse the decomposition of hydrogen peroxide [7]. This also applies to very high concentrations (80% and more), which are achieved in modern plants that use the anthraquinone process.

The resistance of aluminium depends on the concentration of hydrogen peroxide. Below a concentration of 10%, a slight superficial pitting attack with a depth of less than 0.1 mm is observed. The dissolution rate increases up to a concentration of 30% and reaches a value on the order of 0.20 mm per year. For concentrations of 30% or higher, the pitting attack ceases, and the decrease in thickness of the metal drops below 0.01 mm per year [8].

For concentrations above 40%, the formation of a black protective film on the surface is sometimes observed. At lower concentrations, from 40% down to 5%, the surface is rather mat grey. Natural light has no influence on the resistance of aluminium in hydrogen peroxide [9].

The presence of chlorides in hydrogen peroxide can significantly modify the resistance of aluminium. For example, 10 ppm of chlorides can lead to pitting attack with a depth of 100–200 μm.

Aluminium and aluminium alloys 1050, 1100, 3003, 5254, 5454, 5754, 5052, 5086, A-S13 (44100), etc., with the exclusion of copper-containing alloys of the 2000 and 7000 series, are widely used in the hydrogen peroxide industry. Large production units (20 000 t per year using the anthraquinone process) have been manufactured entirely in 1100: reaction vessels, tubing, heat exchangers, storage tanks, etc. Many tank wagons and tank trucks are made of 5083, 5086, etc.

In order to improve the resistance of aluminium equipment, passivating aluminium has been suggested prior to its first use using a nitric acid solution at 50% [10, 11]. This has at least the advantage of eliminating any steel particles and other metals that may have become embedded in the surface of the metal in the metal workshop. These residues may destabilise hydrogen peroxide.

Tests have shown that very strong galvanic corrosion of aluminium in contact with stainless steel may develop in hydrogen peroxide.

1.9.2.   *Alkali peroxides*

■ **ADR numbers**

–   Sodium peroxide Na_2O_2 [1504]
–   Potassium peroxide K_2O_2 [1491]
–   Barium peroxide BaO_2 [1449]
–   Calcium peroxide $CaO_2 \cdot 8H_2O$ [1457]
–   Lithium peroxide Li_2O_2 [1472]
–   Strontium peroxide $SrO_2 \cdot 8H_2O$ [1509]
–   Magnesium peroxide MgO_2 [1476]

When strictly no trace of humidity is present, alkali peroxides have no action on aluminium. This makes it possible to prepare sodium peroxide by oxidation of sodium at 300 °C in vessels made of 1100. Anhydrous barium peroxide has no action on aluminium either, even at a high temperature (300 °C).

In the presence of humidity, the peroxides will decompose and yield the corresponding base, such as sodium hydroxide, when sodium peroxide is decomposed. This medium thus will become strongly alkaline and attack aluminium.

1.10.   **OXYCHLORIDES**

■ **ADR numbers**

–   Chromium oxychloride CrO_2Cl_2 [1758]
–   Phosphorus oxychloride $POCl_3$ [1810]
–   Nitrosyl chloride $NOCl$ [1069]
–   Sulphuryl chloride SO_2Cl_2 [1834]
–   Thionyl chloride $SOCl_2$ [1836]

■ Action on aluminium

Oxychlorides have the general formula MOCl. They decompose in the presence of water and release hydrochloric acid that attacks aluminium. This applies to thionyl chloride $SOCl_2$, sulfuryl chloride SO_2Cl_2, nitrosyl chloride NOCl, phosphoryl chloride $POCl_3$, chromium oxychloride CrO_2Cl_2, bismuth oxychloride BiOCl and copper oxychloride CuOCl. The latter is, like copper sulphate, widely used as an anticryptogamide for the treatment of certain diseases of fruit trees and vine.

REFERENCES

[1] Fink F.W., White E.L., Corrosion effects of liquid fluoriner and liquid oxygen on materials of construction, *Corrosion*, vol. 17, 1961, p. 58t/60t.

[2] Schaläpfer V.P., Gäumann H., Bukowiecki A., Untersuchungen über Korrosionsangriffe in Kohlendioxyd-Druckgefässen aus Aluminiumlegierungen, *Schweiz. Arch. Angew*, vol. 15, 1949, p. 316/324.

[3] Anonyme, *Corrosion data survey, Section 1: Main tables*, NACE Engineers, 1985, p. 124, 167.

[4] Bollinger J., Über die Korrosion verschiedener Metalle in verflüsigtem Schwefedioxyd, *Schweizer Archiv*, vol. 10, 1952, p. 321.

[5] Shukotin A.M., Lantratova N.Y., Gerasimova V.A., *Corrosion resistance of metallic materials in nitrogen oxides at high temperature and pressures*, 1968, Translated from Russian, Document ORNL-TR-2539.

[6] Waldrep P.G., Trayer D.M., *Nitrogen tetroxide corrosion studies of cryopanel materials for space chamber propulsion testing*, Rapport AEDC TR-68-138, 1968.

[7] Tatu H., Le blanchiment à l'eau oxygénée, *Revue de l'Aluminium*, no. 55, 1933, p. 2077–2089.

[8] Alley C.W., Hayford A.W., Scott H.F., *Nitrogen tetroxyde corrosion studies*, P. B. Report 171 301 U.S. Department of Commerce. 1960.

[9] Vogel H.U.V., Vergleichende untersuchung über das Korrosionverhalten von Al 99,99 und 99, 99 plattierten Blechen mit der normalen Leichtmetall, Werkstoffen des chemischen Apparatebaues, *Aluminium*, vol. 20, 1938, p. 85/94.

[10] Wiederholt W., Das Verhalten von Aluminium und Aluminiumlegierungen in Wasserstoffsu-peroxydlösungen, *Korrosion Metallschutz*, vol. 8, 1932, p. 4/15.

[11] Voir par exemple le brevet U.S. 2 948 392 aug 1960, *Aluminium treatment for resistance to hydrogen peroxyde*. J. H. Young.

Chapter E.2

Hydrogen, Nitrogen, and Noble Gases

Chapter E.2
Hydrogen, Nitrogen, and Noble Gases

2.1. HYDROGEN H$_2$

■ **ADR numbers**

–   Compressed hydrogen [1049]
–   Liquid cooled hydrogen [1966].

■ **Action on aluminium**

Gaseous or liquid hydrogen does not interact with aluminium at any pressure. Due to their very good resistance at low temperatures, aluminium alloys have interesting applications in installations for the cooling, compression and liquefaction of this gas, as well as for the construction of reservoirs for liquid hydrogen in rockets [1].

Hydrogen is highly soluble in liquid aluminium, but insoluble in the solid state (Figure E.2.1).

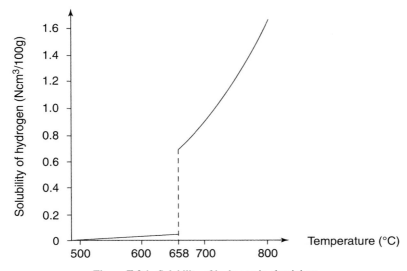

Figure E.2.1. Solubility of hydrogen in aluminium.

369

During TIG or MIG arc welding, traces of lubricants used during transformation, machining and forming, as well as humidity from condensation or from the ambient air, are decomposed by the electric arc. These are sources of hydrogen that may dissolve in the liquid aluminium.

Due to the insolubility in the solid state, small bubbles are formed in the welding seam during cooling. These pores lead to poor welding seams, which is why the zones to be welded must be carefully brushed in order to eliminate any surface contamination. Furthermore, it is necessary to work under controlled temperature and humidity conditions, in order to avoid condensation.

The role of hydrogen in stress corrosion of aluminium alloys is dealt with in Section B.2.5.

2.2. NITROGEN

■ ADR numbers

- Nitrogen, compressed [1066]
- Nitrogen, refrigerated liquid [1977]

■ Action on aluminium

Gaseous or liquid nitrogen does not interact with aluminium at any pressure.

2.3. NOBLE GASES: ARGON Ar, HELIUM He, KRYPTON Kr, NEON Ne, AND XENON Xe

■ ADR numbers

- Argon, compressed [1006]
- Argon, refrigerated liquid [1951]
- Helium, compressed [1046]
- Helium, refrigerated liquid [1963]
- Krypton, compressed [1056]
- Krypton, refrigerated liquid [1970]
- Neon, compressed [1065]
- Neon, refrigerated liquid [1913]
- Xenon, non-refrigerated liquid [2036]
- Xenon, refrigerated liquid [2591].

■ Action on aluminium

These gases are chemically inert. Therefore, they have no action on aluminium, whatever the pressure and the temperature. Arc welding (TIG and MIG) of aluminium and aluminium alloys is done with argon or helium, or a mixture of both.

All these gases can be stored under pressure, at low temperatures, in vessels made of 5083 or 5086. For many decades, gas cylinders made of 2001, 5283, 6061 or 7060 have been used for the storage of industrial gases or respiratory gases for medical use or diving.

REFERENCE

[1] Réservoir de stockage d'hydrogène liquide, *Revue de l'Aluminium*, 1960, no. 290, p. 1048, no. 295, p. 199, 1961, no. 302, p. 1132–1133, 1963, no. 312, p. 874.

Chapter E.3

Metalloids and Halides

Chapter E.3
Metalloids and Halides

3.1. PHOSPHORUS P, ARSENIC As, AND ANTIMONY Sb

■ ADR numbers

- Phosphorus, amorphous P [1338]
- Phosphorus, white, molten P [2447]
- Phosphorus, white or yellow, under water or in solution P [1381]
- Phosphorus, white or yellow, dry P [1381]
- Arsenic As [1558]
- Antimony powder Sb [2871]

■ Action on aluminium

In the absence of humidity, these elements have no action on aluminium. Phosphorus can be stored in aluminium vessels at room temperature and in the absence of air.

In the presence of water or humidity, phosphorus yields phosphoric acid that attacks aluminium.

With antimony, galvanic corrosion may develop on aluminium in the presence of humidity.

3.2. CARBON C, SILICON Si, AND BORON B

■ ADR numbers

- Silicon powder, amorphous Si [1346]
- Calcium carbide CaC_2 [1402]

■ Action on aluminium

□ Carbon and graphite

Carbon in the form of coal, soot and powder has no action on aluminium. Aluminium equipment is widely used in coal mining, for two reasons: aluminium resists corrosion well in this particular environment comprising a high level of humidity and carbon dust, and for safety reasons, because aluminium does not produce sparks upon mechanical impact.

In the presence of water or humidity, contact between graphite and aluminium can lead to strong galvanic corrosion of aluminium (see Section B.3.8).

☐ Calcium carbide

Calcium carbide CaC_2 has no action on aluminium. Sludge residues resulting from the decomposition of carbide by water during the preparation of acetylene mainly consist in lime that leads to a superficial pitting attack of the metal.

☐ Silicon and boron

Silicon and boron have no action on aluminium at any temperature, even in the presence of humidity.

3.3. SULPHUR S

3.3.1. *Sulphur*

■ **ADR numbers**

–   Sulphur [1350]
–   Sulphur, molten [2448]

■ **Action on aluminium**

Unlike other common metals, aluminium does not react with sulphur to yield sulphides, even at high temperatures. Aluminium is thus inert in contact with sulphur over a wide range of temperatures, whether solid or liquid (melting point 119 °C). Tests have confirmed that aluminium is totally inert in contact with vapours of boiling sulphur at 444 °C.

Aluminium equipment is widely used in the sulphur industry, for mining, storage and transportation.

3.3.2. *Hydrogen sulphide H_2S*

■ **ADR number**

–   Hydrogen sulphide H_2S [1053]

Table E.3.1. Dissolution rate in a solution containing H_2S and NH_3 (mm per year)

Alloy	Temperature	
	49 °C	71 °C
3003	0.05	0.01
5052	0.05	0.01
6061	0.05	1.1

■ **Action on aluminium**

Dry gas has no action on aluminium up to 500 °C, and the metal surface in contact with the gas does not become altered or discoloured.

Humidity does not fundamentally change the behaviour of aluminium in contact with hydrogen sulphide. The solubility of hydrogen sulphide in water amounts to 87.7 $g \cdot l^{-1}$ at 20 °C and to 247 $g \cdot l^{-1}$ at 100 °C. These solutions do not attack aluminium at all, even at high temperatures; the annual dissolution rate is below 1 μm [1].

Tests on 1100 and 5754 in a saturated hydrogen sulphide solution at room temperature, with or without the addition of 200 $g \cdot l^{-1}$ of triethanolamine, have shown that these alloys resist well up to 125 °C, with a decrease in thickness on the order of a few micrometres per year.

Aluminium resists mixtures of hydrogen sulphide and ammonia in solution well. As an example, in water containing 15.74 $g \cdot l^{-1}$ NH_3 and 4.82 $g \cdot l^{-1}$ H_2S, the dissolution rate is low [2] (see Table E.3.1). Therefore, aluminium exchangers of the liquid–liquid type can be used in circuits with water containing NH_3 and H_2S from petrochemical processes [3].

The presence of very large quantities of carbon dioxide (30–50% in a mixture $CO_2 + H_2S$) and moisture can very significantly modify the resistance of aluminium to hydrogen sulphide. It is thus necessary to assess a possible risk of corrosion by preliminary tests.

3.4. HALOGENS

3.4.1. *Fluorine F₂*

■ **ADR numbers**

– Fluorine, compressed F_2 [1045]
– Oxygen difluoride, compressed OF_2 [2190]

■ **Action on aluminium**

In the absence of traces of humidity, liquid or gaseous fluorine has no action on aluminium, whether welded or not welded. At very low temperatures, up to − 190 °C, aluminium, especially 2219, is not attacked by gaseous or liquid fluorine, or by the FLOX mixture containing 88% fluorine and 12% oxygen [4].

Prior to its first use, aluminium equipment should be exposed to a gaseous stream of fluorine diluted in an inert gas such as nitrogen or argon; this pre-treatment at room temperature aims at eliminating any superficial grease and metallic particles that could react with liquid fluorine [5].

Up to temperatures of approximately 200 °C, the dissolution rate of 1050A is on the order of 1 μm per year [6].

The good resistance of aluminium in contact with dry gaseous or liquid fluorine results from the formation of a film consisting of a mixture of aluminium fluoride and alumina of the probable composition $2AlF_3 \cdot Al_2O_3$. This film is insoluble in fluorine and protects the surface of the metal.

As in any conductive liquid, galvanic coupling with a less electronegative metal such as stainless steel can lead to a risk of galvanic corrosion of aluminium in the contact areas.

In the presence of humidity, hydrofluoric acid can form; it attacks aluminium very severely.

Tests at a low temperature (-79 °C) have shown that oxygen difluoride OF_2 has no action on 2014 T6, but leads to pitting corrosion on 1100 and 6061 T6 [7].

3.4.2. Chlorine Cl₂

■ **ADR number**

–   Chlorine Cl_2 [1017]

■ **Action on aluminium**

Dry gaseous chlorine at room temperature and liquefied chlorine (point of liquefaction -34.6 °C at 1 bar) has no action on aluminium. With increasing temperature, the risk of a (very) violent reaction strongly increases. As an example, the dissolution rate of aluminium in a stream of dry chlorine amounts to 0.30 mm per year at 130 °C, but is 500 times higher at 160 °C. The presence of air or oxygen does not significantly change the resistance of aluminium to chlorine.

In the presence of humidity, hydrochloric acid will form and lead to a rapid attack of aluminium with localised pitting.

The action of chlorine as an atmospheric pollutant is dealt with in Section C.2.6, and its influence on the behaviour of aluminium in chlorinated water in Section D.1.6.

3.4.3. Bromine Br

■ **ADR number**

– Bromine or bromine solution Br [1744]

■ **Action on aluminium**

Bromine has an action on aluminium that is similar to chlorine. In the absence of humidity, bromine (a liquid at room temperature, with a boiling point of 58.8 °C at 1 bar) does not attack aluminium.

In the presence of humidity, hydrobromic acid [1788] will form, leading to pitting corrosion of aluminium.

3.4.4. Iodine I
Dry iodine crystals do not attack aluminium. Like chlorine and bromine, any trace of humidity will result in pitting of the metal.

3.5. HALOGEN DERIVATIVES

3.5.1. Fluorinated derivatives

■ **ADR numbers**

– Chlorine trifluoride ClF_3 [1749]
– Bromine trifluoride BrF_3 [1746]
– Antimony pentafluoride SbF_5 [1732]
– Bromine pentafluoride BrF_5 [1745]
– Iodine pentafluoride IF_5 [2495]
– Phosphorus pentafluoride PF_5 [2198]
– Chlorine pentafluoride ClF_5 [2548]

■ **Action on aluminium**

Dry fluorine derivatives such as antimony trifluoride SbF_3, antimony pentafluoride SbF_5, bromine pentafluoride BrF_5, and chlorine trifluoride ClF_3 have no action on aluminium, except on alloys with silicon (40000). At 30 °C, the dissolution rate of alloys 1100, 2024, 3003, 5052 and 7079 in chlorine trifluoride is below 1 μm [8].

The same applies to chlorine pentafluoride [9], which can be produced in aluminium alloy vessels. Liquid bromium pentafluoride does not attack aluminium up to 200 °C, and in the vapour phase up to 340 °C.

In the presence of humidity, all these fluorine derivatives decompose and yield hydrofluoric acid that may attack aluminium to greater or lesser degrees.

3.5.2. Chlorinated derivatives

■ ADR numbers

–   Arsenic trichloride $AsCl_3$ [1560]
–   Phosphorus trichloride PCl_3 [1809]
–   Silicon tetrachloride $SiCl_4$ [1818]
–   Antimony pentachloride, liquid $SbCl_5$ [1730]
–   Antimony pentachloride, solution $SbCl_5$ [1731]
–   Molybdenum pentachloride $MoCl_5$ [2508]
–   Phosphorus pentachloride PCl_5 [1806]

■ Action on aluminium

Chlorine derivatives such as arsenic trichloride $AsCl_3$, phosphorus trichloride PCl_3, silicon tetrachloride $SiCl_4$, antimony pentachloride $SbCl_5$ and phosphorus pentachloride PCl_5 have no action on aluminium at room temperature, as long as these products are totally dry. In the presence of humidity, they decompose and form hydrochloric acid that attacks aluminium.

At high temperatures, even in the absence of humidity, these chlorine derivatives can attack aluminium slightly.

3.5.3. Iodine derivatives

Solutions of arsenic triiodide AsI_3 have no action on aluminium at room temperature.

3.6. SULPHIDES

■ ADR number

–   Phosphorus sesquisulphide P_4S_3, free from yellow and white phosphorus [1341]

■ **Action on aluminium**

Arsenic monosulphide AsS (or As_2S_2), phosphorus sulphides, phosphorus sesquisulphide P_4S_3 and phosphorus trisulphide P_4S_6 have no action on aluminium at room temperature.

In the presence of water, phosphorus trisulphide P_4S_6 decomposes under formation of phosphoric acid and thus attacks aluminium. On the other hand, phosphorus sesquisulphide P_4S_3 is insoluble in water and does not attack aluminium even in the presence of humidity.

Aluminium is used for the storage and transportation of these products as well as for the production of matches that contain these phosphorus sulphides.

3.6.1. Carbon disulphide CS_2

■ **ADR number**

–   Carbon disulphide CS_2 [1131]

■ **Action on aluminium**

Aluminium is very resistant in contact with liquid or gaseous carbon disulphide (boiling point at 1 bar: 46 °C).

Equipment in alloys 3003, 6061, A-S13 (44100) is used for the production, storage and transportation of this product.

3.6.2. Sulphur hexafluoride SF_6

■ **ADR number**

–   Sulphur hexafluoride SF_6 [1080]

■ **General information**

Sulphur hexafluoride is a gas at room temperature. It has excellent dielectric properties and is used in gas-insulated electrical devices at high and very high voltage such as switches.

■ **Action on aluminium**

In the absence of humidity, sulphur hexafluoride SF_6 has no action on aluminium. Humidity decomposes it into hydrofluoric acid that attacks aluminium.

Envelopes of gas-insulated electrical switches are in alloys 5754 and 5083, because of the good resistance of aluminium to SF_6 and its amagnetic properties.

REFERENCES

[1] Smrcek K., Sekerka I., Seifert V., Corrosion resistance of aluminium materials to aqueous hydrogen sulfides solutions, *Chem. Prumsyl*, vol. 8, 1958, p. 297–301.
[2] Cronau R.C., Wilde B.E., Procedures for minimizing corrosion in coal-chemical processing facilities, *Materials Performance*, vol. 19, 1980, p. 9.
[3] Demeulenaere R., Tenue d'alliages d'aluminium dans un milieu d'eaux résiduaires chargées en H2S et NH3, *Matériaux et Techniques*, 1976, p. 233–234.
[4] Constantino L.L., Denson J.R., Krishnan C.S., Toy A., *Compatibility testing of space-craft materials and space-storable liquid propellents*, TRW Systems Group, report 23162-6023-RV-00, May 1974.
[5] Fink F.W., White E.L., Corrosion effects of liquid fluorine and liquid on materials of construction, *Corrosion*, vol. 17, 1961, p. 58t–60t.
[6] Hauffe K., *Corrosion Handbook, Fluorine, hydrogen fluoride, hydrofluoric acide*, Vol. 1, Dechema, 1989, p. 106.
[7] Tiner N.A., Corrosion of metals by flowing liquid fluorine compounds, *Advance Cryogenic Engineering*, vol. 12, 1967, p. 771–779.
[8] Grigger J.C., Miller H.C., Effect of chlorine trifluoride and perchloride on construction materials, *Materials Protection*, vol. 3, 1964, p. 53–58.
[9] Hensley W.E., Walter R.J., Compatibility of materials with chlorine pentafluoride, *Journal of Spacecraft Rockets*, vol. 7, 1970, p. 174–180.

Chapter E.4

Inorganic Bases

Chapter E.4
Inorganic Bases

In alkaline media, aluminium dissolves as aluminate AlO_2^- under hydrogen release, according to the reactions:

$$Al \quad + \quad 4OH^- \quad \longrightarrow \quad AlO_2^- \quad + \quad 2H_2O \quad + \quad 3e^-$$

$$3H_2O \quad + \quad 3e^- \quad \longrightarrow \quad 3OH^- \quad + \quad 3/2H_2$$

$$\overline{4Al \quad + \quad H_2O + OH^- \quad \longrightarrow \quad 2Al_2O^- \quad + \quad 3/2H_2}$$

The higher the pH of formed aluminate, the more soluble it is. As the OH^- is introduced by the inorganic base, the stronger the base, the higher the pH value: this is the case of sodium hydroxide NaOH and potassium hydroxide KOH.

Experience shows that at a given pH, the dissolution rate of aluminium depends on the nature of the base. The rate is high with strong bases (such as sodium or potassium hydroxide). The attack will neither slow down nor stop, as long as the volume of the solution is sufficient for the surface under consideration, or if the solution is replenished.

On the other hand, the dissolution rate is low with weak bases such as ammonia, and the attack will stop as soon as a film has formed on the metal surface. For example, at the same pH of 13, a solution with 10 g·l^{-1} of sodium hydroxide attacks aluminium 25 times faster than a solution of 200 g·l^{-1} of ammonia.

In mineral bases, the attack of aluminium is always uniform.

Sodium silicate is a very effective inhibitor for the corrosion of aluminium in alkaline media.

4.1. SODIUM HYDROXIDE NaOH

■ ADR numbers

– Sodium hydroxide, solid [1823]
– Sodium hydroxide, solution [1824]

■ General information

Sodium hydroxide (caustic soda) is highly soluble in water, and sodium hydroxide solutions are strong bases. The annual world production of sodium hydroxide is on the order of 60 million tons. It is universally used as a neutralisation agent in the chemical industry, paper making, etc. Soda lye contains in general 30 wt% of sodium hydroxide.

■ Action on aluminium

The attack of aluminium is uniform and regular. The dissolution rate depends on the concentration (Table E.4.1) and can be very high: in a sodium hydroxide solution with 0.1 $g \cdot l^{-1}$, the annual decrease in thickness would be 7 mm!

The dissolution rate sharply increases with temperature [1] (Table E.4.2). To plunge a piece of aluminium into hot sodium hydroxide solution presents a real hazard of projections, due to the very violent release of hydrogen gas resulting from the attack.

In general, the dissolution rate of aluminium alloys is even higher than that of aluminium grade of the 1000 series.

Sodium hydroxide solutions are widely used for pickling before surface treatments of pieces in aluminium alloy. This process has to be carried out under strictly controlled conditions, always close to the following typical set of parameters:

– NaOH concentration: 50 $g \cdot l^{-1}$
– Temperature: 50–60 °C
– Duration of pickling: 5–10 min
– Rinsing with cold water
– Neutralisation in a solution with 50% nitric acid
– Rinsing with water
– Drying (optional)

This pickling results in a decrease in thickness in the order of 0.05 mm for an immersion of 10 min in a solution at 50 $g \cdot l^{-1}$.

Most pickling baths are based on sodium hydroxide (or sodium carbonate). They contain inhibitors to control the dissolution rate. Sodium silicate efficiently inhibits the corrosion of aluminium in sodium hydroxide solutions. A solution of 10 $g \cdot l^{-1}$ NaOH containing 40 $g \cdot l^{-1}$ sodium silicate does not attack aluminium during an immersion of at least 2 h.

Table E.4.1. Dissolution rate of 1050 in sodium hydroxide at 20 °C

Concentration ($g \cdot l^{-1}$)	pH	Mass loss ($g \cdot m^{-2} \cdot h^{-1}$)	Dissolution rate ($mm \cdot h^{-1}$)
0.01	10.4	0.0	0.0
0.1	11.4	2.2	0.001
1	12.4	8.5	0.003
10	13.2	30.0	0.01
50	13.7	61.5	0.02

Table E.4.2. Dissolution rate of 1050 in sodium hydroxide ($mm \cdot h^{-1}$)

Temperature (°C)	Concentration NaOH ($g \cdot l^{-1}$)				
	20	100	200	300	400
30	0.03	0.07	0.08	0.09	0.07
50	0.10	0.24	0.35	0.36	0.31
60	0.18	0.46	0.64	0.66	0.56
80	0.40	1.27	1.70	1.81	

The addition of sodium silicate to cleaning solutions based on sodium carbonate or sodium hydroxide, which are used in certain food industries, allows cleaning of aluminium equipment with no attack on aluminium.

In the absence of any trace of humidity, sodium hydroxide has no action on aluminium, even above its melting point of 318 °C [2]. However, the slightest trace of humidity provokes a violent attack on aluminium. For example, in an eutectic mixture of sodium hydroxide and potassium hydroxide, the dissolution rate is 2 $\mu m \cdot h^{-1}$ with 1% water, but 0.40 $mm \cdot h^{-1}$ with 8.5% water [3].

4.2. POTASSIUM HYDROXIDE KOH

■ ADR numbers

– Potassium hydroxide, solid [1813]
– Potassium hydroxide, solution [1814]

■ Action on aluminium

Potassium hydroxide, also called caustic potash, is much less used than sodium hydroxide. Anhydrous or in solution, its action is the same as that of sodium hydroxide, and it has the same aggressiveness towards aluminium.

4.3. LITHIUM HYDROXIDE LiOH

■ ADR numbers

– Lithium hydroxide [2680]
– Lithium hydroxide, solution [2679]

■ **Action on aluminium**

Lithium hydroxide solutions severely attack aluminium, even at room temperature. No equipment in aluminium alloy can be used in applications that contain lithium hydroxide [4], such as lithium batteries.

4.4. CALCIUM HYDROXIDE Ca(OH)$_2$

■ **General information**

The word "lime" designates several different products:

–   limestone $CaCO_3$, that occurs as calcite and aragonite,
–   quicklime CaO,
–   slaked lime or hydrated lime $Ca(OH)_2$,
–   lime water, an aqueous solution of $Ca(OH)_2$,
–   milk of lime, a suspension of $Ca(OH)_2$ in water,
–   hydraulic limestone resulting from the calcination of more or mess clayey natural limestone.

All these products contain calcium hydroxide $Ca(OH)_2$, a weak base that has as a saturated solution a pH of 12 at 20 °C.

Lime, in any definition, is very important for many chemical processes. It is the most inexpensive agent for the neutralisation of acidic media and for slag in steel making. It is widely used in agriculture and as the basis for certain types of concrete and plaster. The world production of lime is such that it is ranked in sixth place of chemical product production.

■ **Action on aluminium**

Loose quicklime and slaked lime are transported in dry bulk cargo tankers in aluminium alloy (5083 or 5086).

In the absence of humidity, these products have no corrosive action on aluminium alloys. However, depending on their grain size distribution, they can be abrasive (even in the presence of humidity).

In the presence of water, superficial pitting corrosion can occur. As shown by Bailey in 1920 [5] and confirmed by Takatani [6], corrosion will cease soon due to the formation of an insoluble surface film of calcium aluminate $Ca(AlO_2)_2$ by the reaction of aluminate ions AlO_2^- that are present at pH values above 12, with Ca^{2+} ions originating from the dissociation of $Ca(OH)_2$.

4.5. BARIUM HYDROXIDE Ba(OH)$_2$·8H$_2$O

This hydroxide, also called caustic baryta, is poorly soluble in water. Its solutions attack aluminium superficially by pitting. This attack ceases very quickly, as the metal is covered by an insoluble film. At a high temperature, the attack can be faster and more intense.

4.6. AMMONIA NH$_3$

■ ADR number

– Ammonia, anhydrous NH$_3$ [1005]

■ General information

This is a product of utmost importance, the starting point of nitrogen chemistry for products such as nitric acid, fertilisers (urea, ammonium nitrate, ammonium sulphate, ammonium phosphate), amides, amines and nitriles. The annual world production is approximately 100 million tons, ranking it in fifth place among the chemical products.

Ammonia is a gas that is easy to liquefy at ordinary temperatures (boiling point $-33.5\,°C$). It is highly soluble in water. Its solutions are called ammonia water.

It is used as a cooling liquid in the large refrigerators of warehouses and ships. It has even been considered for use in heat exchangers for the recovery of renewable energy, especially for Ocean Thermal Energy Conversion OTEC (see Section D.3.4).

■ Action on aluminium

With the exception of copper-containing alloys of the 2000 and 7000 series, aluminium resists gaseous or liquid ammonia very well at a very wide range of temperatures, up to $450-500\,°C$ [7]. In the absence of humidity, the dissolution rate of 1050 is below 25 μm per year at 20 °C [8] and below 50 μm per year at 100 °C [9].

In liquid ammonia, at a pressure of 8.75 bar, 3003 and 5454 will develop a very thin, transparent film within a few weeks. Tests have shown that the decrease in the thickness of aluminium in contact with ammonia containing less than 0.04% humidity at 200 °C and 10 bar is in the order of 1 μm per year.

In the presence of up to 5% humidity, the metal can suffer a slight attack that will stop quickly because a protective oxide film will cover the metal.

The presence of carbon dioxide CO$_2$, hydrogen sulphide H$_2$S or sulphur dioxide SO$_2$, three gases that are inert towards aluminium, do not change the behaviour of aluminium in ammonia.

It has been shown that alloys 1100, 3003 and 6063 resist well in a coolant mixture $SrCl_2 +$ NH_3, and poorly in the mixtures $CaBr_2 + NH_3$ and $CaCl_2 + NH_3$ for heat pumps [10].

In liquid ammonia, the difference in potential between aluminium and steels is on the order of 350 mV; in the presence of air, this difference increases to about 600 mV [11]. There is, therefore, a risk of galvanic corrosion of aluminium components if no precaution is taken to insulate aluminium from steel or any other metal or alloy.[1]

With the exception of copper-containing alloys of the 2000 and 7000 series, aluminium and aluminium alloys are very widely used in the storage and transport of gaseous or liquid ammonia, in cooling machines (heat exchangers, etc.) as well as in production plants for ammonia and nitric acid. The ASME Boiler and Pressure Code as well as the German Code AD Merkblatt W/6/1 permit the use of aluminium equipment up to 200 °C.

Internal stress in strain-hardened or cold-rolled alloys of the 5000 series with more than 3% magnesium (5754, 5083, 5086) may give rise to a risk of stress corrosion [12]. The use of temper H111 is thus recommended.

4.7. AMMONIA WATER NH_4OH

■ **ADR numbers**

–   Ammonia solution, relative density less than 0.880 at 15 °C in water, with more than 35% but not more than 50% ammonia [2073]
–   Ammonia solution, relative density between 0.880 and 0.957 at 15 °C in water, with more than 10% but not more than 35% ammonia [2672]

■ **General information**

Gaseous ammonia is highly soluble in water, up to 900 g·l^{-1} at 20 °C, but commonly, only less concentrated solutions, up to 30%, are used. These solutions (ammonia water) are a base that is less strong than sodium or potassium hydroxide.

■ **Action on aluminium**

Aluminium in contact with ammonia water will suffer uniform pickling that ceases after a few hours, as soon as a protective oxide film has reformed on the metal surface. The higher the concentration of ammonia, the weaker the initial attack on 1050 (Table E.4.3 and Figure E.4.1) [13].

---

[1] In ammonia, the ranking in the scale of dissolution potentials is the same as in seawater.

Table E.4.3. Dissolution rate of 1050 in ammonia water (μm·h^{-1}) [13]

Concentration NH$_3$ (%)	Duration of immersion (d)			
	1	2	4	7
0.5	1.4	2.9	1.7	–
2.0	1.9	3.2	3.2	3.0
5.0	2.6	3.9	3.8	3.8
10.0	2.4	3.5	3.7	3.5
21.8	0.9	1.2	1.3	1.1

An increase in temperature has no influence on the resistance of aluminium in ammonia water. In fact, two processes are competing, attack and formation of the oxide layer, and the rate of both processes increases with temperature. These two antagonistic effects have the result that aluminium resists well, if not better, in hot ammonia water than in ammonia water at room temperature.

Aluminium is very widely used in equipment for the transportation and the storage of ammonia water, and in production plants for nitric acid. As in gaseous ammonia, the presence of carbon dioxide CO_2, sulphur dioxide SO_2, hydrogen sulphide H_2S, ammonium carbonate NH_4CO_3 and ammonium sulphide $S(NH_4)_2$ in ammonia does not alter the corrosion resistance of aluminium. On the other hand, the presence of chlorides (and of salts of certain heavy metals such as copper) has a noxious effect.

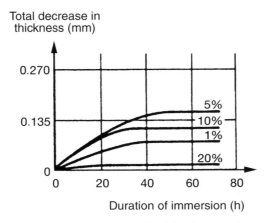

Figure E.4.1. Dissolution rate of aluminium in ammonia water.

4.8. HYDRAZINE N$_2$H$_4$

■ **ADR numbers**

–   Hydrazine, anhydrous [2029]
–   Hydrazine, aqueous solution, with more than 37% hydrazine, by mass [2030]
–   Hydrazine, aqueous solution with not more than 37% hydrazine, by mass [3293]

■ **General information**

Hydrazine is a liquid (melting point at 1.5 °C, boiling point at 113 °C) that is highly soluble in water. It is a weak base, its solutions have a pH close to 11 and are used as deoxidants in water circuits of boilers.

   Concentrated hydrazine is used as a propellant in rockets, the comburant being, for example, dinitrogen tetraoxide N$_2$O$_4$.

■ **Action on aluminium**

Pure hydrazine and its derivatives monomethylhydrazine (MMH) CH$_3$NHNH$_2$ and dimethylhydrazine (DMH) (CH$_3$)$_2$NNH$_2$ have no action on aluminium in a wide range of temperatures between − 196 and +27 °C [14].

   Hydrazine solutions have no action on aluminium; the annual decrease in thickness in solutions at 5 and 10% is below 1 μm.

   Hydrazine and its derivatives are not decomposed or altered in contact with aluminium alloys [15]. This makes it possible to manufacture the tanks of rockets in aluminium alloys such as 6061. Tests have shown that this alloy is not prone to stress corrosion in hydrazine [16].

4.9. UREA NH$_2$CONH$_2$

■ **General information**

Urea is the eleventh most used chemical product. Its world annual production is on the order of 90 million tons, 80% of which is used for fertilisers.

   Urea is highly soluble in water, up to 900 g·l^{-1}. Its solutions at 50% are slightly alkaline (pH 9.5).

■ **Action on aluminium**

Urea powder or granules have no action on aluminium.

Highly concentrated urea solutions, such as that manured as a fertiliser, can provoke a slight pickling of aluminium. As in the case of ammonia, this attack ceases quickly. In a solution of 51% urea, the initial attack is roughly 5 μm.

Aluminium equipment is used in production plants, for the storage and for the transportation of urea.

REFERENCES

[1] Chatterjee B., Thomas R.W., The chemical etching of aluminium in caustic soda solutions, *Transactions of the Institute of Metal Finishing*, vol. 54, 1976, p. 17.

[2] Gregory J.N., Hodge N., Iredale J.V.J., *The static corrosion of nickel and other materials in molten caustic soda*, Atomic Energy Research Establishment Harwell, England, 1956.

[3] Dmitruk B.F., Zarutbitskii O.G., Babich N.N., Kinetics of aluminium corrosion in a hydrated alkaline melt, *Ukraine Khim. Zh*, vol. 49, 1983, p. 690–693.

[4] John K., *Corrosion Handbook*, vol. 3, Dechema, 1988, p. 221.

[5] Bailey G.H., Rate of corrosion of aluminium, *Ind. Soc. Chem. Ind.*, vol. 39, 1920, p. 118.

[6] Takatani Y., Yamakawa K., Yoshizawa S., Corrosion behavior of aluminum alloys in saturated calcium hydroxide solution, *Journal of the Society Material Science, Japan*, vol. 32, 1983, p. 1218–1222.

[7] Tsejtlin K.L., Sorokin Yu.I., Balashova A.A., Babitzkaya S.M., Levin Ya.S., Konyuschenko A.T., Golovkin R.V., Ladyzhenski B.S., High temperature corrosion of metals in gaseous ammonia, *Zashchita Metallov*, vol. 6, 1970, p. 451–454.

[8] Bird D.B., Flournoy R.W., Refinery experience and testing of aluminium condenser tubes, *Materials Protection*, vol. 3, 1964, p. 56.

[9] Nelson G.A., *Corrosion data survey—Table A-6*, 2nd edition, NACE, Houston, 1971.

[10] Wilson D.F., Howell M., DeVan J.H., *Materials corrosion in ammonia/solid heat pump working media*, ORNL, report TM-12004, 1992.

[11] Jones D.A., Wilde B.E., Corrosion performance of some metals and alloys in liquid ammonia, *Corrosion*, vol. 33, 1977, p. 46–50.

[12] Blenkin R., Werdale J.K., Storage vessels, Design, fabrication and use, *Chemical Process & Engineering*, vol. 44, 1963, p. 344.

[13] Röhrig H., Roch J., Über die Einwirkung kalter wässriger Ammoniaklösungen auf Aluminiumblech normaler und höherer Reinheit, *Aluminium*, vol. 21, 1939, p. 128.

[14] Waldrep P.G., Trayer D.M., *Effects of monoethylenehydrazine on cryopanel materials for space simulation chamber propulsion test*, AEDC, report TR-68-194, December 1968.

[15] Van der Wall E.M., Suder J.K., Beegle R.L., Cabeal J.A., *Propellant/Material compatibility study*, AFRPL, report TR-71-741.

[16] Gilbreath P., Adamson M.J., *The stress-corrosion susceptibility of several alloys in hydrazine fuels*, NASA, report TN D-7604, February 1974.

Chapter E.5

Inorganic Acids

Chapter E.5
Inorganic Acids

Aluminium is amphoteric: it is attacked both in acidic and alkaline media. The natural oxide film covering the metal is soluble in acidic media, and its solubility increases very rapidly once the pH has dropped to a value of 5 (Figure B.1.18).

As with inorganic bases, the attack is uniform and regular. In general, the dissolution rate depends on the acid concentration, the temperature and the contact time.

Several of these acids, alone or in combination, are used in more or less diluted solutions as pickling baths for aluminium. These treatments are performed under standard operating conditions with baths of a well-defined concentration. The baths can contain inhibitors for controlling the dissolution rate; the resulting attack is always very slight, in the order of 1 μm on each face.

Several corrosion tests are carried out in acidic media (see Section B.4.4).

5.1. HYDRACIDS

5.1.1. Hydrochloric acid HCl

■ **ADR number**

–   Hydrochloric acid HCl [1789]

■ **General information**

Hydrochloric acid is a gas that is highly soluble in water. It is commonly delivered as a solution containing between 35 and 39 wt% of acid.

■ **Action on aluminium**

This is a strong acid and one of the most aggressive products towards aluminium and even more so towards aluminium alloys [1]. For grades of the 1000 series, the dissolution rate slightly decreases with increasing aluminium content.

The dissolution rate increases greatly with temperature (Table E.5.1).

Aluminium can be used only exceptionally in contact with highly diluted hydrochloric acid: at a concentration of less than 1%, the decrease in thickness remains within rather

Table E.5.1. Dissolution rate in hydrochloride acid (mm per year)

Alloy	Temperature (°C)	HCl concentration				
		1%, 0.3 M	5%, 1.6 M	10%, 3.2 M	15%, 4.8 M	20%, 6 M
1199	20	0.1	0.4	2.6	7.0	27
	50	1.3	3.9	5.2	>50	>50
	98	18.5	>50	>50	>50	>50
1100	20	0.2	7.2	>50	>50	>50
	50	5.2	16.0	>50	>50	>50
	98	>50	>50	>50	>50	>50
3103	20	1.10				
6082	20	2.32				
44100	20	1.13				

acceptable limits: at a concentration of 1%, the decrease is in the order of 0.50 mm per year, and at a concentration of 0.1%, it is in the order of 0.15 mm per year [2].

Cleaning and pickling can be performed with highly diluted hydrochloric acid solutions. Inhibitors must be added to limit the decrease in thickness.

Certain organic products can inhibit the action of diluted hydrochloric acid. Most of them are amines, such as aniline, dibutylamine, thiourea, naphthoquinone, acridine, nicotine. Their effectiveness depends on the acid concentration and on the temperature. Their lifetime is limited. Preliminary tests are recommended in order to determine the performance of an inhibitor in a given medium.

Aqua regia, a mixture of hydrochloric acid (four parts) and nitric acid (one part), is extremely aggressive towards aluminium.

Pure, strictly anhydrous hydrochloric acid in the gaseous or liquid state does not attack aluminium at room temperature [3]. However, in practice it is very difficult, if not impossible, to prevent all traces of humidity. As a consequence, the storage or transportation of anhydrous hydrochloric acid in aluminium vessels is practically excluded, because it presents a substantial hazard.

5.1.2.   *Hydrofluoric acid HF*

■ ADR number

–   Hydrofluoric acid [1790]

■ General information

Like hydrochloric acid, anhydrous hydrofluoric acid is gaseous. It is highly soluble in water.

■ **Action on aluminium**

This is the halogen acid that attacks aluminium most. In an acid at a concentration of 40%, the dissolution rate is 0.5 mm·h^{-1} at room temperature. In anhydrous acid, the dissolution rate is low, in the order of 0.02 mm per year.

Tests at -10 °C have shown that the dissolution rate depends on the acid concentration and on the alloy. 5056 is particularly sensitive as the acid concentration decreases [4] (Table E.5.2).

The very low dissolution rate of aluminium in concentrated hydrofluoric acid is due to the fact that AlF$_3$ resulting from the reaction of aluminium with hydrofluoric acid

$$2Al + 6HF \rightarrow 2AlF_3 + 3H_2$$

is insoluble in concentrated acid. It forms a protective film on the metal's surface. This film is soluble in water, which explains the very high dissolution rate in diluted acid.

Is it nevertheless possible to store and transport anhydrous hydrofluoric acid in aluminium vessels? The same remark as for anhydrous hydrochloric acid applies to hydrofluoric acid.

Fluorosilicon acid H$_2$SiF$_6$ attacks aluminium.

5.1.3.   *Hydrobromic acid HBr and hydroiodic acid HI*

■ **ADR numbers**

– Hydrobromic acid HBr [1788]
– Hydroiodic acid HI [1787]

■ **Action on aluminium**

These acids are highly soluble in water. The dissolution rate of aluminium in their solutions is of the same order as that in hydrochloric acid of the same concentration.

Table E.5.2. Dissolution rate of aluminium alloys in hydrofluoric acid at -10 °C

HF concentration (%)	Dissolution rate (mm per year)		
	1050	3105	5056
99.5	0.13	0.11	0.07
95	0.21	0.22	0.82
90	0.24	0.40	1.86
85	0.33	0.42	17.6
80	0.55	1.15	28.2

5.1.4. Hydrocyanic acid HCN

■ **ADR number**

–   Hydrocyanic acid, aqueous solution HCN [1613]

■ **Action on aluminium**

Whether gaseous or in aqueous solution, hydrocyanic acid has no action on aluminium. This acid and its solution can thus be stored and transported in aluminium alloy vessels. Aluminium equipment can be used in production plants for hydrocyanic acid.

5.2. OXACIDS

5.2.1. Perchloric acid HClO₄

■ **ADR number**

–   Perchloric acid with more than 50% but not more than 72% acid, by mass [1873]
–   Perchloric acid with not more than 50% acid, by mass [1802]

■ **Action on aluminium**

This is the strongest mineral acid. In solution, it is not an oxidant, in spite of the presence of oxygen in its molecule. It, therefore, severely attacks aluminium.

5.2.2. Chloric acid HClO₃

■ **ADR number**

–   Chloric acid HClO₃ [2626]

■ **Action on aluminium**

Chloric acid attacks aluminium.

5.2.3. Hypochloric acid HOCl

Although a weak acid, hypochloric acid attacks aluminium because it decomposes readily into hydrochloric acid.

5.2.4. Sulphuric acid H_2SO_4

■ ADR numbers

- Sulphuric acid, fuming [1831]
- Sulphuric acid with not more than 51% acid or battery fluid, acid [2796]
- Sulphuric acid with more than 51% acid [1830]
- Sulphuric acid, spent [1832]

■ General information

This is the chemical product with the highest tonnage. Its annual world production is in the order of 150 million tons. It has a great number of applications, first of all the production of fertilisers, and also the production of polymers, titanium dioxide, etc. This is a strong acid.

■ Action on aluminium

The dissolution rate of aluminium increases with the acid concentration (Table E.5.3) [5, 6], and very sharply increases with temperature (Table E.5.4) [5]. For a given concentration, it is much smaller than what is observed with hydrochloric acid. As a consequence, there are a few applications of aluminium in contact with diluted sulphuric acid such as cathodes of electrolysis cells for the production of zinc sulphate [7].

Table E.5.3. Dissolution rate in sulphuric acid

Concentration H_2SO_4 (%)	Dissolution rate (mm per year)			
	1199	1100	3005	44100
0.5	0.04	0.06		
1	0.05	0.15	1.12	0.07
2.5	0.06	0.20		
5	0.08	0.25		
10	0.10	0.30	1.07	0.92
20	0.13	0.40		
50	0.35	1.30		
62.5			3.07	3.07
93.5	3.50	4.50		
96			2.94	2.80

Table E.5.4. Dissolution rate of 1050 in sulphuric acid, influence of temperature

Concentration H_2SO_4 (%)	Dissolution rate (mm per year)		
	20 °C	50 °C	98 °C
1	0.15	1.60	3.60
10	0.22	3.20	17.50
25	0.27	6.00	Dissolved
62.5	3.34	Dissolved	Dissolved
78	11.60	Dissolved	Dissolved
96	3.60	1.10	2.94

In spinning factories for rayon, in sulphuric medium at a concentration of 10–15% and at 45 °C, 1199 is used because the corrosion products of aluminium are colourless and thus do not alter the synthetic textile fibre.

Battery fluid is very aggressive towards aluminium. That is why components in aluminium alloys (especially on small crafts) must be protected against the spilling of battery fluid.

5.2.5.   Sulphamic acid H_2NSO_3H

■ **ADR number**

–   Sulphamic acid H_2NSO_3H [1967]

■ **General information**

This acid is also called aminosulphonic acid. It is a sulphonation agent. It is a strong acid: the pH of a solution at 22 $g \cdot l^{-1}$ is 1.5. It is highly soluble in water: 20 $g \cdot l^{-1}$ at 15 °C and 320 $g \cdot l^{-1}$ at 50 °C. Its solutions are used as a descaling agent, especially in household appliances that heat water, such as electric coffee-makers.

■ **Action on aluminium**

Boiling sulphamic acid partly decomposes into sulphuric acid, which explains why the dissolution rate of aluminium rapidly increases with temperature (Table E.5.5) [6].

Experience has shown that hot sulphamic acid solutions can be used for descaling tubes and heat exchangers in aluminium without risk of corrosion. However, concentrated sulphamic acid solutions close to the boiling point should not remain in contact with aluminium equipment for hours, because the risk of pitting corrosion becomes significant.

Table E.5.5. Dissolution rate in sulphamic acid

Alloy	Concentration ($g \cdot l^{-1}$)	Dissolution rate (mm per year)		
		20 °C	55 °C	95 °C
1199	1	0	0.3	0.9
	10	0	0.5	2.5
	50	0	0.4	3.0
1050	1	0	0.7	1.1
	10	0	1.1	7.4
	50	0	1.3	15.0
3003	1	0	0.5	0.7
	10	0	1.3	8.0
	50	0	1.2	16.0
5754	1	0.01	0.7	
	10	0.01	1.6	16.0
	50	0.01	2.1	18.0
6005A	10	0	1.2	7.5
	50	0	1.7	15.0

5.2.6. Fluosulphonic acid $HFSO_3$ and Chlorosulphonic acid $HClSO_3$

■ ADR numbers

– Fluosulphonic acid $HFSO_3$ [1777]
– Chlorosulphonic acid $HClSO_3$ [1754]

■ Action on aluminium

In the absence of any trace of humidity, these acids do not attack aluminium. They can be stored and transported in aluminium equipment.

In the presence of humidity, fluosulphonic acid and chlorosulphonic acid decompose into sulphuric acid and hydrochloric acid or hydrofluoric acid, respectively, and attack aluminium severely.

5.2.7. Hyposelenic acid H_2SeO_3 and Selenic acid H_2SeO_4

■ Action on aluminium

Aluminium cannot be used in contact with these acids, because they decompose in contact with aluminium, leading to a slight, superficial attack.

5.2.8.   *Nitric acid HNO₃*

■ **ADR number**

–   Nitric acid, other than red fuming, with not more than 70% nitric acid [2031]
–   Nitric acid, other than red fuming, with more than 70% nitric acid [2031]
–   Nitric acid, red fuming [2032]

■ **General information**

The annual world production of nitric acid is on the order of 60 million tons. Its main uses are the production of ammonium nitrate NH_4NO_3 [1942], which is a nitrogen fertiliser, of dye-stuffs, explosives, as well as for the synthesis of adipic acid, an intermediate for the synthesis of nylon.

■ **Action on aluminium**

Diluted nitric acid is a strong acid that reacts with almost all materials. It is also an oxidant: the higher its concentration, the more powerful it is. The behaviour of aluminium in contact with nitric acid solution thus depends on whether the acidic or the oxidant character predominates.

In all cases, whatever the concentration, the temperature and the nature of the alloy, the dissolution is always regular and uniform. Pitting is never observed; this also applies to welding seams and heat-affected zones.

☐ **Influence of the concentration**

Nitric acid concentration has been the subject of many studies [8]. In weakly concentrated solutions, the acidic character is dominant. However, the attack of aluminium has nothing to do with what happens in hydrochloric or sulphuric acid. In a solution at 5%, the dissolution rate is 0.6 mm per year, while in hydrochloric acid of the same concentration, it amounts to 7.2 mm per year.

In concentrated solution, on the other hand, the oxidising character is dominant, and aluminium is only very slightly dissolved.

At room temperature (Figure E.5.1), the dissolution rate of aluminium increases steadily with the concentration up to 40–50%, and then decreases very quickly, as explained above: beyond that maximum, the oxidising character clearly dominates (Table E.5.6).

If all other parameters are kept constant, the dissolution is found to be higher in the gas phase (0.03 mm per year) than in the liquid (0.001 mm per year), and of course higher in zones that are submitted to alternate wet and dry cycles (0.05 mm per year).

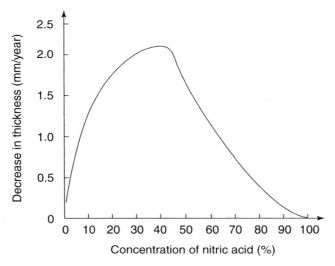

Figure E.5.1. Influence of the concentration of nitric acid on the dissolution of 1050.

Table E.5.6. Dissolution rate in nitric acid

Concentration HNO$_3$ (%)	Dissolution rate (mm per year)		
	1050	1060	1199
1	0.20	0.20	0.20
5	0.65	0.60	0.50
10	1.20	1.00	0.80
15	1.60		
20	1.75	1.65	1.35
25	1.90	1.80	1.60
30	2.00	1.75	1.60
40	2.10		1.30
50	1.60	1.60	
65	0.85		
70	0.45		
80	0.45		
90	0.15	0.15	0.15
93.5		0.10	0.10
96	0.10		
99.6	0.01	0.01	0.01

Corrosion of Aluminium

Table E.5.7. Dissolution rate of 1050 in nitric acid, influence of the temperature

Concentration HNO$_3$ (%)	Dissolution rate (mm per year)			
	24 °C	35 °C	57 °C	79 °C
5	0.80	2.40	13.3	41.5
54	1.30	4.50	20.0	>50
93.5	0.06	0.13	0.4	5.2

☐ Influence of the temperature

The increase in temperature increases the dissolution rate at low concentration very sharply, especially above 50 °C (Table E.5.7). The temperature thus increases the concentration above which the dissolution remains below a given value [9] (Table E.5.8). An increase in temperature leads to a shift of the maximum of the dissolution curves towards lower concentration [10] (Figure E.5.2). In practice, the temperature should be controlled in order to avoid an excessive increase in the dissolution rate.

☐ Influence of the duration of contact

Tests [8] performed over 1000 h have shown that the dissolution rate tends to increase for a concentration of 20% and to decrease in the same proportions for higher concentrations (Figure E.5.3). At high concentrations, the oxidising character of nitric acid tends to reinforce the natural oxide film, while at low concentrations, the acidic character is dominant. In practice, this means that the determination of the behaviour of aluminium should be based on tests that are sufficiently long and representative.

☐ Crevice corrosion

In an assembly, the gap between two aluminium plates or between aluminium and a plastic layer such as Teflon can promote dissolution. This has been shown by tests [11] with a gap of 0.06 μm in boiling (82.6 °C) nitric acid with 1060 for 200 h (Table E.5.9).

Table E.5.8. Dissolution rate of 1050 in nitric acid, influence of the temperature on the concentration

Temperature (°C)	Concentration HNO$_3$ (%)	Dissolution rate (mm per year)
0	66	0.01
10	76	0.01
20	85	0.01
30	95	0.01

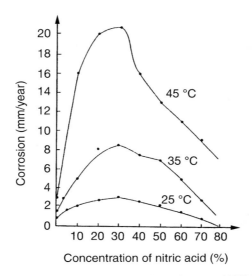

Figure E.5.2. Influence of the temperature on the dissolution rate of 3003 in nitric acid [10].

☐ Galvanic corrosion

As in any strongly conductive medium, aluminium is prone to galvanic corrosion in contact with other metals (see Chapter B.3) such as stainless steel. Tests have shown that in boiling nitric acid, the dissolution rate of aluminium is multiplied by five while that of stainless steel is divided by two.

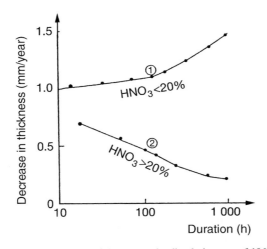

Figure E.5.3. Influence of the duration of the test on the dissolution rate of 1060 in nitric acid [10].

Table E.5.9. Influence of a recess of 0.06 mm on the dissolution rate in boiling nitric acid (duration 200 h)

Width of recess (mm)	Dissolution rate (mm per year)
0.02	0.97
0.05	1.30
0.25	3.89
0.50	6.49
1.00	6.10
2.00	9.21

□ **Addition of hydrofluoric acid to fuming nitric acid**

In highly concentrated nitric acid such as fuming nitric acid used for the preparation of certain explosives and for the propulsion of missiles and rockets, the addition of 0.1–0.5% hydrofluoric acid inhibits the dissolution of aluminium. As an example, between 20 and 50 °C, the dissolution rate of 6061 T6 is 0.5 mm per year in fuming acid, and less than 0.01 mm in the presence of 0.2–0.5% hydrofluoric acid [12].

Under these conditions, a mixed film of composition $2AlF_3 \cdot Al_2O_3$ grows on the surface of the metal, this film is insoluble in nitric acid.

□ **Influence of dinitrogen tetroxide (N_2O_4)**

The dissolution rate of 1050 in nitric acid at a concentration of 99.8% at 10 °C increases from 0.002 mm per year to 0.014 mm per year if the acid contains 20 wt% of dinitrogen tetroxide (Figure E.5.4).

5.2.9. Nitrating acid mixtures

■ **ADR number**

–   Nitrating acid mixture [1796]

■ **Action on aluminium**

Nitrating acid mixtures with more than 50% nitric acid are used for the preparation of dye-stuff and explosives. The behaviour of aluminium mainly depends on the concentration of nitric acid [13] (Table E.5.6): only above a concentration of 85% does the dissolution rate

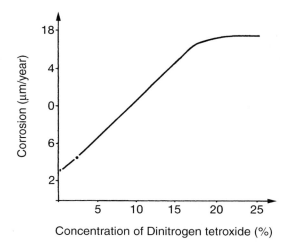

Figure E.5.4. Influence of dinitrogen tetroxide on the dissolution rate of 1050 in HNO$_3$ (99.8%) at 10 °C [19].

drop. The dissolution rate of 1100 in mixtures containing 85% HNO$_3$, 13.5% H$_2$SO$_4$ and 1.5% H$_2$O at 43 °C is in the order of 0.025 mm per year, as shown by tests over 168 h [14].

■ **Use of aluminium alloys in nitric media**

Equipment in aluminium and aluminium alloys (with the exception of alloys of the 2000 and 7000 series) is widely used in the chemistry of concentrated nitric acid: reaction vessels, storage tanks, shipping containers, etc. [15, 16].

In the 1000 series, the aluminium content has no important influence on the dissolution rate, which is the same for 1050, 1080 and 1199, as shown in Table E.5.10. The dissolution rate of unalloyed aluminium (1000 series) in nitric acid has been the subject of numerous

Table E.5.10. Dissolution rate of 1050 in nitrating acid mixtures at 20 °C

Concentrations (%)			
H$_2$SO$_4$	HNO$_3$	H$_2$O	Dissolution rate (mm per year)
11.8	82.2	6.0	0.78
22.9	71.4	5.7	1.87
24.0	53.3	22.7	2.48
55.2	30.6	14.2	3.76
73.5	17.5	9.0	3.82
90.0	5.5	4.5	3.60

studies since 1920 and is well known (probably less for other alloys). For this reason, equipment for storing and transporting nitric acid used to be manufactured in 1080 welded with 1080 Ti.

Due to the rather low mechanical strength (R_m = 90 MPa) of this material, alloys have been considered for this use. In general, the resistance of aluminium alloys at room temperature is slightly less good than that of unalloyed aluminium. Alloying elements such as silicon, copper, zinc or magnesium slightly increase the dissolution rate, especially at low acid concentrations.

Casting alloys with silicon (series 40000), such as aluminium and 5% silicon and A-S7G03 (42100) have a very good resistance to nitric acid at room temperature [17].

Test results on different wrought alloys [18, 19] show that at room temperature, the dissolution rate in concentrated nitric acid (98–99%) remains at a very acceptable level, even for alloys with over 3% magnesium (Table E.5.11).

For alloys, an annealing treatment at high temperatures (400 and 550 °C) for 4–24 h tends to reduce the dissolution rate (Figure E.5.5). This could explain the good resistance of welding seams and of the thermally affected zones of welded structures.

Welding seams are no problem if an appropriate filler metal is chosen. For example, 1080 can be welded with 1050, 1080 Ti or with 1100 as filler metal. No preferential corrosion is observed, neither in the seam nor in the heat-affected zone, except for a narrow strip at the limit between seam and plate on assemblies of 1080 with 1080 Ti (Ti 0.25%), which has been reported after tests at 50 °C in nitric acid at 99% [20].

Plastic deformation, and thus a certain degree of strain hardening, has no influence on the dissolution rate.

Table E.5.11. Dissolution rate of aluminium alloys in nitric acid (98–99%) at room temperature

Alloys	Dissolution rate (mm per year)	Reference
3103	0.0006	[4]
3003	0.085	[19]
3005	0.10	[19]
5005	0.070	[19]
5049	0.09	[19]
5454	0.115	[19]
5052	0.15	[19]
AlMg3	0.001	[4]
AlMg5	0.003	[4]
AlMg6	0.001	[4]
6082	0.08	[19]

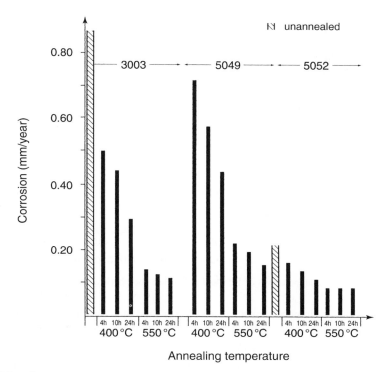

Figure E.5.5. Influence of an annealing treatment on the dissolution rate in nitric acid (99.5%) at 30 °C, for a 306-hour test [19].

5.2.10. *Phosphoric acid H₃PO₄*

■ ADR number

– Phosphoric acid H_3PO_4 [1805]

■ Action on aluminium

Phosphoric acid strongly attacks aluminium. In a solution with 72% phosphoric acid, the annual dissolution at 20 °C amounts to

– 5 mm for 1050,
– 8 mm for 5754,
– 10 mm for 5083.

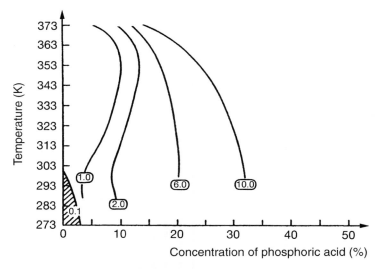

Figure E.5.6. Isodissolution curves of 1050 and A-SG13 in phosphoric acid, without aeration [22].

The dissolution rate in phosphorous acid is lower than in hydrochloric or sulphuric acid, but still too high to use aluminium alloys in contact with phosphoric acid solutions [21], except in a very narrow range of concentrations and temperatures [22] (Figure E.5.6).

Phosphoric acid solutions are used for pickling of metals, especially aluminium alloys. The dissolution rate can be controlled by adding chromates, which slow down the attack, depending on the acid concentration [23] (Table E.5.12).

Commonly, the following phosphochromic mixture is used:

– chromic acid CrO_3: 20 g·l^{-1}
– phosphoric acid H_3PO_4: 50 g·l^{-1}

for cold pickling (and even at 80 °C) of aluminium within a few minutes.

Table E.5.12. Inhibition efficiency of sodium chromate on the dissolution rate of phosphoric acid

H$_3$PO$_4$ (%)	Na$_2$CrO$_4$ (%)			
	0.1	0.5	1.0	1.5
1	99		100	
5	98			100
10	99		94	
20		97		
88		50		90

5.2.11. *Fluorophosphoric acids H$_2$FPO$_3$, HF$_2$PO$_2$, and HF$_6$PF*

■ **ADR numbers**

– Fluorophosphoric acids H$_2$FPO$_3$, HF$_2$PO$_2$, and HF$_6$PF [1776]

■ **Action on aluminium**

They have no action on aluminium if anhydrous. Under these conditions they can be stored in aluminium vessels.

5.2.12. *Boric acid H$_3$BO$_3$*

Boric acid is available as white flakes. It is rather soluble in water (63 g·l^{-1} at 20 °C, and 276 g·l^{-1} at 90 °C). Its solutions, even hot, only lead to slight pickling of aluminium.

Aluminium equipment is used for the production, drying, transportation and storage of boric acid. Tests on 1050 with a solution containing:

– boric acid H$_3$BO$_3$: 15 g·l^{-1}
– sodium hydroxide NaOH: 6 g·l^{-1}

have shown that 1050 does not suffer corrosion in this medium.

5.2.13. *Chromic acid CrO$_3$*

■ **ADR number**

– Chromic acid CrO$_3$ [1755]

■ **Action on aluminium**

Chromic acid CrO$_3$, also called chromic acid anhydride, pure or in solution, has a very moderate action on aluminium. For example, in a solution of 5% chromic acid, the decrease in thickness of 1050, 3103, 6082, A-S13 (44100) is on the order of 0.1 mm per year.

The sulphochromic mixture

– CrO$_3$: 20 g·l^{-1}
– H$_2$SO$_4$: 50 g·l^{-1}

is used as a pickling bath for aluminium (see Section B.5.2).

REFERENCES

[1] Vogel H.U., Vergleichende Untersuchung über das Korrosionverhalten von Al 99,99 und Al 99, 97 plattierten Blechen mit den normalen Leichmetall. Werkstoffen des chemischen Apparatebaues, *Aluminium*, vol. 20, 1938, p. 85–94.

[2] McKee A.B., Brown R.H., Resistance of aluminium to corrosion in solutions containing various anions and cathions, *Corrosion*, vol. 3, 1947, p. 595–612.

[3] Chemische Beständigkeit von Aluminium gegenüber verschiedenen Stoffen, *Werkstoffe und Korrosion*, vol. 14, 1963, p. 913.

[4] Zotikov V.S., Bakhmutova G.B., Corrosion of aluminium and its alloys in hydrofluoric acid, *Zaschita Metallov*, vol. 10, 1974, p. 164–166.

[5] Rabald E., *Corrosion Guide*, 2nd revised edition, Elsevier, Amsterdam, 1968.

[6] Junière P., Sigwalt M., *Les applications de l'aluminium dans les industries chimiques et alimentaires*, Eyrolles, Paris, 1962.

[7] Zhurin A.I., Kosmynin A.I., Vlasenko O.I., Vyssh I., Corrosion of aluminium cathodes during the electrodeposition of zinc, *Ucheb. Zaved. Tsvet. Met.*, vol. 16, 1973, p. 71–75.

[8] Hauffe K., *Corrosion Hanbook*, Dechema, vol. 10, 1991, p. 117–129.

[9] Horn E.M., Schoeller K., Dölling H., Zur Korrosion von Aluminium-Werkstoffen in Salpetersäure, *Werkstoffe und Korrosion*, vol. 41, 1990, p. 308.

[10] Singh D.D., Chaudarry R.S., Agarwall C.V., Corrosion characteristics of some aluminium alloys in nitric acid, *Journal of the Electrochemical Society*, vol. 129, 1982, p. 1869.

[11] Malakhova E.K., Zinchenko R.G., Crevice corrosion of aluminium in strong nitric acid, *Khim Prom*, 1980, p. 358.

[12] Mason D.M., Rittenhouse J.B., Mechanism of inhibiting effect of hydrofluoric acid in fuming nitric acid on liquid phase corrosion of aluminium and steel alloys, *Corrosion*, vol. 14, 1958, p. 345t–347t.

[13] Rabald E., *Ullmanns Encyclopädie der technischen Chemie*, 3e édition, Urban & Schwarzenberg, München, 1951.

[14] Dillon C.P., Corrosion of type 347 stainlesss steel and 1100 aluminium in strong and mixed nitric sulfuric acid, *Corrosion*, vol. 12, 1956, p. 623t–626t.

[15] Chevillotte R., Les emplois de l'aluminium dans l'industrie de l'acide nitrique, *Revue de l'Aluminium*, vol. 88, 1954, p. 314–318.

[16] Junière P., Les réservoirs de Bergerac, *Revue de l'Aluminium*, vol. 214, 1937, p. 559–566.

[17] Everhart J.L., Aluminium alloys casting, *Materials Engineering*, vol. 47, 1958, p. 125.

[18] Zhuravleva L.V., Rubrenov V.P., Aluminium alloys for nitric acid citerns, *Zaschita Metallov*, 1970, p. 224.

[19] Horn E.M., Diekmann H.W., *Corrosion of aluminium and aluminum alloys in nitric acid*, ASTM, STP 1134, 1990, p. 60–69.

[20] Vargel C., *Tenue à 50 °C de l'A8 et de l'A-M1G plaqué A8 soudés dans l'acide nitrique concentré à 99%*, rapport Pechiney CRV 796, 1973.

[21] Kosting P.R., Heins C., Corrosion of metals by phosphoric acid, *Industrial and Engineering Chemistry*, vol. 23, 1931, p. 140–150.

[22] Berg F.F., *Korrosionsschaubilder*, 2e édition, VD-Verlag, GmBH Dusseldorf, 1969.

[23] *Werkstofftabelle: Phosphorsäure*, Dechema, Frankfurt am Main, 1969.

Chapter E.6

Inorganic Salts

Chapter E.6
Inorganic Salts

Inorganic salts result from the action of an acid on a base according to the reaction

$$AH + BOH \rightarrow AB + H_2O$$

or from the action of an acid on a metal (accompanied by hydrogen release) according to the reaction:

$$AH + M \rightarrow \frac{1}{2}H_2 + AM$$

At room temperature, inorganic salts are usually solids, powders or granules, often white, sometimes coloured, more or less deliquescent depending on their capacity to absorb humidity from the air when stored at the open air.[1]

In general, the action of certain inorganic salts on aluminium can be foreseen.

■ Salts of heavy metals

Salts of heavy metals such as copper, gold, silver, mercury, tin, lead, nickel, and cobalt (to cite only the most common ones) all are aggressive towards aluminium, whatever the associated anion, such as chloride, sulphate or fluoride.

In the presence of humidity, most soluble salts of these metals are reduced in contact with aluminium, taking into account the respective potentials of reduction and oxidation. This leads to more or less severe corrosion, which appears as pitting. The size and density of pitting depends on that of the reduced metal particles.

The same applies to certain metalloids such as antimony, and more generally to all metals having a dissolution potential in the medium under consideration, such as freshwater or saltwater, which is more electronegative than that of aluminium (see Table B.1.3).

■ Halide salts

Halide salts (chlorides, fluorides, bromides, and iodides) are generally more aggressive towards aluminium than sulphates and phosphates. Their aggressiveness decreases in the order given.

---

[1] The term "salt", used for these compounds since the 14th century, first by alchemists, stems from the analogy of their appearance and certain properties, such as their solubility in water, with "salt" NaCl.

■ Oxidising salts

Oxidising salts such as chlorates, persulphates, nitrates, and perchlorates generally have no action, especially if anhydrous.

■ The influence of water

As a matter of principle, in the absence of humidity, a dry powder has no action on aluminium, because corrosion as an electrochemical reaction can develop only in an aqueous medium. Salts that are very insoluble in water, therefore, generally have no action on aluminium.

A distinction must be made between

–   *trapped water* (also called crystal water) that forms a part of the molecule's crystal structure, for example, that of magnesium sulphate $Mg_2SO_4 \cdot 7H_2O$. This water is bonded to the molecule and does not modify the behaviour of aluminium in contact with these products, except if dehydration takes place at a high temperature.[2]
–   *humidity*, part of which is water fixed on the surface of the product. It can have several sources such as incomplete drying, voluntary addition of water, and condensation of humidity contained in the air. Unlike crystal water, this water may dissolve the wet product(s), depending on its solubility in water, which generally depends on the temperature. If the product is aggressive towards aluminium, corrosion may occur.[3]

Salts of strong acids or strong bases will react with water: this so-called hydrolysis will release the corresponding acid or base. As an example, aluminium sulphate will hydrolyse in water to sulphuric acid.

As a consequence, depending on the nature of the salts, these solutions can be more or less aggressive (or not aggressive at all) and lead to localised corrosion by pitting, or to uniform corrosion.

Salts of strong acids and strong bases, if highly soluble, can significantly modify the behaviour of aluminium alloys.

Remark
In this chapter, highly diluted solutions with a concentration generally below $1 \ g \cdot l^{-1}$ are not considered. These dilutions can sometimes be found in drinking water. The influence

---

[2] The temperatures of dehydration have been tabulated. They are above room temperature.

[3] Experience shows that condensation on a surface of aluminium alloy such as the surface of a road tanker or a shipping container can lead to pitting corrosion, as well as to more or less pronounced tarnishing of the wall surfaces. In practice, this means that load should not be exposed to variations in temperature that could lead to evaporation of water from the load.

of inorganic salts (chlorides, carbonates, etc.), which are commonly dissolved in these solutions forms a part of the corrosion of water, discussed in Part D.

Since 1950, many electrochemical tests have been performed in very diluted solutions of certain inorganic salts in order to study the anodic dissolution of aluminium. Their interpretation is beyond the scope of the present book and is part of the discussion of electrochemical test methods (see Section B.4.5).

6.1. FLUORIDES

■ ADR numbers

- Sodium fluoride NaF [1690]
- Potassium fluoride KF [1812]
- Ammonium fluoride NH_4F [2505]

■ General information

Fluorides have many uses, for example, as mordants in dyeing, as wood preservatives, in plant-care products, in ceramics, and as frosting agents in glass-making. They are part of the formulation of certain welding fluxes. Most fluorides are more or less water-soluble, some decompose in water.

■ Action on aluminium

Fluorides that are insoluble in water such as calcium fluoride CaF_2, magnesium fluoride MgF_2, barium fluoride BaF_2 and zinc fluoride ZnF_2 have no action on aluminium.

Cryolithe, a double fluoride of aluminium and sodium Na_3AlF_6, has a very low solubility in water. In the absence of humidity, it has no action on aluminium. It is stored and transported in aluminium vessels. In the presence of humidity, aluminium can suffer a superficial pitting attack.

Solutions of sodium fluoride NaF, potassium fluoride KF and ammonium fluoride NH_4F at room temperature only lead to a very moderate attack as superficial pitting; its deepening rate decreases rapidly. For example, the dissolution rate of AMg-6 in a solution of 10% potassium fluoride KF at room temperature is 0.02 mm per year [1].

The decomposition of fluorides in water yields free hydrofluoric acid, leading to a more or less important attack on aluminium. This is observed with ammonium fluoride NH_4F, which begins decomposing at 70 °C.

Welding flux comprises equimolar mixtures of lithium chloride and potassium chloride with an addition of lithium fluoride, sodium fluoride, potassium fluoride or ammonium fluoride. At 580 °C, the intensity of etching shows a linear decrease with the fluoride content. It may reach $3-4$ mm·h^{-1} with 5052, if the fluoride content exceeds 10% [2].

6.1.1. Uranium hexafluoride UF$_6$

■ **Action on aluminium**

Uranium hexafluoride UF$_6$ has no action on aluminium up to about $130-150$ °C (and maybe even higher) if the aluminium does not have a high silicon content [3]. Casting alloys of the 40000 series are thus excluded. 5754 has been widely used for the fabrication of nozzles in a gas diffusion isotopic separation plant, and A-G3T (51000) has been used for the fabrication of compressors. 7075 can be used up to 140 °C [4].

6.1.2. Fluorosilicates

■ **ADR numbers**

– Sodium fluorosilicate Na$_2$SiF$_6$ [2674]
– Potassium fluorosilicate K$_2$SiF$_6$ [2655]

■ **Action on aluminium**

Sodium fluorosilicate Na$_2$SiF$_6$ and potassium fluorosilicate K$_2$SiF$_6$ are poorly water soluble (in cold water about 0.6% for the sodium salt, and only 0.12% for the potassium salt). Their solutions provoke superficial pitting corrosion. In saturated solutions, and at a temperature close to the boiling point, a black protective film that adheres well to the surface is formed.

Sodium fluorosilicate is used as an inhibitor in sodium carbonate solutions.

Latex coagulation is achieved in the presence of sodium fluorosilicate. This operation can be carried out in aluminium vessels; they will turn black, but will not be attacked.

6.1.3. Fluoroborates

■ **Action on aluminium**

Cadmium fluoroborate Cd(BF$_4$)$_2$, chromium fluoroborate Cr(BF$_4$)$_2$ and nickel fluoroborate Ni(BF$_4$)$_2$ are used in electroplating baths. Like all salts of heavy metals, they are aggressive towards aluminium.

6.2. CHLORIDES

Chlorides are ubiquitous. Seawater contains between 25 and 30 $g \cdot l^{-1}$ of sodium chloride NaCl, brackish water between 3 and 4 $g \cdot l^{-1}$. For most common metals, they are the most corrosive compounds.

6.2.1. Sodium chloride NaCl

■ General information

Sodium chloride NaCl (salt) is the chloride that is produced in the highest quantity. It is extracted from seawater by evaporation, or from salt mines (so-called rock salt). It has many uses in the food and chemical industries.

■ Action on aluminium

The behaviour of aluminium in seawater and water in general has been dealt with in Part D. Here, only dry or humid salt and concentrated brine solutions will be discussed.

Many wrought and cast alloys have very good resistance to dry or humid solid sodium chloride present as powder or crystals, as well as to brine. This is not the case with copper-containing casting alloys of the 20000 series such as A-U5GT (21000), of the 40000 series (AlSi5Cu, AlSi9Cu, etc.) and with wrought alloys of the 2000 and 7000 series. Among all common metals and alloys, aluminium has the best resistance to sodium chloride (Figure E.6.1).

Experience has shown that during the first months in contact with brine or with the salt itself, which may be more or less humid, aluminium will tarnish and suffer a very dense pitting corrosion: between 50 and 100 per dm^2, with a maximum depth between 100 and 300 μm depending on the alloys and their strain-hardening temper. However, experience also shows that these pits, once formed, will not deepen after an initiation period of a few months.

■ Action of brine

The brine concentration does not have a marked influence on the penetration rate and the density of pitting. Pitting density rather decreases as the brine concentration increases. At room temperature, aluminium and some aluminium alloys such as 3003, 5083, A-S7G (42000) resist as well in a solution containing 1 $g \cdot l^{-1}$ chloride as in a saturated solution with 357 $g \cdot l^{-1}$.

Nor does temperature have a marked effect: at a given concentration, the behaviour is the same at room temperature and at the boiling point.

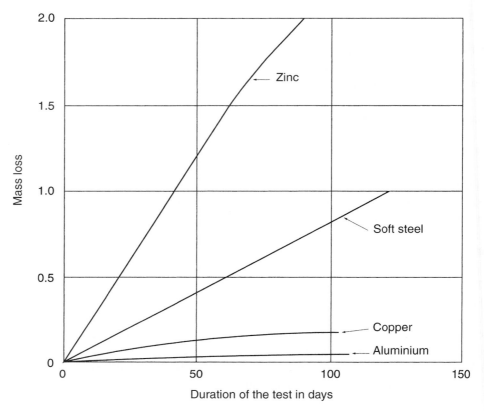

Figure E.6.1. Corrosion resistance of some common metals in NaCl solution (from Evans).

■ Use of aluminium in the presence of NaCl

There is a wide range of applications of aluminium alloys such as 3003, 5754, 5083, 5086, 6060, 6061, A-S7G (42000), A-S13 (44100) in the presence of sodium chloride. They are used in the food industry (see Chapter G.6), for example, as saucepans for cooking food in salt water containing about 10 g·l^{-1} of NaCl (which may contain small impurities such as sodium sulphate and nitrate), trays for the production of salted food and in the canning industry. Aluminium does not alter the taste or colour of salt solutions.

In the production of salt (salterns, rock salt mining), carts, elevator buckets, wagons [5, 6], etc. aluminium alloys are commonly used.

It has been shown that in cooling circuits, brine at a concentration of 260 g·l^{-1} has no action on 1100 or 5754 at low temperatures. Pitting can be inhibited by adding 2% sodium chromate to the brine.

6.2.2. *Potassium chloride KCl*

■ General information

Potassium chloride is mainly produced from mixtures of sodium chloride and potassium chloride extracted from salt mines. It is used in fertilisers. Aluminium alloys are very widely used in mining equipment, where the conditions are particularly aggressive for a large number of materials, and for the transport of the minerals.

■ Action on aluminium

Potassium chloride, solid or in solution, has an action similar to sodium chloride [7].

6.2.3. *Calcium chloride CaCl₂*

■ General information

Anhydrous calcium chloride $CaCl_2$ is a white, very hygroscopic powder. At room temperature, at a relative humidity of 60%, 1 kg of calcium chloride can absorb 1.6 kg of water. It is, therefore, widely used for desiccating and dehydrating gases. Solutions of the hydrate $CaCl_2 \cdot 6H_2O$ have a melting point of $-55\,°C$ and are used as a coolant liquid.

The addition of about 2% of calcium chloride to concrete accelerates the setting and avoids freezing of the concrete before setting under cold weather conditions.

■ Action on aluminium

In the absence of humidity, dry calcium chloride has no action on aluminium alloys. Solutions have a slight action: the dissolution rate is low [8], in the order of a few tenths of a millimetre per year (Table E.6.1). Concentration and temperature have no clear influence. If pitting occurs, its depth normally does not exceed 0.1–0.2 mm. The addition of 2% sodium chromate to calcium chloride solution inhibits possible pitting corrosion.

Table E.6.1. Dissolution rate of 1050 and A-S13 (44100) in calcium chloride solutions (mm per year)

	1050			A-S13 (44100)		
Concentration (%) CaCl₂	20 °C	50 °C	102 °C	20 °C	50 °C	102 °C
1	0.20	0.06	0.02	0.29	0.16	0.00
10	0.02	0.07	0.01	0.06	0.15	0.29
60	0.07	0.16				

Cooling coils in aluminium have been widely used with calcium chloride solution at a concentration of 250 g·l^{-1} or more. Calcium chloride is used in winter as an antifreeze on roads. Its effect on car components in aluminium is very comparable to that of sodium chloride.

Tests with mixtures in gaseous or liquid ammonia and calcium chloride at various temperatures (15, 70 and 130 °C) have shown that 1050 and 5754 show neither corrosion nor any alteration of their surface [9].

6.2.4. Other chlorides

■ **ADR numbers**

- Aluminium chloride AlCl$_3$, anhydrous [1726]
- Aluminium chloride AlCl$_3$ [2581]
- Zinc chloride ZnCl$_2$ [2331]
- Zinc chloride ZnCl$_2$, solution [1840]
- Copper chloride CuCl [2802]
- Copper dichloride CuCl$_2$ [2802]
- Mercuric chloride HgCl$_2$ [1624]
- Tin chloride SnCl$_4$ [1827]
- Ferric chloride FeCl$_3$, anhydrous [1773]
- Ferric chloride FeCl$_3$, solution [2582]

■ **Action on aluminium**

Ammonium chloride solutions (NH$_4$Cl) lead to slight pitting corrosion at room temperature. Corrosion is much stronger in concentrated, boiling solutions.

Lithium chloride LiCl, whether in the solid state or in solution, has an action similar to that of sodium chloride.

Barium chloride solutions (BaCl$_2$) have only a slight action, comparable to that of calcium chloride. Tests in solutions at 1–10% on 1050 have shown that up to 50 °C, the dissolution rate is below 0.03 mm per year; at 98 °C, it is in the order of 0.15 mm per year, and superficial micro-pitting is observed [10].

Magnesium chloride solutions (MgCl$_2$) lead to pitting, especially hot solutions. The use of aluminium in contact with solutions containing more than 1% MgCl$_2$ should be avoided.

Anhydrous beryllium chloride BeCl$_2$ has no action on aluminium. In concentrated solutions, the formation of a film protects aluminium even at 100 °C. Anhydrous beryllium chloride is stored and transported in aluminium alloy vessels.

Anhydrous bismuth and antimony chlorides (BiCl$_3$ and SbCl$_3$) have no action on aluminium.

Anhydrous aluminium chloride $AlCl_3$ has no action on aluminium. It can be stored or transported in aluminium vessels. In the presence of humidity, it hydrolyses, the medium becomes highly acidic and attacks aluminium. The same applies to its solutions.

Anhydrous zinc chloride $ZnCl_2$ has no action. The higher the concentration and temperature, the more aggressive its solutions are. At 98 °C, the dissolution rate of 1050 in a solution at 10% is in the order of 1.5 mm per year, with a strong tendency to pitting corrosion.

Chlorides of heavy metals:

- Copper (I) chloride $CuCl$,
- Copper (II) chloride $CuCl_2$,
- Mercuric (I) chloride Hg_2Cl_2 (calomel),
- Mercuric (II) chloride $HgCl_2$,
- Tin (II) chloride $SnCl_2$,
- Tin (IV) chloride $SnCl_4$,
- Chromium (II) chloride $CrCl_2$,
- Chromium (III) chloride Cr_2Cl_6,
- Iron (III) chloride $FeCl_3$,
- Cadmium chloride $CdCl_2$.

These chlorides, whether in solution at any concentration or in the presence of humidity, can be reduced in contact with aluminium (see Sections D.1.6.6 and E.1.4), which leads to pitting corrosion.

6.3. CHLORATES AND PERCHLORATES

■ ADR numbers

- Barium chlorate $Ba(ClO_3)_2$ [1445]
- Calcium chlorate $Ca(ClO_3)_2$ [1452]
- Calcium chlorate $Ca(ClO_3)_2$, aqueous solution [2429]
- Potassium chlorate $KClO_3$ [1485]
- Potassium chlorate $KClO_3$, aqueous solution [2427]
- Sodium chlorate $NaClO_3$ [1495]
- Sodium chlorate $NaClO_3$, aqueous solution [2428]
- Ammonium perchlorate NH_4ClO_4 [0402] and [1442]
- Potassium perchlorate $KClO_4$ [1489]
- Sodium perchlorate $NaClO_4$ [1502]

■ **General information**

Chlorates and perchlorates are strong oxidants. They are used for the fabrication of explosives, solid propergols, pyrotechnical products, and as bleaching agents and herbicides.

The most common ones are

- ammonium chlorate NH_4ClO_3,
- barium chlorate $Ba(ClO_3)_2$,
- calcium chlorate $Ca(ClO_3)_2$,
- potassium chlorate $KClO_3$,
- sodium chlorate $NaClO_3$,
- ammonium perchlorate NH_4ClO_4,
- potassium perchlorate $KClO_4$,
- sodium perchlorate $NaClO_4$.

■ **Action on aluminium**

Their anhydrous salts have no action on aluminium. Aluminium equipment is used for the production of chlorates and perchlorates, and especially for crystallisation stages for pure salts.

Aqueous solutions of chlorates and perchlorates have very little action on aluminium, even at 100 °C. The presence of impurities, especially sodium chloride, can lead to superficial pitting corrosion in contact with humid salts or their solutions.

Experience has shown that the behaviour of aluminium in ammonium perchlorate solutions, even at 85 °C, is not modified by the addition of up to 16% sodium chloride.

The addition of sodium sulphate to an ammonium perchlorate solution leads to pitting corrosion that is so severe that the use of aluminium cannot be envisaged. This has been shown with solutions containing 200 g·l^{-1} sodium sulphate and 300 g·l^{-1} ammonium perchlorate.

6.4. HYPOCHLORITES

■ **ADR numbers**

Hypochlorite solution [1791].

■ **General information**

Sodium hypochlorite (NaOCl) solutions (bleach), potassium hypochlorite solutions KOCl and ammonium hypochlorite solutions NH_4OCl, diluted to 0.5–1%, are widely used as bactericides for doing the laundry or for cleaning domestic or agricultural premises.

■ **Action on aluminium**

Aluminium withstands cleaning with these solutions very well at room temperature, because the contact is short, intermittent and often followed by rinsing with water.

For prolonged contact, the addition of a corrosion inhibitor such as 2% sodium silicate is recommended.

Concentrated solutions with a pH of 11 lead to a uniform, regular attack of the metal, for a decrease in thickness of 0.5 mm per year with 1100 and 5754.

Copper containing alloys of the 2000 series are much more sensitive to the action of sodium hypochlorite solutions. The decrease in thickness of 2024 is 30 times higher than that of 1100.

6.5. BROMIDES AND IODIDES

■ **General information**

Bromides and iodides are mainly used in photography and for drugs.

The main bromides are

- Ammonium bromide NH_4Br,
- Potassium bromide KBr,
- Sodium bromide NaBr.

The main iodides are

- Ammonium bromide NH_4I,
- Potassium bromide KI,
- Sodium bromide NaI.

They are white crystals, highly soluble in water.

■ **Action on aluminium**

Dry crystals have no action and can be transported or stored in aluminium vessels. Sodium bromide, even in a relative humidity of 90%, does not attack aluminium [11].

Their solutions lead to pitting corrosion, which is more intense with bromides than with iodides, and more intense with hot solutions than with cold solutions. In a solution containing 750 g·l^{-1} of ammonium bromide at 80 °C, pitting corrosion of aluminium is observed with a decrease in thickness in the order of 2 mm per year.

6.6. SULPHATES

6.6.1. *Ammonium sulphate (NH$_4$)$_2$SO$_4$, sodium sulphate Na$_2$SO$_4$·10H$_2$O, and potassium sulphate K$_2$SO$_4$*

■ **General information**

Sodium sulphate Na$_2$SO$_4$ is used for paper and glass making, and in detergents. Potassium sulphate K$_2$SO$_4$ and ammonium sulphate (NH$_4$)$_2$SO$_4$ are almost exclusively used in ternary fertilisers of the NKP type (see Section G.3.1). They are colourless crystals or white powders and are water soluble.

■ **Action on aluminium**

When dry, they have no action on aluminium.

In solution, they can provoke a slight, uniform attack of aluminium at room temperature. The dissolution rate is in the order of 0.01 mm per year in a solution of 10% sodium sulphate. The intensity of the attack does not increase with temperature, but superficial pitting occurs at a temperature of 50 °C and higher [10].

Tests have shown that a solution with 36 wt% of ammonium sulphate leads to a decrease in the thickness in the order of 0.2 mm per year.

In solution, the acidic hydrosulphates NaHSO$_4$ and KHSO$_4$ have a slightly stronger action on aluminium than the neutral sulphates.

6.6.2. *Beryllium sulphate BeSO$_4$, magnesium sulphate MgSO$_4$·7H$_2$O, calcium sulphide CaSO$_4$·2H$_2$O, barium sulphate BaSO$_4$, and cadmium sulphate CdSO$_4$·15H$_2$O*

Whether anhydrous or in solution (beryllium or magnesium sulphate), these sulphates have no action on aluminium. Calcium and barium sulphate are insoluble. It is well known that plaster, which mainly contains calcinated calcium sulphate (gypsum) does not attack aluminium (see Section G.4.2).

Solutions of magnesium sulphate, in a concentration of less than 50%, have no action on aluminium up to about 50–70 °C.

6.6.3. *Aluminium sulphate* $Al_2(SO_4)_3 \cdot 18H_2O$

■ General information

Aluminium sulphate is used in water treatment plants, for paper making, dyeing, and in fire extinguishers.

■ Action on aluminium

When dry, it does not attack aluminium, even at high temperatures. Anhydrous aluminium sulphate can be stored and transported in aluminium alloy vessels.

In solution, hydrolysis of the sulphate yields an acidic medium that attacks aluminium in a uniform manner, and the higher the concentration and temperature the faster this will occur: in a 10% solution, the dissolution rate is on the order of 0.05 mm per year at 20 °C, 0.6 mm per year at 50 °C, and about 10 mm per year at the boiling point [12].

6.6.4. *Zinc sulphate* $ZnSO_4 \cdot 7H_2O$

Zinc sulphate solutions attack aluminium by pitting. Nevertheless, cathodes in 1050 are used in electrolysis cells for the production of zinc sulphate, because in this medium, the resistance of aluminium in superior to that of other common metals.

6.6.5. *Copper sulphate* $CuSO_4 \cdot 5H_2O$

■ General information

Copper sulphate is widely used as a fungicide, either alone or in association with other fungicides (maneb, zineb), for the treatment of orchards and vineyards. It is sprinkled as a solution at about 1%, and in much more concentrated solutions if spread by plane.

■ Action on aluminium

Like all copper salts, copper sulphate solutions are very aggressive towards aluminium, even at very low concentrations: pitting corrosion on aluminium begins at a copper concentration of about 0.5 g·l^{-1}. In solutions at 1% copper sulphate, the pitting attack is generalised and corresponds to a decrease in thickness on the order of 0.5 mm per year. Aluminium equipment for storing, transporting or spreading this product can, therefore, not be used.

6.6.6. Other sulphates

–   Antimony sulphate $Sb_2(SO_4)_3$
–   Chromium (II) sulphate $CrSO_4$
–   Chromium (III) sulphate $Cr_2(SO_4)_3$
–   Iron (III) sulphate $Fe_2(SO_4)_3$
–   Nickel sulphate $NiSO_4$
–   Lead sulphate $PbSO_4$
–   Mercuric (II) sulphate $HgSO_4$
–   Cobalt sulphate $CoSO_4$.

These lead to more or less severe corrosion of aluminium, depending on the nature of the salt and its concentration.

Iron (II) sulphate $FeSO_4 \cdot 7H_2O$ in saturated solution has only a slight action in aluminium at room temperature, and it is, therefore, possible to transport and to store ferrous sulphate solutions in aluminium vessels. However, as soon as ferrous sulphate oxidises to ferric sulphate, the medium becomes very aggressive towards aluminium.

6.6.7. Alums

–   Potassium alum $Al_2(SO_4)_3K_2SO_4 \cdot 24H_2O$ (double sulphate of potassium and aluminium)
–   Chromium alum $Cr_2(SO_4)_3K_2SO_4 \cdot 24H_2O$ (double sulphate of potassium and chromium)

■ General information

Alums are used especially as mordants for dyeing.

■ Action on aluminium

At room temperature and in solutions at less than 10% (which is practically the solubility limit of potassium alum), these alums attack aluminium only superficially and uniformly. The dissolution rate is below 0.1 mm per year.

At high temperatures, and especially in boiling solutions, the metal becomes covered by a rather adherent film that provides protection.

In very concentrated alum solutions and at high temperatures, pitting corrosion can occur.

6.7. SULPHITES AND HYDROGEN SULPHITES

- Ammonium sulphite $(NH_4)_2SO_3 \cdot H_2O$
- Sodium sulphite $Na_2SO_3 \cdot 7H_2O$
- Potassium sulphite $K_2SO_3 \cdot 3H_2O$
- Magnesium sulphite $MgSO_3 \cdot 6H_2O$
- Calcium sulphite $CaSO_3 \cdot 6H_2O$
- Ammonium hydrogen sulphite NH_4HSO_3
- Sodium hydrogen sulphite $NaHSO_3$
- Potassium hydrogen sulphite $KHSO_3$
- Calcium dihydrogen sulphite $Ca(HSO_3)_2$

■ **General information**

Sulphites and hydrogen sulphites are used as bleaching agents for textiles and wood pulp, as antiseptics and disinfectants.

■ **Action on aluminium**

In general, sulphites and hydrogen sulphites of alkali metals have a slight action on aluminium at room temperature. In a 10% sodium sulphite solution, the decrease in thickness is in the order of 0.05 mm per year at room temperature, and 0.25 mm per year at 98 °C. Ammonium sulphite and hydrogen sulphite have a more moderate action than those of sodium and potassium under the same conditions.

Aluminium equipment has occasionally been used for the treatment of synthetic fibres with magnesium sulphite (solution of 50% $MgSO_3$).

6.8. SODIUM THIOSULPHATE $Na_2S_2O_3 \cdot 5H_2O$

Whether in diluted or in concentrated solution, sodium thiosulphate (also called sodium hyposulphate) does not attack aluminium. A protective, iridescent film forms at the surface of the metal, and the decrease in thickness in 10% solution is in the order of 0.01 mm per year between 20 and 98 °C.

6.9. SULPHIDES

Solutions of ammonium sulphide $(NH_4)_2S$ have no action on aluminium, whatever their concentration. Only in boiling solutions and at concentrations above 40%, aluminium will suffer from a uniform pickling, which can be inhibited by adding 0.5–1% sodium silicate. Ammonium sulphide can be stored and transported in aluminium vessels.

Potassium sulphide K_2S, sodium sulphide Na_2S, barium sulphide BaS and magnesium sulphide MgS attack aluminium. Calcium sulphide has a more moderate action; aluminium vessels have been used for the transport and the spreading of mixtures containing up to 20% calcium sulphate and some iron sulphate.

Calcium hydrogensulphide $Ca(HS)_2$ has no action on aluminium.

Molten sodium polysulphide Na_2S_3 attacks aluminium at 350 °C [13].

6.10. PERSULPHATES

■ ADR numbers

-   Ammonium persulphate $(NH_4)_2S_2O_8$ [1444]
-   Potassium persulphate $K_2S_2O_8$ [1492]
-   Sodium persulphate $Na_2S_2O_8$ [1505]

■ Action on aluminium

Persulphate solutions lead to pitting corrosion of aluminium.

6.11. NITRATES

■ ADR numbers

-   Aluminium nitrate $Al(NO_3)_3 \cdot 9H_2O$ [1438]
-   Ammonium nitrate:
    -   hot concentrated solution, in a concentration of more than 80% but not more than 93% [2426]
    -   with more than 0.2% combustible substances, including any organic substance calculated as carbon, to the exclusion of any other added substance [0222]
    -   with not more than 0.2% total combustible material, including any organic substance calculated as carbon, to the exclusion of any other added substance [1942]
-   Silver nitrate $AgNO_3$ [1493]
-   Barium nitrate $Ba(NO_3)_2$ [1446]
-   Calcium nitrate $Ca(NO_3)_2 \cdot 4H_2O$ [1454]

- Chromium nitrate $Cr(NO_3)_2 \cdot 9H_2O$ [2720]
- Magnesium nitrate $Mg(NO_3)_2 \cdot 6H_2O$ [1474]
- Mercuric nitrate $Hg_2(NO_3)_2 \cdot H_2O$ [2725]
- Mercurous nitrate $Hg(NO_3)_2 \cdot H_2O$ [1627]
- Nickel nitrate $Ni(NO_3)_2 \cdot 6H_2O$ [1725]
- Lead nitrate $Pb(NO_3)_2$ [1469]
- Sodium nitrate $NaNO_3$ [1498]
- Zinc nitrate $Zn(NO_3)_2$ [1514]

6.11.1. Ammonium nitrate NH_4NO_3, potassium nitrate KNO_3, and sodium nitrate $NaNO_3$

■ General information

Solid ammonium, potassium and sodium nitrates are white powders. They are highly soluble in water. The solubility of ammonium nitrate is $1180 \text{ g} \cdot \text{l}^{-1}$ at 0 °C and that of ammonium nitrate $921 \text{ g} \cdot \text{l}^{-1}$, while that of potassium nitrate is lower, only $133 \text{ g} \cdot \text{l}^{-1}$. They are used in nitrogen fertilisers. Ammonium nitrate ranks among the top 20 chemical products in the world.

■ Action on aluminium

Dry or in solution at any concentration, these nitrates have no action on aluminium.
The annual decrease in thickness of 3003 at room temperature is

- $< 50 \ \mu\text{m}$ in a solution of $440 \text{ g} \cdot \text{l}^{-1}$ $NaNO_3$,
- $< 20 \ \mu\text{m}$ in a solution of $150 \text{ g} \cdot \text{l}^{-1}$ KNO_3,
- $< 10 \ \mu\text{m}$ in a very concentrated solution of $960 \text{ g} \cdot \text{l}^{-1}$ ammonium nitrate.

However, heating should be avoided since it results in evaporation of ammonia, leading to a decrease in the pH towards more acidic values. In order to use aluminium alloys under proper conditions, the pH of solutions of these nitrates should be maintained above 6 by adding ammonia [14].

The experience with storage tanks in aluminium alloys for concentrated solutions of ammonium nitrate at $830 \text{ g} \cdot \text{l}^{-1}$ shows the very good resistance of alloys 1100, 3003, 5052, 5454 and 6061; this also applies to welded zones [15].

Due to their good resistance to nitric acid, ammonia and nitrates, aluminium alloys are very widely used in nitrate plants. Fertiliser solutions of potassium nitrate and ammonia are transported in aluminium alloy tanks, as are fertiliser granules.

In molten form, these nitrates have no action on aluminium, either. The inner walls of granulation towers for the production of ammonium nitrates are often in 3003 or 5754. Aluminium alloys resist this medium well, in which the temperature reaches 160 °C at the tower's summit.

Salt baths that are used for certain thermal treatments of aluminium alloys such as homogenisation and solution heat treatment are mixtures of sodium nitrate (melting point, 307 °C) and potassium nitrate (melting point, 334 °C).[4] The mass loss of aluminium in a nitrate bath is 20 $mg\cdot m^{-2}\cdot h^{-1}$ of immersion, which corresponds to a decrease in thickness of less than 0.01 $\mu m\cdot h^{-1}$. This is very acceptable for thermal treatments that do not exceed a few hours.

The alkalinity of these baths should be monitored in order to avoid superficial pickling of the metal during rinsing after the heat treatment. The addition of sodium dichromate to the salt bath is recommended to neutralise effects of possible alkalinisation [16].

6.11.2.   *Aluminium nitrate Al(NO3)3*

Dry aluminium nitrate does not attack aluminium. Its solutions have a slight action at room temperature: the dissolution rate is in the order of 0.4 mm per year at 20 °C. At higher temperatures, the hydrolysis of aluminium nitrate releases nitric acid, which, due to its low concentration, may attack aluminium. The dissolution rate is 10 times higher at 50 °C than at 20 °C, and in addition, pitting corrosion is observed. Aluminium nitrate can be produced (from calcium nitrate and aluminium sulphate), stored and transported in aluminium equipment.

6.11.3.   *Other nitrates*

Calcium nitrate $Ca(NO_3)_2$, magnesium nitrate $Mg(NO_3)_2$, barium nitrate $Ba(NO_3)_2$ and bismuth nitrate $Bi(NO_3)_2$, whether dry or in solution, have no action on aluminium. In a solution of 600 $g\cdot l^{-1}$ calcium nitrate $Ca(NO_3)_2$, the decrease in thickness of 3003 is below 0.1 mm per year.

While zinc nitrate $Zn(NO_3)_2$ has no action on aluminium, the nitrates of other heavy metals such as copper, chromium, cobalt, mercury, nickel, and lead provoke severe pitting corrosion of aluminium.

6.12.   NITRITES

■ ADR numbers

–   Potassium nitrite KNO_2 [1488]
–   Sodium nitrite $NaNO_2$ [1500]

---

[4] To prevent explosions, the pieces to be treated should be carefully cleaned from any turnings and casting scrap. Alloys with more than 5 % magnesium should not be treated in these baths.

■ **Action on aluminium**

Solutions of sodium nitrite $NaNO_2$, potassium nitrite KNO_2, and ammonium nitrite NH_4NO_2 practically have no action on aluminium, even if boiling. Molten salts may lead to superficial tarnishing at the metal's surface. The use of nitrite salt baths for the thermal treatment of aluminium alloys is not recommended.

6.13. PHOSPHATES

Usually, the term "phosphates" is used both for the mineral that contains phosphoric anhydride and for fertilisers rich in phosphoric acid. In the following, we will deal only with phosphates in the chemical meaning of salt derivatives of phosphoric acid H_3PO_4.

■ **General information**

Ammonium and sodium phosphates are mainly used in fertilisers and washing powders. They are also used in cattle feed.

Solutions of ammonium hydrogenphosphate $NH_4H_2PO_4$ are less aggressive towards aluminium than solutions of ammonium dihydrogenphosphate $(NH_4)_2HPO_4$. The uniform attack of aluminium increases with concentration and temperature. Solutions of ammonium phosphate $(NH_4)_3PO_4$ are considered as only slightly aggressive.

The action of sodium hydrogen phosphate Na_2HPO_4 and sodium phosphate Na_3PO_4 is better known, because these products are used in the formulation of washing powders. Na_3PO_4 solutions are clearly alkaline (a solution at 10% has a pH above 13), which explains their action on aluminium. Sodium hydrogenphosphate solutions, having a pH close to neutral, are far less aggressive than sodium phosphate solutions (Table E.6.2). The attack is always uniform.

The addition of 0.1% sodium silicate strongly reduces the attack in sodium phosphate solutions.

Table E.6.2. Dissolution rate of 1050 in sodium phosphate solution (mm per year) [9]

Concentration (%)	Sodium hydrogenphosphate Na_2HPO_4			Sodium phosphate Na_3PO_4		
	20 °C	50 °C	98 °C	20 °C	50 °C	98 °C
1	0	0	0.1	0.7	1.8	6.0
10	0	0	0.1	80		

The action of the phosphates of

–   magnesium: $Mg_3(PO_4)_2$ and $MgHPO_4$,
–   calcium: $Ca_3(PO_4)_2$ and $Ca(H_2PO_4)_2$,
–   barium: $Ba_3(PO_4)_2$, $BaHPO_4$ and $BaH_4(PO_4)_2$,

depends on their solubility, on the acidity of their solutions, and on the temperature.

The higher their solubility and acidity and the higher the temperature, the more they attack aluminium.

6.14. ARSENATES AND ARSENITES

■ **ADR numbers**

–   Calcium arsenate $Ca_2(AsO_4)_2$ [1573]
–   Potassium arsenate $K_2(AsO_4)_2$ [1677]
–   Lead arsenates $PbH(AsO_4)$ and $Pb_3(AsO_4)_3$ [1617]
–   Sodium arsenates $Na_3AsO_4 \cdot 12H_2O$, $Na_2HAsO_4 \cdot 7H_2O$, $NaH_2AsO_4 \cdot H_2O$ [1685]
–   Zinc arsenate $Zn_3(AsO_4)_2 \cdot 8H_2O$ [1712]
–   Lead arsenite $Pb_3(AsO_4)_2 \cdot x\,H_2O$ [1618]
–   Sodium arsenite $Na_2As_2O_4$:
    •   solid [2027]
    •   aqueous solution [1686]

■ **General information**

These products enter into the formulation of certain insecticides and are used as wood preservatives.

■ **Action on aluminium**

The water soluble salts such as sodium arsenate sodium arsenite and potassium arsenite, attack aluminium [17]. On the other hand, calcium arsenate and lead arsenate, both with a very low solubility in water, only have a slight action on aluminium.

6.15. CARBONATES

6.15.1. *Ammonium carbonate (NH₄)HCO₃, potassium carbonate K₂CO₃, and sodium carbonate Na₂CO₃*

Solutions of ammonium carbonate $(NH_4)_2CO_3$ and ammonium hydrogencarbonate (also called bicarbonate) $(NH_4)HCO_3$, even at high concentrations up to 60%, have only a slight action on aluminium at a temperature up to 100 °C. Aluminium does not alter the aspect of these products. Equipment in aluminium alloys (except those of the 2000 and copper containing alloys of the 7000 series) is used for producing, drying, transporting and storing these carbonates.

Solutions of sodium carbonate Na_2CO_3 and potassium carbonate K_2CO_3 are very alkaline (the pH at 1% Na_2CO_3 is 11.2, and 11.7 at 10%). The dissolution rate is very high, at least during the first hours of immersion, until a grey protective layer forms on the metal that slows down the attack (Table E.6.3). The attack can be prevented completely by simply adding a corrosion inhibitor such as sodium silicate in a proportion of 0.2–1%, depending on the carbonate concentration.

Solutions of bicarbonate have practically no action: they lead at most to superficial pickling.

6.15.2. *Other carbonates*

- Magnesium carbonate $MgCO_3$
- Calcium carbonate $CaCO_3$
- Barium carbonate $BaCO_3$

■ General information

These salts are insoluble in water. They have many industrial applications: marble, limestone, chalk (calcium carbonate), fire bricks, thermal insulation (magnesium carbonate). They are also used in toothpaste.

Table E.6.3. Dissolution rate of 1050 in sodium carbonate solutions (mm per year) [10]

Concentration (%)	Sodium carbonate Na₂CO₃			Sodium bicarbonate NaHCO₃		
	20 °C	50 °C	98 °C	20 °C	50 °C	98 °C
1	− 0.4	+2.8	− 0.6			
5				0	0	0
10	− 0.8	− 5.1	–	+0.01	+0.01	+0.01
25	+0.8	− 5.8	− 15			

■ Action on aluminium

They have no action on aluminium, whatever their concentration and temperature. Toothpaste is sold in flexible aluminium tubes. Magnesium carbonate is stored in silos with trays in 5754 and 3003.

6.16. SILICATES AND METASILICATES

The behaviour of aluminium in sodium silicate $nSiO_2 \cdot Na_2O$ depends on the alkalinity of the products, i.e. on the mass ratio SiO_2/Na_2O, that usually ranges from 2 to 4. The higher that ratio, the less alkaline the product is, and the less the action of sodium silicate is on aluminium. In any case, the attack is very weak and limited to superficial etching. Silicate (as well as metasilicate SiO_3Na_2) is an excellent corrosion inhibitor of aluminium in alkaline media.

Potassium silicate $nSiO_2 \cdot K_2O$, the ratio SiO_2/K_2O of which ranges only from 2 to 2.5, has an action similar to that of sodium silicate.

Magnesium silicate $xSiO_2yMnO \cdot n\,H_2O$, a white powder that is very insoluble in water, has no action on aluminium. Talcum, a variety of magnesium silicate, suffers no alteration of its appearance in contact with aluminium.

Solutions of sodium silicate $3CaO \cdot SiO_2$ lead to a practically immediate, very superficial dissolution of aluminium, which remains below 5 μm.

6.17. BORATES, PERBORATES AND TETRABORATES

■ General information

Sodium tetraborate $Na_2B_4O_7$ (borax) and sodium perborate $NaBO_3 \cdot 4H_2O$, both rather soluble in water, are part of the formulation of washing powders. They are also used in glass making, for wax emulsions and gums.

■ Action on aluminium

Whether dry or in solution, these products have practically no action on aluminium, even at high temperatures (80 °C and above). At 20 °C, the decrease in thickness of 5754 in a solution of 5% tetraborate is 10 μm per year. These products are stored and transported in aluminium alloy equipment.

Aluminium borate $2Al_2O_3 \cdot B_2O_3 \cdot 3H_2O$, even very humid, does not attack aluminium.

6.18. CYANIDES, CYANATES AND THIOCYANATES

■ **ADR numbers**

– Silver cyanide AgCN [1684]
– Calcium cyanide Ca(CN)$_2$ [1575]
– Copper cyanide Cu(CN)$_2$
– Potassium cyanide KCN [1680]
– Sodium cyanide NaCN [1689]

6.18.1. *Cyanides*

Solutions of ammonium cyanide NH$_4$CN, sodium cyanide NaCN, potassium cyanide KCN, calcium cyanide Ca(CN)$_2$ and barium cyanide Ba(CN)$_2$ are alkaline (due to the hydrolysis of cyanides) and attack aluminium. The addition of 0.5% sodium silicate inhibits this attack.

Cyanides such as silver cyanide AgCN, copper cyanide Cu(CN)$_2$, and gold cyanide AuCN, that are used in galvanoplastic processes, lead to pitting corrosion of aluminium due to the presence of cations of heavy metals (silver, copper, and gold).

6.18.2. *Thiocyanates*

– Ammonium thiocyanate NH$_4$SCN
– Potassium thiocyanate KSCN
– Sodium thiocyanate NaSCN.

Aluminium equipment is used for the fabrication of these different salts. They have no action on aluminium, even at high temperatures.

6.18.3. *Ferrocyanides*

Sodium ferrocyanide Fe(CN)$_6$Na$_4$·12H$_2$O and potassium ferrocyanide Fe(CN)$_6$K$_4$·3H$_2$O, as well as sodium ferricyanide Fe(CN)$_6$Na$_3$·H$_2$O and potassium ferricyanide Fe(CN)$_6$K$_3$, dry or in solution, have no action on aluminum, even at 100 °C. In a solution at 10% ferrocyanide, the decrease in thickness is 10 μm per year for 1100 and 5754.

6.19. CHROMATES AND DICHROMATES

Sodium chromate Na$_2$CrO$_4$, potassium chromate K$_2$CrO$_4$, ammonium chromate (NH$_4$)$_2$CrO$_4$, calcium chromate CaCrO$_4$, zinc chromate ZnCrO$_4$, sodium dichromate

$Na_2Cr_2O_7$, potassium dichromate $K_2Cr_2O_7$, and ammonium dichromate $(NH_4)_2Cr_2O_7$, whether dry or in solution, have no effect on aluminium at low or high temperatures.

These are excellent corrosion inhibitors for aluminium (see Section B.5.8) and are widely used as such: zinc chromate is a pigment for primer coatings, potassium chromate is used (as solutions at $3-7$ g·l^{-1} at 100 °C for 30 min) as a sealing bath after anodising that confers to aluminium components an outstanding corrosion resistance. However, their use is more and more restricted due to their toxicity.

6.20. PERMANGANATES

■ ADR numbers

- Potassium permanganate $KMnO_4$ [1490]
- Sodium permanganate $NaMnO_4$ [1503]

■ Action on aluminium

Solutions of potassium permanganate $KMnO_4$ and sodium permanganate $NaMnO_4$ have no action on aluminium, whatever their concentration and temperature. They have no passivating effect on aluminium, as they have on other metals.

6.21. CARBAMATES AND SULPHAMATES

Solutions of ammonium carbamate $NH_4SO_3NH_2$ have no action on aluminium, even at 125 °C.

Cold ammonium sulphamate solutions $NH_4SO_3NH_2$ do not attack aluminium. Hot sulphamate decomposes, releasing sulphuric acid that renders these solutions aggressive towards aluminium.

REFERENCES

[1] Zotikov V.S., Semenyuk E.Y., Corrosion of metals in ammonium fluoride solutions, *Tr Gos. In-Ta Prikl. Khim*, vol. 67, 1971, p. 199.
[2] Kawakatsu I., Osawa T., Erosion of aluminium based alloys by molten brazing flux, *Journal of Japan Institute of Metals*, vol. 37, 1977, p. 435.
[3] Herenguel J., Scheidecker M., Lelong P., Biais R., Boghen R., Whitwham D., Aluminium et magnésium en énergie nucléaire, *Revue de l'Aluminium*, no. 334, 1965, p. 989–997.
[4] Saninger J.M., Alba F., Garrido S.J., Avendano E., The kinetics of aluminium-7075 corrosion by uranium hexafluoride, *Corrosion Science*, vol. 30, 1990, p. 903–913.

[5] Chevillotte R., L'aluminium dans l'industrie du sel, *Revue de l'Aluminium*, no. 94, 1937, p. 849–852.

[6] Bandet P., Les wagons Pechiney pour le transport du sel, *Revue de l'Aluminium*, no. 251, 1958, p. 189–192.

[7] Hasenberg L., *Potassium Chloride. Corrosion Handbook*, Dechema, vol. 7, 1990.

[8] Rabald E., *Corrosion Guide*, 2nd edition, Elsevier, Amsterdam, 1968, p. 84.

[9] Kohl H.K., The corrosion of steel and aluminium in calcium chloride, ammonia, magnesium chloride-methylamine and magnesium chloride-methylaminedecane, *Werkstoffe und Korrosion*, vol. 30, 1979, p. 171.

[10] Vogel H.U.V., Korrosionsverhalten von Aluminium und Aluminiumlegierungen gegen wäßrige Lösungen verschiedener Temperaturen, *Korrosion und Metallschutz*, vol. 16, 1940, p. 259–278.

[11] John K., *Bromides. Corrosion Handbook*, Dechema, vol. 3, 1988.

[12] Bohner H. von, Übersicht über das Verhalten des Aluminiums und seiner Legierungen gegenüber Stoffen der chemischen und der Nahrungsmittel-Industrie, *Korrosion und Metallschutz*, vol. 9, 1933, p. 86–92, see also page 113–122.

[13] Brown A.P., Battles J.E., The corrosion of metals and alloys by sodium polysulfides melts at 350 °C, *Proceedings of the Electrochemical Society*, 1987, p. 237–246.

[14] Horst R.L., Binger W.W., Corrosion studies of aluminium in chemical process operations, *Corrosion*, vol. 17, 1961, p. 25t–30t.

[15] Un réservoir de nitrate d'ammonium de 8000 mètres cubes, *Revue de l'Aluminium*, no. 232, mai 1956, p. 513.

[16] Lesage P., Goetz M., Bains de sels: bains de nitrites-nitrates et de nitrates-nitrites, *Traitements thermiques*, vol. 114, 1977, p. 29–36.

[17] Cook G.S., Dickinson N., Corrosion of metals by insecticidal solutions, *Corrosion*, vol. 6, 1950, p. 137–139.

Part F

The Action of Organic Products

Chapter F.1

Hydrocarbons

Chapter F.1
Hydrocarbons

Hydrocarbons are the main constituents of crude oil and natural gas. They are the largest class of compounds in organic chemistry, to which fuels, polymers etc. belong, and are very important in today's chemical industry.

Three families of hydrocarbons can be distinguished:

- *acyclic, saturated alkanes* correspond to the general formula C_nH_{2n+2}. At room temperature, the first four members, from C_1 to C_4 (methane, ethane, propane and butane), are gases. The next members, from C_5 to C_{16}, are liquids; the others are solids, with a melting point generally below 100 °C. Alkanes occur naturally as complex mixtures in crude oil. Natural gas mainly consists of methane.
- *Acyclic alkenes with a double bond C=C* have the general formula C_nH_{2n}. At room temperature, the first three members, from C_2 to C_4, are gases. The next members are liquids, and the heaviest ones are solids. The four most important products of this family are ethylene, propene, butene and isobutene.
- *Acyclic alkynes with a triple bond C≡C* have the general formula C_nH_{2n-2}. Acetylene being the first member of that family, they are often called acetylenic carbides. The first three members, from C_2 to C_4, are gases as room temperature, the others are liquids. Acetylene is the only alkyne to be produced industrially.

Hydrocarbons are chemically rather inert and, therefore, have no action on aluminium. They are insoluble in water.

Gasolines for explosion engines and turbines are mixtures of liquid acyclic and benzenic hydrocarbons.

1.1. ALKANES

■ ADR numbers

- Methane, compressed CH_4 [1971]
- Methane, refrigerated liquid CH_4 [1972]
- Ethane C_2H_6 [1035]
- Ethane, refrigerated liquid C_2H_6 [1961]
- Propane C_3H_8 [1978]
- Butane C_4H_{10} [1011]

- Isobutane C_4H_{10} [1969]
- Pentanes, liquid C_5H_{12} [1265]
- Hexanes C_6H_{14} [1208]
- Heptanes C_7H_{16} [1206]
- Octanes C_8H_{18} [1262]

■ Action on aluminium

Alkanes have no action on aluminium in a very wide temperature range, from $-200\ °C$ up to the highest temperatures possible for aluminium alloys.

Because of their cryogenic properties, i.e. their very good mechanical resistance at low temperatures, their weldability [1] and their excellent corrosion resistance in marine atmosphere, the alloys of the 5000 series (5083, 5086, 5754) and 6000 series (6082, 6061) are widely used in plants for gasification and regasification of liquid natural gas (LNG), as well as for the transportation (tanks of methane cargos in 5083) and the land storage (tanks in 5083) of LNG [2].

Tubing for domestic gas distribution grids have been manufactured in 6060 tubes. Burners of gas cookers are in aluminium casting alloys.

At elevated temperatures, i.e. in the range from 100 to 150 °C, gases such as methane, ethane and mixtures thereof, possibly containing a few percent of carbon dioxide, water vapour and hydrogen sulphide H_2S, have no action on aluminium.

The liquid alkanes C_5 to C_{16} have no action on aluminium.

Paraffins, i.e. saturated hydrocarbons higher than C_{15}, have no action. Aluminium is used for the construction of equipment used for the refining, reheating, distillation and casting of these products, which are not discoloured in contact with aluminium alloys.

The same applies to mineral oils and greases that are based on crude oil. Certain greases that are deposited on outdoor cables in aluminium during their fabrication provide effective protection against possible atmospheric corrosion due to moisture and pollution for a very long time (several decades).

1.2. ALKENES

■ ADR numbers

- Ethylene C_2H_4 [1962]
- Ethylene, refrigerated liquid C_2H_4 [1038]
- Ethylene oxide $(CH_2)_2O$ [1040]
- Propylene C_3H_6 [1077]
- Butylenes, mixture or 1-butylene or *cis*-2-butylene or *trans*-2-butylene C_4H_6 [1012]

- Isobutylene C_4H_8 [1055]
- Butadienes, mixtures of 1,2-butadiene and 1,3-butadiene C_4H_8 [1010]
- Isoprene, stabilised $C_6H_4(CO_2C_4H_6)_2$ [1218]

■ Action on aluminium

Ethylene C_2H_4 and ethylene oxide $(CH_2)_2O$, propylene C_3H_6 and butylene (also called butene) C_4H_8, which are gases at room temperature, have no action on aluminium. The same applies to higher members that are liquids. Butadiene C_4H_8, the basis of a great number of synthetic rubbers also called Buna, and isoprene also have no action on aluminium.

The same is true of their polymers and copolymers. Equipment in aluminium alloy is used in plants producing these polymers, as well as for their storage and transportation. Polymer granules are stored in silos made in 5083 or 5086.

The plastic and rubber transformation industry increasingly makes use of moulds in aluminium alloys for blowing, thermoforming and injection. The influence of polymers on the complexes used in the building sector and on packaging is discussed in Section G.4.4.

1.3. ALKYNES

■ ADR number

- Acetylene, dissolved C_2H_2 [1001]

■ Action on aluminium

Tests and experience with acetylene generators have shown that this gas has no action on aluminium, even in the presence of humidity. This also applies to gas that contains the classic impurities due to its preparation from calcium carbide, namely hydrogen sulphide and phosphine. The dissolution rate is very low, in the order of 0.10 mm per year [3].

Acetylene gas cylinders for portable oxyacetylene torches, as well as welding equipment and tubes for torches can be made in aluminium alloys.

1.4. ARENES OR AROMATIC HYDROCARBONS

■ ADR numbers

- Benzene C_6H_6 [1114]
- Toluene $C_6H_5CH_3$ [1294]
- Xylenes $C_6H_4(CH_3)_2$ [1307]

- Styrene monomer, stabilised $C_6H_5C_2H_3$ [2055]
- Ethylbenzene $C_6H_5C_2H_5$ [1175]
- Naphthalene, crude or refined $C_{10}H_8$ [1334]
- Naphthalene, molten $C_{10}H_8$ [2304]

■ General information

Arenes are cyclic hydrocarbons derived from benzene. At room temperature, they are liquids or solids. Benzene is by far the most important arene and is among the twenty most important chemical products. Its annual world production is in excess of 6 million tons. It forms the basis for a very large number of products: plastics, resins, dye stuffs, explosives, detergents, insecticides, and synthetic fibres.

■ Action on aluminium

Benzene C_6H_6 has no action on aluminium, even as a mixture with alcohol. Aluminium equipment is used for the distillation (boiling point 80 °C), storage and transportation of benzene and its derivatives, such as toluene $C_6H_5CH_3$, xylene $C_6H_4(CH_3)_2$, ethylbenzene $C_6H_5C_2H_5$, styrene $C_6H_5C_2H_3$ [4].

Aromatic hydrocarbons with condensed benzene rings such as naphthalene $C_{10}H_8$, anthracene $C_{14}H_{10}$ and their derivatives have no action on aluminium alloys that are used in equipment for their distillation, condensation and storage.

Biphenyl $C_6H_5C_6H_5$ has no action on aluminium.

1.5. TERPHENYLS

Terphenyls $C_6H_5C_6H_4C_6H_5$ are liquids with a boiling point close to 350 °C. They are chemically very stable (their flame point is in excess of 500 °C). Their thermal conductivity is sufficiently high to allow their use as heat exchanger fluid in certain applications such as thermal solar energy plants and heat exchangers.

Tests performed at 450 °C over 200 h with sheet in 1199, 1050, 5754 and 7049 have shown that these alloys are not attacked, but only stained in anhydrous (<15 ppm water), degassed terphenyl.

1.6. CYCLIC, NON-BENZENIC HYDROCARBONS

■ ADR numbers

- Cyclopropane C_3H_6 [1027]

–   Cyclobutane C_4H_8 [2601]
–   Cyclopentane C_5H_{10} [1146]
–   Cyclohexane C_6H_{12} [1145]

■ Action on aluminium

The cyclic hydrocarbons cyclopropane C_3H_6, cyclobutane C_4H_8, cyclopentane C_5H_{10}, and cyclohexane C_6H_{12} have no action on aluminium.

Their higher homologues and their derivatives such as naphtenes, cyclenes, and terpenes, which occur naturally in certain crude oils, essences of flowers and in the roots of many plants and trees, also have no action on aluminium. The most common ones are menthadiene, pinene, terebenthine, which makes up the bulk of turpentine, and camphor. Aluminium equipment is widely used for the extraction, purification and storage of these products.

1.7.   FUEL FOR ENGINES

■ ADR numbers

–   Motor spirit or gasoline or petrol [1203]
–   Fuel, aviation, turbine engine [1863]

■ General information

Higher alkanes, from C_5 (pentane C_5H_{12}, hexane C_6H_{12}, heptane C_7H_{14}, octane C_6H_{18}) up to C_{15}, are liquid at room temperature. These products, in mixture with other unsaturated or benzenic hydrocarbons (which can also contain common additives such as tetraethyl lead), are fuels for explosion engines and reaction engines: gasoline, gasoil, and kerosene.

■ Action on aluminium

Tests with alloys such as 1050, 3003, 5754, and 5082 over a very long period of time have shown that fuel does not attack aluminium alloys at all, whether fuel for motor vehicles or for reaction turbines.[1]

If fuel contains traces of water of whatever origin, such as freshwater or seawater, this water may accumulate at the bottom of the tank, due to its higher density. This may lead to pitting corrosion.

---

[1] This also applies to leaded fuel.

Under certain conditions of humidity and temperature, especially in tropical zones, bacteriological corrosion may develop in kerosene tanks of aircraft (see Section B.2.11). Many uses of aluminium with fuel are known:

–   tanks for gasoline or fuel oil of vehicles;
–   road transport of hydrocarbons: in Europe, most road tankers for gasoline, fuel oil and heating oil are made in aluminium alloys 5754, 5083 and 5086. Pitch and tar, even hot (240 °C), are transported in specifically designed aluminium tankers;
–   equipment for distribution: petrol pumps, meters, etc.
–   engine components: pistons, carburettors, fuel pumps, etc.

1.8. PETROLEUM INDUSTRY

■ ADR number

–   Petroleum crude oil [1267]

■ General information

Petroleum is a mixture of liquid hydrocarbons of variable composition depending on its origin: alkanes, cyclic hydrocarbons (cyclopentane, cyclohexane), and aromatic hydrocarbons. Crude oil always contains light gases (methane, ethane), particles in suspension, more or less brackish fossil water, and sulphur compounds. The sulphur content ranges from 0.5 to 3 %, depending on its origin.

■ Action on aluminium

The idea of using aluminium in the petroleum industry goes back to the 1920s, especially as screens in tanks in order to limit evaporation [5, 6].

Aluminium has several advantages in the petroleum industry. It resists atmospheres containing carbon dioxide CO_2, hydrogen sulphide H_2S, sulphur dioxide SO_2 and ammonia NH_3 very well, as well as the marine environment in the case of offshore installations [7]. Aluminium alloys do not alter the aspect of polymer granules, because the corrosion product alumina gel is white.

However, for heat exchangers, the use of aluminium alloys is limited to temperatures below 100–150 °C. The drop in the mechanical strength at higher temperatures must be taken into account, and certain reservations as to fire resistance must be overcome (see Chapter G.7).

The resistance of aluminium in contact with petroleum and naphtha (a mixture of hydrocarbons in C_5 and C_6) is excellent. Tests have shown that crude oil from the Sahara does not attack aluminium alloys such as 6060, 5083, 5086, and 7020 at all. This applies to other crude oils, too, even as their sulphur content reaches 3 %, an occasional occurrence. Widespread experience has been accumulated with embedded pipelines in 3003, 6063, 6351, etc. in Canada during the 1950s. Not all of these tubes were protected externally, and a number of cases of corrosion have been reported [8, 9] (see Chapter G.1).

The very good resistance of aluminium in atmospheres charged with sulphur-containing gases allows its use at any stage of the petroleum industry [10]: drilling [11], refining, petrochemistry, in heat exchangers, in exterior heat-insulation shafts [12], fins of air coolers, etc.

1.9. DRILLING SLUDGE

The resistance of aluminium alloys depends on the composition of the sludge: the more alkaline, the quicker the metal's dissolution. In lime sludge (pH 11.9), the dissolution rate is 5 mm per year, while in betonitic sludge (pH 9.8) it amounts to 0.20 mm per year.

Since brine sludge has a less alkaline pH, it gives the best results both at room temperature and at 100 °C. In seawater sludge (pH 8.3–8.6) or in sludge containing saturated brine (density 1.212 and pH 9.5), the decrease in thickness is about 0.10 mm per year, and a slight pitting corrosion is observed.

REFERENCES

[1] Gadeau R., Les alliages d'aluminium en cryogénie, *Revue de l'Aluminium*, no 304, 1962, p. 1355–1368.

[2] Tournon A., Arzew, usine de liquéfaction du gaz d'Hassi R'Mel, 850 tonnes d'aluminium, *Revue de l'Aluminium*, nos 329, 1965, p. 327–340.

[3] Lichtenberg H., Behavior of light metal alloys toward moist acetylene, *Aluminium*, vol. 20, 1938, p. 465–466.

[4] Hauffe K., *Benzene and benzene homologous. Corrosion Handbook*, vol. 6, Dechema. 1990.

[5] Wagner P., Aluminum stock tanks compared with steel in corrosion tests, *National Petroleum News*, vol. 20, 1928, p. 50.

[6] Gill S., Aluminum foil tank coatings resist corrosion and reduce evaporation, *National Petroleum News*, vol. 21, 1929, p. 42–47.

[7] Lancaster J.F., Materials for the petrochemical industry, *International Metals Review*, vol. 23, 1978, p. 101–149.

[8] Whiting J.F., Wright T.E., Tests to five years indicate aluminium alloys pipe gives economical service, *Materials Protection*, vol. 1, 1962, p. 36–46.

[9] Ellis A., Some corrosion experiences with aluminium crud-oil, *Corrosion*, vol. 8, 1952, p. 289–291.

[10] Gastoué D.Y., L'aluminium dans l'industrie du pétrole, *Revue de l'Aluminium*, no 243, 1957, p. 531−534.

[11] Frossard M., Lamotte J.-P., Les tiges de forage en aluminium, *Revenue de l'Aluminium*, no 320, 1964, p. 554−558.

[12] Les revêtements de réservoirs de pétrole, *Revue de l'Aluminium*, No. 315, 1963, p. 1243.

Chapter F.2
Halogen Derivatives

Chapter F.2
Halogen Derivatives

Halogen derivatives, also called halides, result from the replacement of one or more hydrogen atoms of a hydrocarbon by a halogen X, which can be chlorine, bromine, iodine and, less frequently, fluorine. They, therefore, contain at least one carbon–halogen bond C–X.

There is a great variety of halogen derivatives, and their chemical reactivity depends on whether they are saturated or unsaturated, acyclic or benzenic, monohalogenated or polyhalogenated.

Fluorine derivatives form a separate class. Unlike chlorinated, brominated or iodinated derivatives, they are chemically inert under normal conditions of temperature and pressure. They are very stable up to temperatures much higher than their homologous chlorine, bromine or iodine compounds. This behaviour is due to the peculiar nature of the C–F bond.

Halogen derivatives of aliphatic and aromatic hydrocarbons form a class of compounds with many applications, because they can dissolve many different products: fats, varnish, cellulose, polymers, etc. They are intermediates in the synthesis of numerous compounds, such as dye stuffs or pharmaceuticals, plant care products, and insecticides. Certain halogen derivatives are used as coolants in refrigerators.

■ General information

In general, the higher the chemical stability of a compound, the lower the risk that it will attack aluminium.

The behaviour of aluminium in contact with halogen compounds depends on several factors that are discussed below.

□ Nature of the halide

Fluorine derivatives are very stable, even if they contain chlorine or bromine, and practically never pose a problem of compatibility with aluminium.

On the other hand, chlorine and bromine derivatives are less stable. For a given number of carbon atoms, the higher the number of chlorine or bromine atoms, the higher the reactivity of the product.

457

☐ **Structure of the molecule**

Halogen derivatives are increasingly reactive towards aluminium, the lower the molecular mass of the product. Aluminium will be less resistant in contact with methane derivatives (C_1) and ethane derivatives (C_2) than with higher homologues (C_5 and higher).

Ring-substituted aromatic halogen derivatives are much less reactive than acyclic halogen derivatives. Side-chain substituted aromatic compounds such as toluene derivatives, are as reactive as acyclic derivatives.

☐ **Presence of humidity**

In contact with humidity, some halogen derivatives can decompose, yielding the corresponding acid according to the reaction:
$$RX + H_2O \rightarrow HX + ROH.$$
Chlorinated derivatives will thus form hydrochloride acid, brominated derivatives hydrobromic acid, iodinated compounds hydriodic acid. Hydrolysis of a chlorinated or brominated derivative will lead to a more intense attack of aluminium than hydrolysis of an iodine derivative.

While the presence of humidity favours the corrosion of aluminium, certain halogen derivatives may also react with aluminium in the absence of humidity [1].

☐ **Temperature**

The stability of certain derivatives, especially those with a low molecular mass, decreases rapidly with increasing temperature. In contact with aluminium (and other metals), boiling chlorine and bromine derivatives may suffer discoloration as well as a sudden, more or less explosive decomposition that is catalysed by the contact with a metal [2]. This catalytic effect is stronger with finely divided metal, such as powder or turnings [3].

This decomposition is exothermal and releases the corresponding halogen acid. The metal will be attacked. Light may catalyse the decomposition of certain halides such as carbone tetrachloride.

The attack of aluminium by halogen derivatives may start after an incubation period (an unusual situation in the field of aluminium corrosion), the duration of which is difficult to determine, and the mechanism of which is poorly understood [4].

2.1. DERIVATIVES OF ACYCLIC HYDROCARBONS

2.1.1. *Chlorinated derivatives*

■ **Methane derivatives**

□ ADR numbers

–   Methyl chloride (chloromethane) CH_3Cl [1063]
–   Methylene chloride (dichloromethane) CH_2Cl_2 [1593]
–   Chloroform $CHCl_3$ [1888]
–   Carbon tetrachloride (tetrachloromethane) CCl_4 [1846]

□ Action on aluminium

–   Methyl chloride

The use of aluminium in contact with gaseous or liquid methyl chloride is not recommended. Unstable organometallic compounds may form, which may explode in contact with air [5]. In the presence of moisture, the decomposition of methyl chloride into hydrochloric acid leads to pitting corrosion.

–   Methylene chloride

At room temperature, in the absence of moisture, its action on aluminium is weak, with a decrease in thickness in the order of 0.1 mm per year. At 120 °C, the decrease in thickness has been found to be more than 14 mm per year in the liquid phase, and approximately 3 mm per year in methylene chloride vapours [6].

–   Chloroform

In the presence of moisture, pitting corrosion will occur. Inhibition of the attack can be achieved by adding an alkaline reserve (amines, such as dimethylamine) to the product.

–   Carbon tetrachloride

At room temperature, in the presence of moisture, even in the dark, carbon tetrachloride will decompose slowly into hydrochloric acid and carbon dioxide. In the light, air will oxidise carbon tetrachloride, releasing chlorine and phosgene.

The resistance of aluminium will depend on the temperature and on the possible presence of humidity.

At room temperature and up to 40 °C, aluminium will not be attacked by dry carbon tetrachloride: the decrease in thickness is below 10 μm per year.

In the presence of humidity, carbon tetrachloride will lead to a slight attack by pitting. The presence of air or carbon sulphide will also lead to slight pitting. Carbon tetrachloride vapours up to a concentration of 700 g·m^{-3} of air do not attack aluminium, even at a relative humidity of 50%.

At 45 °C, the decomposition of carbon tetrachloride begins to increase with temperature. Pitting corrosion is observed.

At the boiling point (76 °C), the decomposition, preceded by a pink colouring of the liquid, can increase sharply within a few minutes, leading to possible explosion.

The decomposition of carbon tetrachloride in contact with aluminium follows the reaction [7]

$$2Al + 6CCl_4 \rightarrow 2AlCl_3 + 3C_2Cl_6.$$

It is catalysed by aluminium chloride [8], which, the higher the temperature, the more soluble it is in carbon tetrachloride. Once initiated, this reaction will not stop. The lower the iron content of the aluminium, the more severe the attack is, because insoluble iron chloride forms a layer on the metal, thus impeding the attack. Alloys such as 3003 and 5052 should, therefore, be preferred to unalloyed aluminium 1050A and, of course, to refined aluminium 1199.

■ Ethane derivatives

□ ADR numbers

- Monochloroethane (ethyl chloride) CH_3CH_2Cl [1037]
- 1,1-Dichloroethane (ethylene chloride) CH_3CHCl_2 [2362]
- 1,1,1-Trichloroethane CCl_3CH_3 [2831]
- 1,1,2,2-Tetrachloroethane $CHCl_2CHCl_2$ [1702]
- Pentachloroethane CCl_3CHCl_2 [1669]

□ Action on aluminium

At room temperature and in the absence of moisture, these products have no action on aluminium. Some of them are stored or transported in aluminium vessels. Aluminium drums are used for drying hexachloroethane C_2Cl_6.

In the presence of humidity, hydrochloric acid will form by hydrolysis, leading to pitting corrosion.

As soon as the temperature exceeds 50 °C, the risk of decomposition of these solvents increases with temperature and with the number of chlorine atoms. The decrease in thickness may reach 1 mm per year in dichloromethane, and 1–10 mm per year in trichloromethane, depending on the alloy [9]. Aluminium is attacked in contact with hexachloromethane at 60 °C and will not resist this product in liquid form (melting point 180 °C).

The reaction of these chlorine derivatives with aluminium is exothermic and may be explosive. It takes off after an incubation period, the duration of which cannot be predicted and may vary from a few minutes to a few hundred or even thousand hours [10]. Daylight can catalyse the reaction with aluminium and thus lead to severe damage to storage containers made in aluminium [11].

Hence, the use of aluminium equipment in contact with these products is not recommended above 50 °C.

Vapours of these derivatives are not aggressive, even in the presence of humidity. For example, humid air at 50% relative humidity containing 300 $g \cdot m^{-3}$ ethylene chloride has no action on aluminium.

These solvents need to be stabilised [12] when degreasing parts in aluminium alloys. Any metallic particles (such as turnings) must be regularly eliminated from the degreasing bath.

■ Ethylene derivatives

□ ADR numbers

- Dichloroethylene CHClCHCl [1150]
- 1,1-Dichloroethylene (vinylidene chloride), stabilised CCl_2CH_2 [1303]
- Trichloroethylene $CHClCCl_2$ [1710]
- Tetrachloroethylene CCl_2CCl_2 [1897]

□ Action on aluminium

Tests performed over 3 months at room temperature have shown that aluminium is not attacked in these solvents, even in the presence of 0.1% humidity. The decrease in thickness is below 1 μm per year, irrespective of whether the samples are totally immersed, partially immersed or exposed to the vapour phase only. On welded samples, the behaviour of the seam and the heat-affected zone is identical to that of unwelded metal.

At the boiling point of the solvent (47 °C for dichloroethylene, 87 °C for trichloroethylene, and 121 °C for perchloroethylene), the chlorine derivative may decompose, release hydrochloric acid and thus attack the metal.

This is especially true when the metal is immersed in boiling trichlororethylene or exposed to its vapours at boiling temperature; the decrease in thickness is in the order of 1 mm per year.

Under the same conditions, dichloroethylene and perchloroethylene (as well as their vapours) are much less aggressive; the decrease in thickness is in the order of 10 μm per year.

The contact of aluminium with common metals such as stainless steel, steel, copper, and copper alloys has no influence on the resistance of aluminium to these solvents.

Aluminium pieces assembled with fittings made in ordinary steel, stainless steel or copper are commonly degreased in trichloroethylene or perchloroethylene vapours in surface treatment shops. When performed in a stabilised solvent (containing additives that neutralise the acid originating from decomposition of the solvent in contact with ambient moisture), and under the condition that the temperature does not exceed 60–70 °C, this operation does not lead to any problems. The contact with the solvent vapours does not exceed the short period of roughly half an hour.

In installations for the extraction and purification of fats, vessels, heat exchangers and heat exchangers in aluminium alloy are used. Their operating temperature should not exceed 60–70 °C with trichloroethylene and perchloroethylene.

These solvents can be stored in aluminium vessels at ambient temperature. However, any presence of aluminium powder or turnings should be avoided, because finely divided aluminium could be attacked by these solvents, even anhydrous, and trigger a strong reaction in these degreasing units.

■ Other ethylene derivatives

□ ADR numbers

−   Vinyl chloride, stabilised CH_2CHCl [1086]
−   Ethyl chloride (chloroethane) CH_3CH_2Cl [1037]
−   1,2-Dichloropropane $CH_3CHClCH_2Cl$ [1279]

□ Action on aluminium

Vinyl chloride CH_2CHCl, a gas at room temperature and a liquid under moderate pressure, has no action on aluminium. Both the monomer and the polymer, polyvinylchloride (PVC), are stored and transported in containers and tanks in aluminium alloys 5754, 5083, 5086, 6061, etc., and are not altered in any way.

Ethylene chloride CH_3CH_2Cl, 2-chlorodibutadiene $CH_2CHCClCH_2$ or chloroprene, and propylene dichloride (1,2-dichloropropane) $CH_3CHClCH_2$ have no action on aluminium at room temperature. In the presence of humidity, hydrochloric acid that attacks aluminium may form. The addition of certain stabilisers such as formamide to ethylchloride prevents the attack of the metal.

Neoprene, the polymer of butadiene chloride, is a synthetic rubber. Like its monomer, is has no action on aluminium.

■ Other derivatives of acyclic hydrocarbons

□ ADR numbers

−   Dichloropentane $C_5H_{10}Cl_2$ [1152]

☐ Action on aluminium

Butylchloride C_4H_9Cl and dichlorobutane $C_4H_8Cl_2$ have no action on aluminium up to about 50 °C.

Higher chlorine derivatives:

– amylchloride (chloropentane) $C_5H_{11}Cl$,
– hexylchloride (chlorohexane) $C_6H_{13}Cl$,
– octylchloride (chlorooctane) $C_8H_{17}Cl$,
– laurylchloride (chlorododecane) $C_{12}H_{25}Cl$,

have no action on aluminium at room temperature and in the absence of humidity. Many insecticides are chlorine derivatives such as

– chlorodane $C_{10}H_6Cl_8$,
– chlorocamphene (texaphene) $C_{10}H_{10}Cl_8$,
– heptachlor $C_{10}H_5Cl_7$,
– polychlorocamphenes $C_{10}H_{12}Cl_6$ and $C_{10}H_{16}Cl_4$.

Either in concentrated form or diluted in water (or other solvents such as kerosene or xylene) for manuring, their action on aluminium is only very slight.

They may be packed in aluminium, and their solutions are stored in aluminium alloy vessels, and loaded on aircraft or helicopter for manuring.

2.1.2. *Bromine derivatives*

■ ADR numbers

– Bromoform (tribromomethane) $CHBr_3$ [2515]
– Methyl bromide (bromomethane) CH_3Br [1062]
– Methylene dibromide (dibromomethane) CH_2Br_2 [2664]
– Ethylbromide (bromoethane) CH_3CH_2Br [1891]
– Ethylene dibromide (1,2-dibromoethane) CH_2BrCH_2Br [1605]

■ Action on aluminium

In general, as for chlorine derivatives, the presence of humidity or the increase in the temperature will increase the corrosion of aluminium in contact with bromine derivatives.

Methyl bromide CH_3Br, a nematicide, attacks aluminium; it may violently decompose in contact with aluminium.

In the presence of humidity, bromoform $CHBr_3$ will also attack aluminium. Adding an alkaline buffer (amines such as dimethylamine) to the product may inhibit the attack. Hot bromoform, even anhydrous, will attack aluminium.

In the absence of humidity, aluminium resists dibromomethane CH_2Br_2. The attack in the presence of humidity can be inhibited by adding amines.

Ethylbromide CH_3CH_2Br will attack aluminium, even cold and exempt of humidity.

1,1-Dibromoethane CH_3CHBr_2 has no action.

2.1.3. Iodine derivatives

■ **ADR number**

–   Methyl iodide CH_3I [2644]

■ **Action on aluminium**

Aluminium resists methyl iodide CH_3I at room temperature and in the absence of humidity. In the presence of water, methyl iodide releases hydriodic acid HI, which attacks aluminium. The addition of an alkaline buffer such as dimethylamine or butylamine, which neutralises the released acid, will prevent this attack.

Ethyl iodide CH_3CH_2I and ethylene iodide CH_2ICH_2I have no action on aluminium.

Iodoform CHI_3, a solid at room temperature and insoluble in water, has no action on aluminium.

2.1.4. Fluorine derivatives

■ **ADR numbers**

See Table F.2.1.

■ **General information**

These are derivatives of methane and ethane in which some or all hydrogen atoms are replaced by fluorine, chlorine or even bromine. Because fluorine is present in the molecule, these compounds are very stable, both chemically and thermally. Some are stable up to 400 °C. They are not decomposed in water, and will, therefore, not release hydrochloric and hydrofluoric acid.

Most of these compounds are gases at room temperature and under normal pressure. Because of their thermodynamical characteristics, they are widely used as coolants in

Table F.2.1. Fluorine derivatives: ADR numbers

Designation	Formula	Commercial designation [1]	ADR number
Dichlorodifluoromethane	CCl_2F_2	12	1028
Chlorodifluorobromomethane	$CClF_2Br$	12 B1, 1201	1974
Chlorotrifluoromethane	$CClF_3$	13	1022
Bromotrifluoromethane	CF_3Br	13 B1, 1301	1009
Tetrafluoromethane	CF_4	14	1982
Dichlorofluormethane	$CHCl_2F$	21	1029
Chlorodifluormethane	$CHClF_2$	22	1018
Trifluoromethane	CHF_3	23	1984
1,1-Dichloro-1,1,2,2-tetrafluoroethane	$CClF_2CClF_2$	114	1958
Chloropentafluoroethane	$CClF_2CF_3$	115	1020
1,1,1,2-Tetrafluoroethane	CF_3CH_2F	134a	3159

[1]This designation does not depend on the manufacturer.

refrigerating systems: household refrigerators, freezers, and refrigerating systems. They are used as propellants in aerosols and as fire extinguishing agents for certain highly flammable products such as hydrocarbons and ketones.

Chlorine-containing derivatives are being phased out because they destroy the ozone layer. They are being replaced progressively by hydrofluorocarbons (HFC), without chlorine.

■ Action on aluminium

Aluminium resists fluorinated hydrocarbons well. At 50 °C, the corrosion rate of aluminium in contact with them is minute, and is estimated at 0.01 μm per year. The temperature limit for using aluminium in contact with fluorinated hydrocarbons certainly exceeds 200–250 °C; some have estimated it at more than 300 °C [13].

The use of copper-containing alloys (alloys of the 2000 series, and copper-containing alloys of the 7000 series) as well as alloys of the 5000 series containing more than 2.5% manganese in contact with these fluorine derivatives is not recommended.

The enormous stability of fluorohydrocarbons with respect to water and a great number of inorganic and organic products, as well as their high volatility at room temperature, have led to their widespread use as propellants in aerosol cans [14].

2.2. DERIVATIVES OF AROMATIC HYDROCARBONS

■ ADR numbers

– Chlorobenzene C_6H_5Cl [1134]

- Dichlorobenzene $C_6H_4Cl_2$ [1591]
- Trichlorobenzene, liquid $C_6H_3Cl_3$ [2321]
- Hexachlorobenzene C_6Cl_6 [2729]
- Benzylchloride $C_6H_5CH_2Cl$ [1738]
- Benzylidene chloride $C_6H_5CHCl_2$ [1886]

■ General information

Chlorine atoms bonded to the benzene ring (as in chlorobenzene C_6H_5C) are much less reactive than those bonded to a side-chain (as in benzylchloride $C_6H_5CH_2Cl$).

Chlorine derivatives of benzene such as chlorobenzene C_6H_5Cl, dichlorobenzene $C_6H_4Cl_2$, trichlorobenzene $C_6H_3Cl_3$, and hexachlorobenzene C_6Cl_6 are solvents, intermediates for synthesis or insecticides.

■ Action on aluminium

In general, for a given number of chlorine atoms, chlorinated benzene derivatives are much less reactive towards aluminium than chlorinated aliphatic derivatives.

They have no action on aluminium, even at intermediate temperatures: for example, aluminium resists dichlorobenzene very well, even at 80 °C.

Tests performed over 1 month at 20 °C have shown that 1050A and 3003, immersed in trichlorobenzene, suffer only very superficial pickling, corresponding to a decrease in thickness of less than 1 μm per year.

These products can the stored, treated and transported in aluminium alloy equipment. They are not altered in contact with aluminium.

Benzylchloride $C_6H_5CH_2Cl$ and benzylidene chloride $C_6H_5CHCl_2$ have no action on aluminium at room temperature and in the absence of humidity, except for alloys of the 2000 series. Like acyclic derivatives, they hydrolyse in the presence of humidity, forming hydrochloric acid. When boiling, these compounds decompose in contact with aluminium.

Insecticides such as

- lindane (also called HCH) $C_6H_6Cl_6$,
- aldrine $C_{12}H_8Cl_6$,
- dichlorodiphenylchloroethane (also called DDT) $Cl_2(C_6H_4)CHCCl_3$

can provoke a slight and very superficial pitting corrosion of aluminium, whether concentrated or diluted as solutions for manuring and sprinkling.

These products and their ready-to-use solutions can be stored or transported in aluminium tanks.

2.3. HALOGEN DERIVATIVES OF ACIDS AND PHENOLS

2.3.1. Acyl halides

■ **ADR numbers**

– Phosgene $COCl_2$ [1076]
– Acetyl chloride CH_3COCl [1717]
– Propionyl chloride CH_3CH_2COCl [1815]
– Butyryl chloride $CH_3(CH_3)_3COCl$ [2353]
– Chloroacetyl chloride $CH_2ClCOCl$ [1752]
– Chloroacetic acid, solid $CH_2ClCOOH$ [1751]
– Chloroacetic acid, solution $CH_2ClCOOH$ [1750]
– Chloroacetic acid, molten $CH_2ClCOOH$ [3250]
– Dichloroacetic acid $CHCl_2COOH$ [1764]
– Trichloroacetic acid CCl_3COOH [1839]
– Trichlorocetic acid, solution CCl_3COOH [2564]

■ **Action on aluminium**

In the absence of humidity and in the dark, phosgene $COCl_2$ does not attack aluminium. In the presence of humidity, released hydrochloric acid attacks aluminium [15].

– Chloroacetyl chloride $CH_2ClCOCl$,
– acetyl chloride CH_3COCl,
– propionyl chloride CH_3CH_2COCl,
– butyryl chloride $CH_3(CH_2)_2COCl$,
– lauryl chloride $CH_3(CH_2)_{10}COCl$

are liquids that decompose in the presence of water in hydrochloric acid and the corresponding organic acid. These derivatives are very aggressive towards aluminium.

– Chloroacetic acid $CH_2ClCOOH$,
– dichloroacetic acid $CHCl_2COOH$,
– trichloroacetic acid CCl_3COOH

are water soluble and form very acidic solutions. Whether pure or diluted, they attack aluminium very violently.

The same applies to sodium trichloroacetate CCl_3COONa, an herbicide known under the reference TCA, the solutions of which strongly attack aluminium.

2.3.2. Haldides of phenols

■ **ADR numbers**

- Chlorophenols, liquid ClC_6H_4OH [2021]
- Chlorophenols, solid ClC_6H_4OH [2020]
- Pentachlorophenol C_6Cl_5OH [3155]
- Hexachlorophene $(C_6HCl_3OH)_2CH_2$ [2875]

■ **Action on aluminium**

Pentachlorophenol C_6Cl_5OH and its derivative sodium pentachlorophenate C_6Cl_5ONa, both used as wood preservatives, have no action on aluminium, whether diluted in water or in hydrocarbons.

The same applies to the other chlorine derivatives of phenol that are used as fungicides, bactericides or disinfectants:

- dichlorophene $(C_6H_3ClOH)_2CH_2$,
- hexachlorophene $(C_6HCl_3OH)_2CH_2$,
- chlorophenol C_6Cl_4ClOH,
- dichlorophenol $C_6H_3Cl_2OH$,
- trichlorophenol $C_6H_2Cl_3OH$.

Chloroamphenicole $C_{11}H_{12}Cl_2N_2O_5$ has no action on aluminium that is used for equipment for the fermentation, storage and packing of this antibiotic.

Chloronitrobenzene $NO_2C_6H_4Cl$ and nitrobenzoyle chloride $NO_2C_6H_4COCl$ have no action on aluminium. These products are stored and transported in aluminium containers.

2.3.3. Other halogenides

■ **ADR numbers**

- Chloral, anhydrous, stabilised CCl_3CHO [2075]
- Methyl chloroformate $ClCOOCH_3$ [1238]
- Ethyl chloroformate $ClCOOC_2H_5$ [1182]
- *n*-Propyl chloroformate $ClCOOC_3H_7$ [2740]
- *n*-Butyl chloroformate $ClCOOC_4H_9$ [2743]
- Methyl chloroacetate $CH_2ClCOOCH_3$ [2295]
- Ethyl chloroacetate $CH_2ClCOOC_2H_5$ [1181]

- Isopropyl chloroacetate $CH_2ClCOOC_3H_7$ [2947]
- Vinyl chloroacetate $CH_2ClCOOC_2H_3$ [2589]

■ Action on aluminium

Chloral CCl_3CHO and chloroacetaldehyde CH_2ClCHO are aggressive towards aluminium in the presence of water or even humidity, because they release hydrochloric acid upon decomposition. Chloroethanol CH_2ClCH_2OH, on the other hand, has no action on aluminium. Aluminium equipment is used for storing and transporting this product, as well as in the production of plastics made from chloroethanol.

- Methyl chloroformate $ClCOOCH_3$,
- ethyl chloroformate $ClCOOC_2H_5$,
- propyl chloroformate $ClCOOC_3H_7$,
- butyl chloroformate $ClCOOC_4H_9$

decompose in contact with water and form acids. They attack aluminium.

- Methyl chloroacetate $CH_2ClCOOCH_3$,
- ethyl chloroacetate $CH_2ClCOOC_2H_5$,
- butyl chloroacetate $CH_2ClCOOC_4H_9$,
- cyclohexyl chloroacetate $CH_2ClCOOC_6H_{11}$,
- vinyl chloroacetate $CH_2ClCOOC_2H_3$

attack aluminium.

In the absence of humidity and at room temperature, aluminium resists in contact with benzoyl chloride C_6H_5COCl. In the presence of humidity, this compound decomposes and releases hydrochloric acid that attacks aluminium. At elevated temperatures, and especially close to its boiling point (198 °C), benzoyl chloride attacks aluminium.

REFERENCES

[1] Cutler D.P., Reactions between halogenated hydrocarbons and metals. A literature review, *Journal of Hazardous Materials*, vol. 17, 1987, p. 99–108.
[2] Hamstead A.C., Elder B.B., Canterbury J.C., Reaction of certain chlorinated hydrocarbons with aluminum, *Corrosion*, vol. 14, 1957, p. 189t–190t.
[3] Spencer J.F., Wallace M.L., Interaction of metals of the aluminium group and organic halogens derivatives, *Journal of the Chemical Society*, vol. 93, 1909, p. 1827.
[4] Kyriasis V.C., Heitz E., Zur Kinetik der Korrosion von Metallen in Halogenkohlenwasser-stoffen, *Werkstoffe und Korrosion*, vol. 31, 1980, p. 197–207.

[5] Walker W.O., Wilson K.S., The action of methyl chloride on aluminium, *Refrigeration Engineering*, vol. 34, 1937, p. 89–90.

[6] Borisova L.G., Stability and corrosion of dichloromethane, *Materials Korrosion Svarka, Mashinostroeniya*, Moscou, 1975, p. 162–166.

[7] Stern M., Uhlig H.H., Effect of oxide films on the reaction of aluminium with carbon tetrachloride, *Journal of Electrochemical Society*, vol. 99, 1952, p. 389–392.

[8] Spilker H.G., *Chlorinated hydrocarbons, Chloroethanes. Corrosion Handbook*, vol. 12, Dechema, 1990.

[9] Dölling H., Corrosion tests on aluminium in halogenated hydrocarbons of practical importance, *Werkstoffe and Korrosion*, vol. 312, 1980, p. 173–178.

[10] Kyriasis V.C., Heitz E., Zur Kinetik der Korrosion von Metallen in Halogenkohlenwasserstoffen, *Werkstoffe und Korrosion*, vol. 31, 1980, p. 197.

[11] Brookmann K., Untersuchungen über das Verhalten von Aluminium gegenüber organischen Lösungsmitteln, *Aluminium*, vol. 34, 1958, p. 30.

[12] Dilla W., Koser H.J., Ein Breitag zur Beurteilung der Stabilität chlorierter Kohlenwasserstoffe in der Leichtmetallentfettung, *Chemiker Zeitung*, vol. 107, 1983, p. 227.

[13] Moroni V., Macchi E., Giglioli G., Investigation on thermal stability and corrosion effects of dichloro-difluoro-methane in view of its possible application as working fluid in a power plant, *La Termotecnica*, vol. 28, 1974, p. 209–221.

[14] Minford J.D., Compatibility studies of aluminium with propellant and solvents for use in aerosol, *Journal of the Society of Cosmetic Chemists*, vol. 15, 1964, p. 311–326.

[15] Helsner G., *Acid Halides. Corrosion Handbook*, vol. 3, Dechema, 1988.

Chapter F.3

Alcohols, Ethers, Thiols and Phenols

Chapter F.3
Alcohols, Ethers, Thiols and Phenols

Alcohols are compounds of the general formula R–OH in which the hydroxyl group –OH is bonded to a saturated carbon atom. Three classes of alcohols can be distinguished:

- primary alcohols RCH_2OH
- secondary alcohols R_1R_2CHOH
- tertiary alcohols $R_1R_2R_3COH$

according to the degree of substitution of the carbon atoms that bears the OH group. R_1, R_2 and R_3 are organic groups that, in the case of secondary or tertiary alcohols, can be different or identical.

$$R_1\text{—}CH_2OH \qquad R_1\text{—}\underset{\underset{\textstyle OH}{|}}{CH}\text{—}R_2 \qquad R_1\text{—}\overset{\overset{\textstyle R_2}{|}}{\underset{\underset{\textstyle OH}{|}}{C}}\text{—}R_3$$

Primary alcohol   Secondary alcohol   Tertiary alcohol

Primary alcohols are the most common ones, and the most frequently used. They occur naturally in liquor, wine, beer, and in many natural essences of fruits and flowers. They are very important for the chemistry of fragrances and detergents.

Several alcohols, and especially methyl alcohol and ethyl alcohol, are produced industrially in large quantities.

At room temperature, none of them is gaseous. Primary alcohols are liquids up to C_{11} (undecyl alcohol); beyond, they are solids. Their boiling point is always higher than that of the homologous hydrocarbon having the same number of carbon atoms. As an example, ethane C_2H_6 boils at $-88.5\,°C$, while ethyl alcohol boils at $+78.2\,°C$.

Only the first alcohols are water soluble; their solubility decreases rapidly with increasing molecular weight. Methyl alcohol and ethyl alcohol are water soluble in any proportion, while the solubility of amyl alcohol (C_5) is only 2.7%. They are excellent solvents for a very large number of organic substances.

The reactivity of alcohols decreases with increasing molecular weight and, at a given number of carbon atoms, from primary to tertiary. They are amphoteric: with strong acids, they form esters, and with strong bases they form alcoholates RONa.

Their dehydration leads to ethers of the general formula ROR, if symmetrical, and R_1OR_2 if unsymmetrical. Their reactivity being low, they are used as solvents.

Phenols are compounds in which the OH group is bonded to a carbon atom that belongs to a benzene ring. Almost all of them are solids at room temperature. They generally have a strong, pungent smell. Only the first ones are soluble in water. They react with sodium hydroxide to phenolates.

Thiols, also called mercaptans, have the general formula RSH, and thioethers, also called alkyl sulphides, have the general formula R_1SR_2.

Remark
In this chapter, the traditional designation "X alcohol", for example, "ethyl alcohol" is used rather than the systematic designation with the suffix "ol", for example, "ethanol".

3.1. THE ACTION OF BOILING ALCOHOLS AND DEHYDRATED PHENOLS ON ALUMINIUM

At high temperatures, i.e. close to the boiling point of alcohols and phenols, and in a totally dehydrated environment, the OH group can react with aluminium according to the reaction

$$3ROH + Al \rightarrow \frac{3}{2}H_2 + Al(RO)_3$$

$$3ArOH + Al \rightarrow \frac{3}{2}H_2 + Al(ArO)_3.$$

These reactions take place because alcohol (or phenol) dehydrates the protective natural oxide film and modifies its structure. Moreover, since the resulting alcoholate (or the phenolate) is soluble in this medium, aluminium is no longer protected and will, therefore, be continuously attacked. The attack will proceed as pitting.

Experience shows that the addition of a small amount of water, 0.05–0.1%, depending on the nature of the alcohol or phenol, will completely inhibit the attack of aluminium by phenols, alcohols or mixtures thereof at the boiling point. These traces of water are sufficient to form the natural oxide layer [1].

3.2. ACYCLIC SATURATED ALCOHOLS

3.2.1.   *Methyl alcohol CH_3OH*

■ **General information**

Like the first alcohol, methyl alcohol (methanol) CH_3OH enjoys great industrial importance: the annual world production is about 15 million metric tons. It is used as an intermediate in the synthesis of methacrylates, and enters into the formulation of gasoline.

■ **Action on aluminium**

At room temperature, it has no action on aluminium. In methyl alcohol of technical purity, the decrease in thickness is below 0.1 mm per year. The resistance in mixtures of methyl alcohol and water depends on the proportion of alcohol, and the dissolution rate increases with temperature (Table F.3.1). The results are the same irrespective of whether the samples are completely immersed in the alcohol, partially immersed or exposed to the vapour.

The action of methyl alcohol depends on its additives or impurities [3]. As an example, the presence of formaldehyde, up to 2%, does not modify the behaviour of aluminium in contact with methyl alcohol; however, the presence of up to 1% formic acid leads to a superficial attack on the order of 30 μm per year.

The addition of methyl alcohol to classic gasoline for engines can modify the resistance of aluminium, if water, chlorides or formic acid are present; pitting corrosion may occur [4−6].

3.2.2. Ethyl alcohol C_2H_5OH

■ **ADR number**

Ethyl alcohol (ethanol) or ethanol solution C_2H_5OH [1170]

■ **Action on aluminium**

In contact with ethyl alcohol, pure or in solution, the behaviour of aluminium is very similar to that observed with methyl alcohol; it is even generally better, all the other parameters (concentration, temperature, etc.) being identical.

The addition of other products such as benzene or methyl alcohol, in order to transform ethyl alcohol into fuel or to render it unsuitable for consumption (denatured alcohol), does not modify at all the resistance of aluminium.

In practice, and for a long time, equipment in aluminium alloy is widely used for the extraction of alcohols by distillation (heat exchangers), as well as for the storage and transportation of these alcohols or mixtures thereof.

Table F.3.1. Methyl alcohol−water mixtures, yearly decrease in thickness of 3003 (mm per year) [2]

| Temperature (°C) | Proportion of methyl alcohol (%) | | | | | | | | | | |
	5	10	20	30	40	50	60	70	80	90	95
25	0.17	0.18	0.19	0.19	0.19	0.19	0.19	0.19	0.18	0.14	0.10
60	0.31	0.62	0.75	0.77	0.78	0.77	0.75	0.68	0.66	0.53	0.44

3.2.3. *Higher alcohols*

■ **ADR numbers**

– Propyl alcohol, normal (*n*-propanol) C_3H_7OH [1274]
– Isopropyl alcohol (isopropanol) C_3H_7OH [1219]
– Isobutyl alcohol (isobutanol) C_4H_9OH [1212]
– Pentanols (amyl alcohols) $C_5H_{11}OH$ [1105]

■ **Action on aluminium**

The most common alcohols, which are liquids up to C_{11} or solids beyond:

– propyl alcohol C_3H_7OH,
– butyl alcohol C_4H_9OH,
– pentyl alcohol (or amyl alcohol) $C_5H_{11}OH$,
– hexyl alcohol $C_6H_{13}OH$,
– heptyl alcohol $C_7H_{15}OH$,
– octyl alcohol (or capryl alcohol) $C_8H_{17}OH$,
– nonyl alcohol $C_9H_{19}OH$,
– decyl alcohol $C_{10}H_{21}OH$,
– undecyl alcohol $C_{11}H_{23}OH$,
– dodecyl alcohol (or lauryl alcohol) $C_{12}H_{25}OH$,
– tetradecyl alcohol (or myristyl alcohol) $C_{14}H_{29}OH$,
– pentadecyl alcohol $C_{15}H_{31}OH$,
– hexadecyl alchol (or cetyl alcohol) $C_{16}H_{33}OH$,
– heptadecylalcohol $C_{17}H_{35}OH$,
– octadecyl alcohol (or stearyl alcohol) $C_{18}H_{37}OH$,

pure, as mixtures or in solution in organic solvents such as ethyl alcohol, benzene, ether, and acetone have no action on aluminium at room temperature.

At their boiling point, the complete dehydration of the alcohol has to be prevented, so that aluminium is not attacked.

Aluminium equipment is used for the fabrication of these alcohols, for their transportation and their storage. Neither their aspect nor their properties are altered in contact with aluminium.

3.3. ACYCLIC UNSATURATED ALCOHOLS

■ **ADR number**

– Allyl alcohol C_3H_5OH [1098]

■ **Action on aluminium**

Less important than saturated alcohols, the most common ones are allyl alcohol C_3H_5OH and oleic alcohol $C_{18}H_{34}OH$. They have no action on aluminium.

3.4. AROMATIC ALCOHOLS

■ **ADR number**

– Alpha-methylbenzyl alcohol C_8H_9OH [2937]

■ **General information**

The most common ones are

– Benzyl alcohol $C_6H_5CH_2OH$,
– phenylethyl alcohol $C_6H_5C_2H_4OH$,
– phenylbutyl alcohol $C_6H_5C_2H_8OH$,
– phenylpropyl alcohol $C_6H_5C_3H_6OH$,
– cinnamyl alcohol $C_6H_5C_3H_4OH$,
– veratryl alcohol $(CH_3O)_2C_6H_3CH_2OH$,
– anisic alcohol $(CH_3O)C_6H_4CH_2OH$,
– xylenols $C_6H_3H_6OH$.

■ **Action on aluminium**

Like saturated acyclic alcohols, aromatic alcohols, which are widely used in perfumery, have no action on aluminium, neither pure nor as mixtures or in solution with other organic solvents. They can be stored and transported in aluminium vessels.

3.5. OTHER ALCOHOLS

■ **ADR number**

– Furfuryl alcohol $C_4H_3OCH_2OH$ [2874]

■ **Action on aluminium**

Menthol $C_{10}H_{19}OH$, camphols (also called borneols) $C_{10}H_{17}OH$, and furfuryl alcohol $C_4H_3OCH_2OH$ have no action on aluminium.

The same applies to sulphonated alcohols such as sulphocetyl alcohol $C_{16}H_{32}OH(SO_3H)$, sulpholauryl alcohol $C_{12}H_{24}OH(SO_3H)$, and sulphostearyl alcohol $C_{18}H_{36}OH(SO_3H)$, which are surfactants used as emulsifiers, foaming agents, wetting agents or detergents.

3.6. POLYALCOHOLS

■ **General information**

Polyalcohols (diols, triols, etc.), in spite of their high molecular weight, are highly soluble in water, due to the multiplicity of hydroxyl groups.

Dialcohols, also called glycols, have two alcohol groups. They are liquids in a wide temperature range from -40 to $+200$ °C. They are very hygroscopic and miscible with water in any proportion.

The most common ones are ethylene glycol (1,2-ethanediol) CH_2OHCH_2OH and propylene glycol (1,2-propanediol) $CH_3CHOHCH_2OH$. Because of their thermal properties, namely their thermal conductivity, their very low melting point (down to -40 °C) and their high boiling point (above 190 °C), glycol–water mixtures with up to 30% glycol are used as cooling liquids in car and truck engines.

Glycols are part of the formulation of anti-icing agents that are sprayed on aircraft in winter before takeoff, as well as brake fluid and fluids for hydraulic circuits. They are a basis for the production of polymers. They are also used as heat exchange fluids in cooling systems, and in systems for the recovery of thermal solar energy.

In the presence of oxygen, at high temperatures, glycols decompose to form organic acids: ethylene glycol forms oxalic acid, glycolic acid, and formic acid, while propylene glycol forms pyruvic acid, lactic acid, formic acid, and acetic acid [7]. In this situation, water–glycol mixtures become aggressive to common metals: cast iron, steel, copper and copper alloys, and aluminium alloys [8].

■ **Action on aluminium**

Cold glycols and polyalcohols in general have no action on aluminium, whatever their concentration. The dissolution rate increases at alkaline pH values (Table F.3.2).

Polyalcohols can be stored and transported in equipment in aluminium alloy. Polyalkene–glycol mixtures are used as a quenching fluid for forgings made in aluminium alloys with high mechanical strength such as 7075 [10].

Table F.3.2. Influence of ethylene glycol on the dissolution rate (mm per year) [9]

pH	Concentration (%)		
	20	40	60
4	0.005	0.006	0.004
7	0.0006	0.0003	0.0001
11	0.06	0.03	0.09

Mixtures of water and polyalcohols (with up to 30% glycol), which are used as coolants in engines and heat exchanger fluid in systems for the recovery of thermal energy (solar energy, etc.), have to be modified in order to avoid corrosion. Therefore, these mixtures contain corrosion inhibitors to protect all the metals in the circuit, especially copper and cuprous alloys, which upon corrosion release Cu^{2+} ions, leading to corrosion of steel and aluminium and to degradation of elastomers and polymers in contact with these fluids [11] (see Section D.1.1.6).

The presence of chlorides, even in small amounts such as 10 ppm, promotes corrosion [12], which is why cooling fluids used in circuits comprising heat exchangers (or radiators in the case of cooling systems for car engines) are specially conditioned for this use.

That is why these fluids must not be replaced or replenished with mixtures of tap water and antifreeze; otherwise, there will be a risk of pitting, especially since heat exchanger tubes in automotive engines are only a few tenths of millimetres thick.

☐ Glycerine

Glycerine is a trialcohol $CH_2OHCHOHCH_2OH$. It is miscible with water in any proportion. Pure or diluted in water, it does not attack aluminium, even at elevated temperatures. The decrease in thickness in a mixture glycerine (95%) and water (5%) at a temperature close to 100 °C is below 1 μm per year.

The presence of sodium chloride can lead to a slight pitting attack. For example, in a mixture containing

– glycerin: 81%
– sodium chloride: 10%
– water: 7%

superficial pitting corrosion with a maximum pitting depth of 0.2 mm has been observed at 100 °C.

Distillation, transportation and storage of glycerine can be carried out in equipment made of aluminium alloy. Glycerine is not altered by a prolonged contact with aluminium.

3.7. ETHERS

■ **General information**

Ethers of the general formula R_1-O-R_2 are chemically rather inactive compounds that have many applications: solvents, plasticisers, paints, lubricants, brake fluids, dielectric fluids, drugs, perfumes. Most ethers are liquid at room temperature. The first ethers are gaseous.

3.7.1. *Acyclic ethers*

■ **ADR numbers**

- Dimethyl ether (dimethyl oxide) $(CH_3)_2O$ [1033]
- Diethyl ether (diethyl oxide) $(C_2H_5)_2O$ [1155]
- Di-*n*-propyl ether (dipropyl oxide) $(C_3H_7)_2O$ [2384]
- Diisopropyl ether (diisopropyl oxide) $[(CH_3)_2CH]_2O$ [1159]
- Dibutyl ether (dibutyl oxide) $(C_4H_9)_2O$ [1149]
- Ethyl methyl ether (methoxyethane) $CH_3OC_2H_5$ [1039]
- Butyl methyl ether (methoxybutane) $CH_3OC_4H_9$ [2350]
- Methyl propyl ether (methoxypropane) $CH_3OC_3H_7$ [2612]
- Ethyl propyl ether (ethoxypropane) $C_2H_5OC_3H_7$ [2615]
- Ethyl butyl ether (ethoxybutane) $C_2H_5OC_4H_9$ [1179]

■ **Action on aluminium**

Methyl ether, a gas at room temperature, is stored and transported under pressure in vessels made of aluminium alloy. The presence of humidity can lead to slight pitting of aluminium, certainly due to traces of sulphuric acid that it may contain (it is produced from methyl ether using sulphuric acid).

Ethyl ether, commonly called simply ether, has no action on aluminium. In the presence of humidity, a slight, superficial attack of aluminium can occur, because ether may contain traces of sulphuric acid and acetic acid.

Higher ethers

- Propyl ether $(C_3H_7)_2O$,
- butyl ether $(C_4H_9)_2O$,
- amyl ether $(C_5H_{11})_2O$,
- hexyl ether $(C_6H_{11})_2O$,

- heptyl ether $(C_7H_{15})_2O$,
- octyl ether $(C_8H_{17})_2O$,
- decyl ether $(C_{10}H_{21})_2O$,
- lauryl ether $(C_{12}H_{25})_2O$

are solvents, lubricants, or dielectric fluids and have no action on aluminium.

3.7.2. *Vinyl ethers and cellulose ethers*

■ ADR numbers

- Vinyl ether $(C_2H_4)_2O$ [1167]
- Vinyl methyl ether $CH_3OC_2H_3$ [1087]
- Vinyl isobutyl ether $C_4H_9OC_2H_3$ [1304]
- Vinyl ethyl ether $C_2H_5OC_2H_3$ [1302]
- Butyl vinyl ether $C_4H_9OC_2H_3$ [2352]

■ General information

The vinyl ethers listed above and mentioned below:

- Vinyl ethyl ether $C_2H_5OC_2H_3$,
- propyl vinyl ether $C_3H_7OC_2H_3$,
- polyvinyl ethers $(CH_2CHO-R)_n$

are polymerisation agents. They are part of the formulation of certain glues, adhesives, varnishes, resins, etc.

■ Action on aluminium

They have no action on aluminium. A large number of glues are sold in soft aluminium tubes. Cellulose ethers also have no action on aluminium.

3.7.3. *Glycol ethers*

■ ADR numbers

- Ethylene glycol diethyl ether $C_2H_5OCH_2CH_2OC_2H_5$ [1153]
- Ethylene glycol monomethyl ether $CH_3OCH_2CH_2OH$ [1188]
- Ethylene glycol monoethyl ether $C_2H_5OCH_2CH_2OH$ [1171]

■ **General information**

The main glycol ethers are

–   Ethylene glycol monomethyl ether $CH_3OCH_2CH_2OH$,
–   ethylene glycol monoethyl ether (also called ethylglycol) $C_2H_5OCH_2CH_2OH$,
–   ethylene glycol propyl ether $C_3H_7OCH_2CH_2OH$,
–   ethylene glycol butyl ether $C_4H_9OCH_2CH_2OH$.

These products are solvents or intermediates for syntheses. They are part of the formulation of paints, varnishes, resins, etc.

Brake fluids mainly consist of a mixture of glycol ether (75–90%), a lubricant base (5–20%) that ensures the lubrication of moving parts in the brake circuit, and specific additives, the nature and concentration of which depends on the brands. They aim at protecting the brake circuitry against corrosion and preventing the oxidation of the brake fluid.

■ **Action on aluminium**

Glycol ethers have no action on aluminium.

Aluminium alloys are used as components for hydraulic braking systems in cars: pistons of master cylinders are in 2011, 6060, etc., and the master cylinders in 21000 (A-U5GT), 44100 (A-S13), etc.

Experience has shown that parts in aluminium alloys can corrode superficially in contact with elastomer joints of brake circuits, if these joints contain graphite or are altered by the brake fluid.

Producers of brake fluids and brake components have developed formulations that prevent the alteration of the elastomers and the corrosion of metallic components in the circuit.

3.7.4. *Aromatic ethers*
Aromatic ethers such as

–   Phenyl ether (or phenyl oxide) $(C_6H_5)_2O$,
–   benzyl oxide $(C_6H_5CH_2)_2O$,
–   benzyl cresyl oxide $C_6H_5CH_2OC_6H_5CH_3$,
–   anisole $C_6H_5OCH_3$,
–   anethole $CH_3OC_6H_4C_3H_5$

are used as plasticisers. They have no action on aluminium, even at high temperatures (200–250 °C).

3.8. THIOLS AND THEIR DERIVATIVES

■ ADR numbers

- Amyl mercaptan [1111]
- Butyl mercaptan [2347]
- Cyclohexyl mercaptan [3054]
- Ethyl mercaptan [2363]
- Methyl mercaptan [1064]
- Phenyl mercaptan [2337]
- Maneb [2210]
- Maneb, stabilised [2968]

■ General information

Thioalcohols, also called mercaptans, correspond to the general formula $R-S-H$. The most common ones are

- Ethyl mercaptan C_2H_5SH,
- propyl mercaptan C_3H_7SH,
- butyl mercaptan C_4H_9SH,
- amyl mercaptan $C_5H_{11}SH$.

■ Action on aluminium

Mercaptans have no action on aluminium, even at high temperatures. Certain of their derivatives are prepared at elevated temperatures in reactor vessels made of aluminium alloy. An example is sulphonal, which does not attack aluminium at 125 °C.

Thiophenols, like mercaptans, have no action on aluminium. Benzyl sulphide $(C_6H_5CH_2)_2S$ is a corrosion inhibitor for aluminium in acid medium.

■ Other derivatives

Thioethers, also called alcoyl sulphides, R_1SR_2, have no action on aluminium, even at high temperatures (100–120 °C). This applies to allyle sulphide $C_6H_{10}S$ and to thiodiglycol $(HOC_2H_4)_2S$.

Sodium salts of sulphonic acids $RHSO_3$ as well as acid sulphates of fatty alcohols $ROHSO_3$ are surfactants. These derivatives have no action on aluminium. They can be produced and stored in aluminium equipment.

Paratoluene sulphonic acid $CH_3C_6H_4HSO_3$ is highly soluble in water. Up to 20%, its solutions have no action on aluminium. Above this concentration, a slight attack of the metal can occur, however.

Anthraquinone sulphonic acid $C_6H_4(CO)_2C_6H_3SO_3H$, an intermediate in the synthesis of dyestuffs in solution, has no action on aluminium, even at elevated temperatures.

Sodium sulphonates (sodium salts of sulphonic acids) are surfactants. They have no action on aluminium.

Ethylsulphuric acid $C_2H_5OSO_3H$ and its salts are highly soluble in water. In the presence of water, their hydrolysis creates an acidic medium, leading to a superficial attack of aluminium.

Derivatives of thiocarbamic acid are a family of very important fungicides used in many plant-care products. Manganese salts (maneb), zinc salts (zineb), sodium and iron salts have no action on aluminium, whatever their degree of dilution. On the other hand, copper salts of dithiocarbamic acid may attack aluminium.

3.9. PHENOLS

■ General information

Phenols, of the general formula ArOH (where Ar is a group including one or more benzenic rings), are solids at room temperature. They are water soluble, and their solubility decreases very rapidly as their molecular weight increases. Aqueous phenol solutions are slightly acidic, because of the OH group. For example, a phenol C_6H_5OH solution has a pH close to 6.

They are used as disinfectants, additives to tanning solutions, intermediates in the synthesis of dyestuffs, and in the fabrication of explosives and certain plastics.

■ Action on aluminium

As in the case of alcohols and organic acids, the OH group can, under certain conditions, react with aluminium according to

$$3ArOH + Al \longrightarrow \frac{3}{2}H_2 + Al(ArO)_3.$$

This reaction occurs only at high temperatures, 180–200 °C, which corresponds to the boiling point of phenols. As with alcohols, the reaction results from phenol dehydrating the alumina film on the metal, and thus modifying its structure and sensitising it to its corrosive action. Experience shows that adding a small amount of water, about 0.3%, prevents attack of aluminium by phenols at high temperatures [13].

3.9.1. Phenol C_6H_5OH

■ **ADR numbers**

- Phenol, molten C_6H_5OH [2312]
- Phenol, solid C_6H_5OH [1671]
- Phenol, solution C_6H_5OH [2821]
- Picric acid (trinitrophenol) $(NO_2)_3C_6H_2OH$ [0154]

■ **General information**

Phenol is a solid at room temperature, melts at 43 °C and boils at 182 °C under atmospheric pressure. Phenol is soluble in water: up to 8 wt% at 20 °C. This solution is slightly acidic, and its pH is close to 6.

Phenol, therefore, is a very hygroscopic product and capable of absorbing up to 26% water at room temperature. At 70 °C, water and phenol are totally miscible in any proportion.

■ **Action on aluminium**

At room temperature, anhydrous phenol has no action on aluminium, and phenol solutions only have a slight action on aluminium. The uniform attack is less than 10 μm per year. Phenol is stored and transported in aluminium alloy equipment. Railway tanks in 5754 have been specifically designed for the transportation of this product [14].

At 60–70 °C, aqueous solutions of phenol lead to a slight, very superficial attack. The use of aluminium in contact with phenol and its solutions is possible in this temperature range. The attack becomes negligible when the phenol concentration reaches 75%.

At higher temperatures, phenol has no action on aluminium if it contains traces of humidity (at least 0.3%), necessary to avoid the dehydration of the natural oxide film, which would lead to a strong attack of the metal, as explained above.

Alcoholic solutions of phenol and mixtures of phenol and formaldehyde have no action on aluminium, even at temperatures on the order of 100–110 °C.

Nitrous derivatives of phenol such as picric acid $(NO_2)_3C_6H_2OH$ have no action on aluminium, neither at low nor at high temperatures. Aluminium is used for the production and casting of picric acids and products derived from picric acid, especially explosives.

3.9.2. *Other common phenols*

■ **ADR numbers**

–   Xylenols $C_8H_{10}O$ [2261]

■ **Action on aluminium**

Cresols C_7H_8O, xylenols $C_8H_{10}O$, naphthols $C_{10}H_8O$, pyrocatechol $C_6H_4(OH)_2$, resorcinol $C_6H_4(OH)_2$, hydroquinone $C_8H_{10}(OH)_2$, and pyrogallol $C_6H_3(OH)_2$, all of which are polyphenols, have no action on aluminium, neither as pure compounds nor in solution, whether at room temperature or at elevated temperatures. As in the case of alcohols, at very high temperatures of approximately 180–200 °C, aluminium may undergo severe attack in the absence of humidity.

REFERENCES

[1] Seligman R., Williams P., The inhibitory effects of water on the interaction of aluminium and the fatty acids, phenol, cresol, alpha and beta naphtols, methyl, ethyl, butyl, amyl and benzyl alcohols, *Journal of Society Chemical Industry*, vol. 37, 1918, p. 159–165.
[2] Zaritski B.I., Kostyk A.P., Tagaev O.A., Corrosion activity of methanol–water systems, *Khim. Prom.*, 1986, p. 341.
[3] Spilker H.G., *Methanol. Corrosion Hanbook*, vol. 9, Dechema, 1991.
[4] Poteat L.E., Automotive materials compatibility with methanol fuels blends, *Fourth International Symposium*, University of Miami, February 1978, p. 707.
[5] Wing L., Evarts G.L., Tramontana D.M., Protect gasoline: methanol fuel systems, *Materials Engineering*, vol. 109, 1992, p. 26–27.
[6] Lash R.J. The corrosion behavior of metals, plated metals, and metals coatings in methanol/gasoline fuel mixtures, *Sixth Automotive Corrosion & Prevention Conference*, Deadborne (Etats-Unis), October 1993, p. 153–171.
[7] Rossiter W.J., Brown P.W., Godette M., The determination of acidic degradation products in aqueous glycol and propylene glycol, *Solar Energy Materials*, vol. 9, 1983, p. 267–279.
[8] Jackson J.D., Miller P.D., Fink F.W., Boyd W.K., *Corrosion of materials by ethylene glycol–water*, Battelle, Report DMIC 216, May 1965.
[9] Agnew R.J., Truitt J.K., Robertson W.D., Corrosion behavior of metals in ethylene glycol solutions, *Industry Engineering Chemical*, vol. 50, 1959, p. 894.
[10] Collins J.F., Maducell C.E., Polyalkilene glycol quenching of aluminium alloys, *Materials Performance*, vol. 16, 1977, p. 20.
[11] Elsner G., *Polyols. Corrosion Handbook*, Dechema, 1989.

[12] Sigurdsson H., Tsuvik R., Corrosion of aluminium in water-based heat exchangers systems, *10th Scandinavian Corrosion Congress*, 1986, p. 65–72.

[13] Neufeld P., Chakrabarty A.K., The corrosion of aluminum and its alloys in anhydrous phenol, *Corrosion Science*, vol. 12, 1972, p. 517–525.

[14] Strömmen K.G., Wagons-citernes à quatre essieux en aluminium pour le transport du phénol, *Revue Suisse de l'aluminium,* 1974, p. 120–121.

Chapter F.4

Amines

Chapter F.4
Amines

Amines are compounds in which a nitrogen atom is bonded to one or more carbon atoms. They can be considered as derivatives of ammonia in which one, two or three hydrogen atoms are replaced by hydrocarbon groups R_1, R_2 and R_3. These three groups can be acyclic or benzenic, identical or different.

According to the degree of substitution of the hydrogen atoms of ammonia, three classes of amines can be distinguished:

- primary amines R_1NH_2
- secondary amines R_1NHR_2
- tertiary amines $R_1R_2NR_3$

$$\underset{\text{Ammonia}}{\overset{\displaystyle H-N-H}{\underset{\displaystyle H}{|}}} \qquad \underset{\text{Primary Amine}}{\overset{\displaystyle R_1-N-H}{\underset{\displaystyle H}{|}}} \qquad \underset{\text{Secondary Amine}}{\overset{\displaystyle R_1-N-R_2}{\underset{\displaystyle H}{|}}} \qquad \underset{\text{Tertiary Amine}}{\overset{\displaystyle R_1-N-R_2}{\underset{\displaystyle R_3}{|}}}$$

The first amines, from C_1 to C_5, show physical properties close to ammonia, i.e. a high solubility in water, and a characteristic ammonia smell. Their solubility decreases rapidly with their molecular weight: hexylamine $C_6H_{13}NH_2$ has a very low solubility in water (1.2% at room temperature).

The first acyclic amines: methylamine CH_3NH_2, dimethylamine $(CH_3)_2NH$, trimethylamine $(CH_3)_3N$ and ethylamine $C_2H_5NH_2$ are gases at room temperature; the next ones are liquids or solids, depending on their molecular weight.

Aromatic amines $Ar-NH_2$ are thick liquids or solids, insoluble in water, and have a harsh smell.

Because of the presence of the NH_2 group, solutions of primary amines are strongly alkaline: a solution of 10% methylamine CH_3NH_2 has a pH of 13, as alkaline as an ammonia solution of the same concentration. This alkalinity decreases for secondary and tertiary amines.

Amines have a very large number of uses: intermediates in the synthesis of polymers, certain insecticides, fungicides, dyestuffs (aniline), and pharmaceuticals. They are used as surfactants, vulcanisation promoters, herbicides, corrosion inhibitors, etc.

4.1. ACYCLIC AMINES

■ **ADR numbers**

- Methylamine, anhydrous CH_3NH_2 [1061]
- Methylamine, aqueous solution CH_3NH_2 [1235]
- Dimethylamine, anhydrous $(CH_3)_2NH$ [1032]
- Dimethylamine, aqueous solution $(CH_3)_2NH$ [1160]
- Trimethylamine, anhydrous $(CH_3)_3N$ [1083]
- Trimethylamine, aqueous solution, not more than 50% trimethylamine by mass $(CH_3)_3N$ [1297]
- Ethylamine $C_2H_5NH_2$ [1036]
- Ethylamine, aqueous solution $C_2H_5NH_2$ [2270]
- Diethylamine $(C_2H_5)_2NH$ [1154]
- Triethylamine $(C_2H_5)_3N$ [1296]
- Ethylenediamine $NH_2CH_2CH_2NH_2$ [1604]
- Diethylenetriamine $(H_2NC_2H_4)_2NH$ [2079]
- Tetraethylenepentamine $NH_2(C_2H_4NH)_3C_2H_4NH_2$ [2320]
- Propylamine $CH_3(CH_2)_2NH_2$ [1277]
- Dipropylamine $(C_3H_7)_2NH$ [2383]
- Diisopropylamine $(C_3H_7)_2NH$ [1158]
- Diallylamine $(C_3H_5)_2NH$ [2359]
- *n*-butylamine $C_4H_9NH_2$ [1125]
- Isobutylamine $C_4H_9NH_2$ [1214]
- Di-*n*-butylamine $(C_4H_9)_2NH$ [2248]
- *N*-methylbutylamine $C_4H_9NHCH_3$ [2945]
- Diisobutylamine $(C_4H_9)_2NH$ [2361]
- Amylamine $C_5H_{11}NH_2$ [1106]
- Di-*n*-amylamine $(C_5H_{11})_2NH$ [2841]
- Hexamethylene-diamine, solid $NH_2(CH_2)_6NH_2$ [2280]
- Hexamethylene-diamine, solution $NH_2(CH_2)_6NH_2$ [1783]
- 2-Ethylhexylamine $C_4H_9CH(C_2H_5)CH_2NH_2$ [2276]
- Cyclohexylamine $C_6H_{11}NH_2$ [2357]
- Dicyclohexylamine $(C_6H_{11})_2NH$ [2565]

■ **Action on aluminium**

The most common aliphatic amines are

- Methylamine,

- dimethylamine,
- trimethylamine,
- ethylamine,
- diethylamine,
- propylamine,
- butylamine,
- dibutylamine,
- tributylamine,
- amylamine,
- hexylamine $C_6H_{13}NH_2$,
- octylamine $C_8H_{17}NH_2$,
- decylamine $C_{10}H_{21}NH_2$,
- ethylendiamine.

The resistance of aluminium in amine solutions depends on the nature and the concentration of the amine: the dissolution rate decreases with increasing molecular weight and concentration. Totally anhydrous amines attack aluminium at room temperature [1].

Aluminium immersed in a solution primary amine will be uniformly pickled; this pickling ceases after a few hours, when a protective oxide film has been formed on the surface. The colour of this film depends on the alloy: it is white on 1100, grey on 6060, light brown on 5754.

Concentrated solutions of primary amines with a concentration above 30% in the case of methylamine and acetylamine and 70% in the case of butylamine, have no action on aluminium, although their pH is comprised between 13 and 14 [2] (Figure F.4.1). A sodium hydroxide solution of the same pH (concentration of 5 $g \cdot l^{-1}$) dissolves a thickness of 1 mm within a few hours.

The intensity of pickling is weaker for aluminium alloys than for pure aluminium 1199 (or 1100): in a solution of 10% methylamine, pickling is five times greater with 1100 than with 7075, 2024 or 5754 (Figure F.4.2).

Hot amines have a moderate action on aluminium: the formation of a protective film hinders the attack if the amine contains traces of humidity and is exempt of chlorides.

Higher aliphatic amines, above C_5, have no action at room temperature. The same applies to allylamine $CH_2CHCH_2NH_2$.

Sodium silicate slows down the pickling of aluminium in amine solutions.

4.2. ALCOHOL AMINES

■ ADR numbers

- Ethanolamine $NH_2CH_2CH_2OH$ [2491]

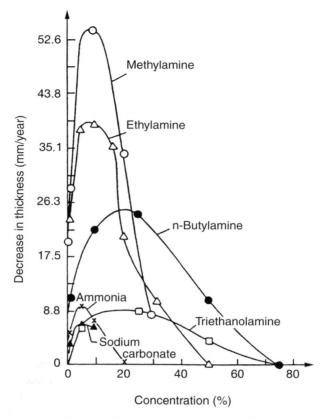

Figure F.4.1. Action of amines on 1100 [2].

■ **General information**

The three following amines

–   ethanolamine $NH_2CH_2CH_2OH$,
–   diethanolamine $NH(CH_2CH_2OH)_2$,
–   triethanolamine $N(CH_2CH_2OH)_3$

have the greatest number of industrial applications. They are used for the extraction of hydrogen sulphide and carbon dioxide from natural gas and gas produced in oil refineries. They are soluble in water in any proportion and form alkaline solutions with a pH close to 11, whatever their concentration.

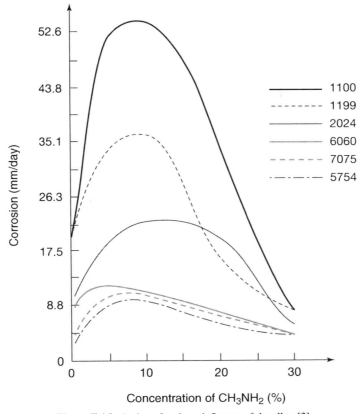

Figure F.4.2. Action of amines: influence of the alloy [2].

■ Action on aluminium

The action on aluminium is much more moderate than that of acyclic amines and ceases as the concentration exceeds 50% (Figure F.4.2). The solutions of glycolamine

- monoglycolamine: 20%
- dimethyleneglycol: 75%
- water: 5%

that are used for the desulphurisation of gases have no action on aluminium up to 180 °C [3].

The presence of carbon dioxide or hydrogen sulphide has no influence on the corrosion resistance of aluminium in ethanolamine solutions. Gas desulphurisation plants have been built with heat exchangers in 3003 [4].

Table F.4.1. Action of MEEDA on 3003: Dissolution rate (mm per year)

MEEDA concentration (%)	Temperature (°C)		
	22	50	75
20	9.9	>50	>50
40	4.5	33	15
50	1.1	4.1	5
60	0.5	1.6	2.1
80	0.1	0.35	0.5
90	0.03	0.2	0.3

At its boiling point at atmospheric pressure (360 °C), dehydrated triethanolamine can attack aluminium.

N-aminoethyl ethanolamine $NH_2(CH_2)_2NHC_2H_5OH$ has no action on aluminium, even up to 200 °C. This product is stored in aluminium containers.

The higher the concentration of solutions of monoethanol ethylenediamine $NH_2C_2H_4$-NHC_2H_4OH (MEEDA), the less aggressive they are to aluminium alloys. At 60% or more, the effect of the temperature is low on 3003 (Table F.4.1) [5].

Boehmite coatings can be formed on aluminium and its alloys in solutions containing 3% triethanolamine (see Section B.5.1).

4.3.   AROMATIC AMINES

■ ADR numbers

- Aniline $C_6H_5NH_2$ [1547]
- N-methylaniline $C_6H_5NHCH_3$ [2294]
- N,N-dimethylaniline $(CH_3)C_6H_5N$ [2253]
- N-ethylaniline $C_6H_5NHC_2H_5$ [2272]
- N,N-diethylaniline $C_6H_5N(C_2H_5)_2$ [2432]
- Toluidine $CH_3C_6H_4NH_2$ [1708]
- Benzidine $(NH_2C_6H_4)_2$ [1885]
- Xylidines $C_6H_3(CH_3)_2NH_2$ [1711]

■ General information

Aniline, the first representative of the aromatic amines, is highly important in the industrial context as the basis for the synthesis of a large number of dyestuffs. Aniline and water show reciprocal miscibility: aniline is soluble in water up to 3.5%, and water is soluble in aniline up to 5%.

■ **Action on aluminium**

Aniline $C_6H_5NH_2$ and the most common aromatic anilines:

- ethylaniline $C_6H_5NHC_2H_5$,
- benzylaniline $C_6H_5CH_2NH_2$,
- toluidine $CH_3C_6H_4NH_2$,
- naphthylamine $C_{10}H_7NH_2$,
- diphenylamine $(C_6H_5)_2NH$,
- benzidine $(NH_2C_6H_4)_2$,
- indole $C_6H_4CHCHNH$,
- azobenzene $C_6H_5N=NC_6H_5$,
- diphenylamine $(C_6H_5)_2NH$,
- indoline $C_6H_4NH(CH_2)_2$

have no action on aluminium up to their boiling points.

Close to their boiling points of approximately 200 °C (aniline boils at 184 °C), certain amines, if totally anhydrous, can severely attack aluminium. As in organic acids, traces of humidity prevent this attack.

Aluminium can be used for the storage and the transport of aniline and other aromatic amines.

4.4. DERIVATIVES OF AROMATIC AMINES

■ **ADR numbers**

- Anisoyl chloride $CH_3OC_6H_4COCl$ [1729]
- Nitroanilines (*o-*,*m-*,*p-*) $C_6H_4NH_3NO_2$ [1661]
- Nitroanisole $C_6H_4(NO_2)(OCH_2)$ [2730]

■ **Action on aluminium**

If anhydrous, these salts, whatever the associated acid, have no action on aluminium.

In aqueous solutions, they decompose and release the corresponding acid. The resistance of aluminium in contact with aniline salts and other aromatic amines depends on the nature of the acid that corresponds to the amine.

Chlorides form hydrochloric acid that attacks aluminium. This also applies to aniline chloride and chlorides of other amines. Sulphates, which are much less soluble in water, also release acid, but to a lesser degree.

On the other hand, nitrates, like aniline nitrate, have no action. Aluminium equipment can be used for the production of aniline nitrate.

4.5. AMINOPHENOLS

■ ADR numbers

–   Anisidines $CH_3OC_6H_4NH_2$ [2431]
–   Aminophenols (*o*-,*m*-,*p*-) $NH_2C_6H_4OH$ and $(NH_2)_2C_6H_3OH$ [2512]

■ Action on aluminium

Aminophenol $NH_2C_6H_4OH$ has no action on aluminium at room temperature. Certain salts of aminophenol such as the sulphate can decompose in contact with aluminium.

At room temperature, anisidine $NH_2C_6H_4OH$ has no action. At elevated temperatures, this product can corrode aluminium.

4.6. AMINO ACIDS

■ General information

Amino acids have an aliphatic or aromatic chain that bears both the amine group NH_2 and the carboxylic acid group COOH. These acids are more or less water-soluble. They are very important in biochemistry.

Protein hydrolysis leads to amino acids. These amino acids, when heated, will decompose into carbon dioxide and ammonia. The aerobic or anaerobic decomposition of proteins and amino acids will always release more or less complex amines, which have a putrid smell, and carbon dioxide and ammonia if the decomposition is pursued further.

■ Action on aluminium

Glycocol (aminoacetic acid) NH_2CH_2COOH, the simplest amino acid, has no action on aluminium.

–   Phenyl glycocol $C_6H_5NHCH_2COOH$,
–   aspartic acid or aminosuccinic acid $COOHCH_2CHNH_2COOH$,

– glutamic acid $COOH(CH_2)_2CH(NH_2)COOH$,
– aminobenzoic acids (anthranilic acids) $C_6H_4(NH_2)COOH$,
– aminosalicylic acid $C_6H_3(NH_2)(OH)(COOH)$

have no action on aluminium, which can be used for the storage of these products or their solutions.

Aluminium resists putrefying organic matter well, which releases large amounts of carbon dioxide, ammoniac and amines.

REFERENCES

[1] *Amine und Aminierung*, Dechema Werkstofftabelle, Mai 1953.
[2] Buckowiecki A., Untersuchungen über das Korrosionsverhalten von Aluminium und Aluminiumlegierungen gegenüber organischen und anorganischen Basen, *Werkstoffe und Korrosion*, vol. 10, 1959, p. 91–105.
[3] Riesenfeld F.C., Hughes C.I., Corrosion in amine gas-treating plants, *Petroleum Refiner*, vol. 30, 1951, p. 97–106.
[4] Riesenfeld F.C., Blohm C.L., Corrosion resistance of alloys in amine gas-treating systems, *Petroleum Refiner*, vol. 30, 1951, p. 107–115.
[5] Rozenboim G.B., Osipov V.N., Corrosion of aluminum alloys in aqueous solutions of monoethanolethylenediamine (MEEDA), Zaschita metallov, vol. 16, no. 1, p. 63–64.

Chapter F.5

Aldehydes and Ketones

Chapter F.5
Aldehydes and Ketones

Aldehydes and ketones contain the same functional group, the so-called carbonyl group:

$$C=O$$

They differ only in the number of alkyl R or aryl groups Ar that surround them, or in the degree of substitution of the carbonyl group:

- aldehydes: $R-CH=O$
- ketones: R_1-CO-R_2 (where R_1 and R_2 can be two identical or different radicals).

5.1. ALIPHATIC ALDEHYDES

■ ADR numbers

- Formaldehyde solution, flammable HCHO [1198]
- Formaldehyde solution HCHO [2209]
- Acetaldehyde CH_3CHO [1089]
- Propionaldehyde C_2H_5CHO [1275]
- Isobutyraldehyde C_3H_7CHO [2045]
- Crotonaldehyde C_3H_5CHO [1143]
- 2-Ethyl-butyraldehyde $C_5H_{11}CHO$ [1178]
- Octylaldehydes C_7H_5CHO [1191]

■ General information

Formaldehyde HCHO is a gas at room temperature, whereas aliphatic aldehydes from C_2 to C_{10} are liquids. Higher aldehydes are crystalline solids. Their solubility in water greatly decreases with increasing molecular weight. The first three aldehydes (formaldehyde, acetaldehyde and propionaldehyde) are miscible with water in any proportion. The solubility of butyraldehyde (C_4) is 750 g·l^{-1}, and that of amylaldehyde (C_5) only 10 g·l^{-1}.

Aldehydes are strong reducing agents and can oxidise in contact with air, yielding the corresponding acid, which is why traces of organic acids can be found in aldehydes after long exposure to air. Certain metals and alloys can accelerate this oxidation. This is not the case of aluminium, however.

Aldehydes have a great number of applications. They are intermediates in the synthesis of certain phenolic resins, acetic acid and detergents. They are also used as vulcanisation accelerators, for papermaking, in perfumery, confectionery, and some are even used as corrosion inhibitors. Some aldehydes are produced in large quantities. Formaldehyde is the twenty-fifth most important chemical product in the United States.

5.1.1. Formaldehyde

Formaldehyde is a gas at room temperature. It is soluble in water in any proportion. Industrially, an aqueous solution (called formol) containing 30–40% of aldehyde together with some traces of methanol and formic acid, is used.

■ Action on aluminium

Aluminium is very resistant to contact with pure formaldehyde, whether at room temperature or as vapour at 60–70 °C. The possible attack corresponds to a decrease in thickness on the order of 25 μm per year [1, 2]. Formaldehyde can be distilled, stored or transported in aluminium equipment.

Solutions of formaldehyde, or formol, only have a very moderate action on aluminium. A protective film is generally formed on the surface, which slows down the dissolution of the metal, whether at room temperature or at high temperatures.

As an example, in a solution containing 37% formaldehyde and 1% methanol, the dissolution rate is 200 μm per year at room temperature. The presence of other organic products such as methanol or formic acid can slightly increase the dissolution rate [3]. Like formaldehyde, formol is stored and transported in equipment made in aluminium alloy.

5.1.2. Other acyclic aldehydes

- Acetaldehyde CH_3CHO
- Propionaldehyde C_2H_5CHO
- Butyraldehyde C_3H_7CHO
- Isobutyraldehyde C_2H_6CHCHO
- Valeraldehyde C_4H_9CHO
- Caproaldehyde $C_5H_{11}CHO$
- *n*-heptylaldehyde $C_6H_{13}CHO$
- Caprylaldehyde $C_7H_{15}CHO$
- Octadecylaldehyde $C_8H_{17}CHO$
- Nonylaldehyde $C_8H_{17}CHO$
- Isodecylaldehyde $C_9H_{19}CHO$
- Decylaldehyde $C_9H_{19}CHO$

- Undecylaldehyde $C_{10}H_{21}CHO$
- Dodecylaldehyde $C_{11}H_{23}CHO$
- Tridecylaldehyde $C_{12}H_{25}CHO$
- Myristic aldehyde $C_{13}H_{27}CHO$
- *n*-pentadecylaldehyde $C_{14}H_{29}CHO$
- Palmitic aldehyde $C_{15}H_{31}CHO$

■ Action on aluminium

These aldehydes, which are widely used in perfumery and cosmetics, either pure or as mixtures, have no action on aluminium. They are neither altered nor modified in contact with aluminium. Aluminium equipment is used for the production, distillation, storage and packaging of many aldehydes [4].

Dialdehydes such as

- adipic aldehyde $OCH(CH_2)_4CHO$,
- glutaric aldehyde (or glutaraldehyde) $CHO(CH_2)_3CHO$,
- glycolic aldehyde $HOCH_2CHO$,

and ethylenic aldehydes such as

- acrylic aldehyde (or acroleine) C_2H_3CHO,
- crotonic aldehyde (or crotonaldehyde) $CH_3C_2H_2CHO$,
- isovaleric aldehyde $(CH)_2CHCH_2CHO$,
- hexenoic aldehyde C_5H_9CHO,

which are the most common ones, have no action on aluminium, even at elevated temperatures, explaining why aluminium is widely used in contact with acrolein for production, storage and transportation equipment [5].

5.2. CYCLIC ALDEHYDES

Cyclic aldehydes such as

- cyclopentane carbaldehyde C_5H_7CHO,
- cyclohexylidene acetic aldehyde $C_6H_{10}CHCHO$,
- cyclohexane carbaldehyde $C_{17}H_{12}CHO$,

and mixed aldehydes with alcohols or ketones such as

- glyceric aldehyde HOCH$_2$CHOHCHO,
- pyruvic aldehyde CH$_3$COCHO

also have no action on aluminium.

5.3. AROMATIC ALDEHYDES

The aromatic aldehydes

- benzoic aldehyde C$_6$H$_5$CHO,
- toluene aldehyde C$_6$H$_4$CH$_3$CHO,
- phenylacetic aldehyde C$_6$H$_5$CH$_2$CHO,
- vanillic aldehyde (vanillin) CH$_3$OC$_6$H$_3$(OH)CHO,
- salicylic aldehyde HOC$_6$H$_4$CHO,
- cinnamic aldehyde C$_6$H$_5$(C$_2$H$_2$)CHO,
- coumaric aldehyde OHC$_6$H$_4$C$_2$H$_2$CHO,
- cumenic aldehyde (CH$_3$)$_2$CHC$_6$H$_4$CHO,
- anisic aldehyde C$_6$H$_4$(OCH$_3$)CHO,
- veratric aldehyde (CH$_3$O)$_2$C$_6$H$_3$CHO,
- terephthalic aldehyde C$_6$H$_4$(CHO)$_2$

are very widely used for cosmetics and perfumery. They have no action on aluminium. Aluminium equipment is used for the production, transportation, storage and packaging of these products or mixtures thereof.

5.4. KETONES

■ ADR numbers

- Acetone CH$_3$COCH$_3$ [1090]
- Butanedione CH$_3$COCOCH$_3$ [2346]
- 3-methylbutan-2-one (CH$_3$)$_3$CHCO [2397]
- Pentane-2,4-dione (acetylacetone) CH$_3$COCH$_2$COCH$_3$ [2310]
- Diethyl ketone C$_2$H$_5$COC$_2$H$_5$ [1156]
- Ethyl methyl ketone (methyl ethyl ketone) CH$_3$COC$_2$H$_5$ [1193]
- Methyl isopropenyl ketone, stabilised CH$_3$COC$_3$H$_7$ [1245]
- Methyl propyl ketone CH$_3$COC$_3$H$_7$ [1249]
- Dipropyl ketone C$_3$H$_7$COC$_3$H$_7$ [2710]

- Methyl isobutyl ketone $CH_3COC_4H_9$ [1245]
- Butyl methyl ketone $CH_3COC_4H_9$ [1224]
- Butyl ethyl ketone $C_2H_5COC_4H_9$ [1224]
- Diisobutyl ketone $(i\text{-}C_4H_9)_2CO$ [1157]
- Ethylamyl ketone $C_5H_{11}COC_2H_5$ [2271]
- 5-methylhexan-2-one $C_5H_{11}COCH_3$ [2302]
- Cyclohexanone $C_6H_{10}O$ [1915]
- Methycyclohexanone $CH_3C_4H_8CO$ [2297]
- Cyclopentanone $C_6H_{12}CO$ [2245]

■ General information

Although they are produced in lesser quantities than alcohols or acids, ketones have many applications. They are intermediates in organic synthesis as well as in the synthesis of dyestuffs, cellulose acetates, etc., and are also used for the production of paints, varnish, glues, perfumes, pharmaceuticals and certain plastics such as methyl methacrylate. They are also excellent solvents.

Ketones are poorly soluble in water, except acetone CH_3COCH_3, which is miscible in any proportion. The solubility decreases rapidly with increasing number of carbon atoms. Up to C_{10}, ketones are liquids at room temperature. Since their carbonyl group is located inside a carbon chain of variable length, ketones are not very reactive. They are stable products and have no action on aluminium.

5.4.1. Aliphatic ketones

The foregoing applies to all common ketones, whether at room temperature or at the boiling point:

- Acetone CH_3COCH_3,
- Butanedione $CH_3COCOCH_3$,
- Ethyl methyl ketone (methyl ethyl ketone) $CH_3COC_2H_5$,
- Diethyl ketone $C_2H_5COC_2H_5$,
- Methyl propyl ketone $CH_3COC_3H_7$,
- Dipropyl ketone $C_3H_7COC_3H_7$,
- Methyl isopropenyl ketone, stabilised $CH_3COC_3H_7$,
- Diisobutyl ketone $(i\text{-}C_4H_9)_2CO$,
- 3-Methylbutan-2-one $(CH_3)_3CHCO$,
- Ethylamyl ketone $C_5H_{11}COC_2H_5$,
- Butyl methyl ketone $CH_3COC_4H_9$,
- Butyl ethyl ketone $C_2H_5COC_4H_9$,

–   Methyl isobutyl ketone $CH_3COC_4H_9$,
–   5-Methylhexan-2-one $C_5H_{11}COCH_3$,
–   Pentane-2,4-dione (acetylacetone) $CH_3COCH_2COCH_3$.

Tests in acetone for 1 month, both at room temperature and at the boiling point, have shown that 3003 has an excellent resistance; the decrease in thickness is below 1 μm per year under these conditions.

Diacetone alcohol $(CH_3)_2COHCH_2COCH_3$ has no action on aluminium.

Equipment in aluminium is used for the production of ketones: reaction vessels, heat exchangers, transportation and storage containers.

5.4.2.   Cyclic ketones

Ketones of the general formula $C_nH_{2n}CO$ are used in perfumery or as solvents. They are liquids or solids (insoluble in water, soluble in alcohol) and have no action on aluminium. These products are stored in aluminium containers. This is the case of the most common cyclic ketones:

–   Cyclobutanone C_3H_6CO,
–   Cyclopentanone C_4H_7CO,
–   Cycohexanone $C_5H_{10}CO$,
–   Methyl cyclohexanone $CH_3C_5H_9CO$,
–   Cycloheptanone $C_6H_{12}CO$,
–   Cyclooctanone $C_7H_{14}CO$,
–   Cyclononanone $C_8H_{16}CO$,
–   Cycloundecanone $C_{10}H_{20}CO$,
–   Cyclotridecanone $C_{12}H_{22}CO$,
–   Cyclotetradecanone $C_{13}H_{26}CO$,
–   Cyclepentadecanone $C_{14}H_{28}CO$,
–   Cyclohexadecanone $C_{15}H_{30}CO$,
–   Cycloheneicosanone $C_{20}H_{40}CO$,
–   Cyclodocosanone $C_{21}H_{42}CO$,
–   Cyclotricosanone $C_{22}H_{44}CO$,
–   Cyclotetracosanone $C_{23}H_{46}CO$.

Camphor $C_{10}H_{16}O$ does not attack aluminium, neither at room temperature nor at its boiling point (180 °C). Solutions in alcohol or camphor oils also do not attack aluminium.

5.4.3.   Aromatic ketones

Aromatic ketones are solid at room temperature. They are used as starting compounds for the synthesis of many dyestuffs.

The most common ones are

- quinone or benzoquinone $C_6H_4O_2$,
- benzophenone or diphenyl ketone $C_6H_5COC_6H_5$,
- anthraquinone $C_6H_4COCOC_6H_4$.

They have no action on aluminium that can be used for the fabrication and transportation of these products.

Anthraquinone is used in a production process for hydrogen peroxide in which a very large number of pieces of equipment such as reaction vessels and heat exchangers are in aluminium (see Section E.1.9).

Sulphonic acids of anthraquinone, when produced in presence of mercury, may lead to a slight attack of aluminium.

REFERENCES

[1] Teeple H.O., Corrosion by some organic acids and related compounds, *Corrosion*, vol. 8, 1952, p. 14–28.

[2] Lingneau E., Das Verhalten der Werkstoffe gegenüber Formaldehyde, *Werkstoffe und Korrosion*, vol. 8, 1957, p. 480–487.

[3] Bally J., Le matériel en aluminium dans la fabrication des produits chimiques dérivés des gaz de four à coke, *Revue de l'Aluminium*, no 100, 1938, p. 1155–1165.

[4] Barkhot H., *Aliphatics ketones. Corrosion Handbook*, Dechema, vol. 7, 1990.

[5] Weigert W.M., *Ullmanns Encyklopädie der technischen Chemie*, 4th edition, vol. 7, Verlag Chemie GmbH, Weinheim, 1974, p. 74.

Chapter F.6

Carboxylic Acids and their Derivatives

Chapter F.6
Carboxylic Acids and their Derivatives

Organic acids are of the general formula

- RCOOH for acyclic acids,
- ArCOOH for aromatic acids,

and form a product family with many applications: they are used in the production of polymers, resins, dyestuffs, detergents, pharmaceuticals, sweeteners, etc.

Several of them belong to the fifty most important chemicals produced in the United States, according to a ranking established in 1990: terephthalic acid (rank 23), acetic acid (rank 34), adipic acid (rank 46), and isopropyl acid (rank 48).

They are weak acids and have only a moderate action on aluminium at room temperature. Their solubility in water and their acidity decrease rapidly as their molecular mass increases. The higher their molecular mass, the weaker the action of organic acids on aluminium.

Like alcohols, they may react violently with aluminium at high temperatures in totally dehydrated media, especially at the boiling point. The OH group can react with aluminium as follows:

$$3RCOOH + Al \rightarrow \frac{3}{2}H_2 + Al(RCOO)_3.$$

This reaction takes place because the acid dehydrates the natural oxide film and alters its structure. Moreover, since the resulting salt $Al(RCOO)_3$ is soluble in the medium, aluminium is no longer protected and will be continuously attacked. This attack often takes place as pitting.

Experience shows that the addition of a small amount of water, between 0.05 and 0.1%, depending on the nature of the acid, totally prevents the attack of aluminium by boiling organic acids or mixtures thereof.

Derivatives of organic acids are also important for the production of polymers, detergents, dyestuffs, explosives, pesticides, insecticides, etc. Many occur in nature.

In this section, we will deal with the behaviour of aluminium in contact with

- *acyclic hydrocarbon acids* of the general formula R−COOH,
- *acid anhydrides* of the general formula R−CO−O−CO−R,
- *salts of acids* of the general formula $(R-COO)_nM$, where M is a metal, most often an alkaline metal (sodium or potassium),
- *esters* of the general formula $R_1-COO-R_2$,

513

-   *amides* of the general formula R–CO–NH$_2$,
-   *nitriles* of the general formula R–C≡N.

6.1. ACYCLIC HYDROCARBON ACIDS

6.1.1. *Saturated hydrocarbon acids*

■ **ADR numbers**

-   Formic acid HCOOH [1779]
-   Acetic acid solution, more than 80% acid, by mass CH$_3$COOH [2789]
-   Acetic acid solution, not less than 50% but more than 80% acid, by mass CH$_3$COOH [2790]
-   Acetic acid solution, more than 10% and less than 50% acid, by mass CH$_3$COOH [2790]
-   Propionic acid C$_2$H$_5$COOH [1848]
-   Butyric acid C$_3$H$_7$COOH [2820]
-   Isobutyric acid *i*-C$_3$H$_7$COOH [2529]

■ **General information**

Acetic acid CH$_3$COOH is used for the production of many important chemicals: synthetic fibres (cellulose acetate), plastics (vinyl acetate), solvents, dyestuffs, etc.

The annual world production in the order of 3 million tons is almost exclusively achieved by synthesis (carbonylation of methanol). Pyrolysis of wood is no more than a very small source of acetic acid (less than 2%).

Propionic acid, less important than formic and acetic acid, is used in the fabrication of synthetic fibres and as an intermediate for synthesis.

■ **Formic acid HCOOH**

Formic acid is miscible with water in any proportion. The maximum concentration commonly used does not exceed 90%.

At ambient temperature, diluted solutions of formic acid have a moderate action on aluminium: the dissolution rate does not exceed 0.2 mm per year (Figure F.6.1), with a tendency to pitting corrosion. Very concentrated solutions are less aggressive than diluted solutions; the maximum attack of aluminium is observed at a concentration of 20% formic acid, whatever the temperature.

An increase in the temperature dramatically increases the dissolution of aluminium in formic acid solutions. At a given concentration, the dissolution rate is 10 times higher at

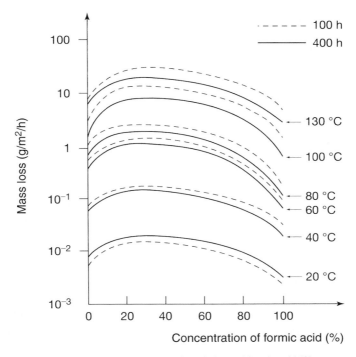

Figure F.6.1. Dissolution rate in solutions of formic acid [2].

40 °C (of the order of 2 mm per year) and 100 times higher at 60 °C than at room temperature. The acid changes its aspect and turns turbid (Table F.6.1).

The possibility of using aluminium equipment in contact with concentrated solutions of formic acid is strictly limited to room temperature, if the acid concentration exceeds 95% and contains no traces of chlorides [1]. Copper-containing alloys of the 2000 and 7000 series are excluded.

■ Acetic acid CH₃COOH

Table F.6.1. Formic acid: dissolution rate (mm per year)

Temperature (°C)	Concentration of formic acid (%)					
	5	10	25	50	75	90
20	0.1	0.1	0.1	0.1	0.08	0.06
50	0.5	0.7	0.7	0.7	0.7	1.2
Boiling[1]	33	40	46	50	36	10

[1]The boiling point of water–formic acid mixtures is close to 100 °C. The pure acid boils at 100.8 °C.

At room temperature, aluminium resists the action of acetic acid and its solutions very well, and the higher the acid concentration, the better the resistance (Figure F.6.2). This has been shown by test results over 3 months at 20 °C on alloy 1100 (Table F.6.2). The dissolution of the metal is uniform, without pitting corrosion.

Up to 50 °C, the increase in temperature does not have a marked effect on the dissolution rate of aluminium. Above 50 °C, and especially close to the boiling point of water–acetic acid mixtures, the dissolution rate increases considerably, except for very high concentrations (Table F.6.3) [3]. Very concentrated or "anhydrous" acid ("glacial" acid) can attack aluminium severely, but traces of water in the order of 0.05–0.20% are sufficient to prevent this attack.

It is generally admitted that 52 °C is the temperature limit not to be exceeded for the use of aluminium in contact with acetic acid for the concentration ranging from 60 to 100%. In this limited range, the dissolution rate is approximately 0.25 mm per year [4]. On welded metal, no specific or preferential corrosion is observed on the welding bead and in the heat-affected zone [5].

The presence of certain products can modify the resistance of aluminium and its alloys at room temperature. The presence of sulphuric acid in highly concentrated acetic acid increases the dissolution rate of aluminium: 1% sulphuric acid leads to a decrease in thickness of 15 μm per year, and 3% to a decrease of 50 μm per year. Halogenated derivatives also accelerate dissolution. The presence of formic acid and chlorides should be carefully avoided. A concentration of 0.2–0.6% of phenol significantly increases the dissolution rate of aluminium.

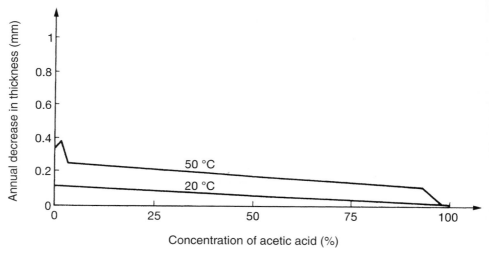

Figure F.6.2. Influence of the concentration of acetic acid and the temperature on the dissolution rate of alloy 1100 [6].

Table F.6.2. Dissolution rate of alloy 1100 in acetic acid

Concentration (%)	pH	Decrease in thickness (mm/year)
1	3.1	0.025
10	2.4	0.015
50	1.7	0.010
95		0.001

Aluminium and alloys of the series 1000, 3000, 5000 and 6000, as well as casting alloys with silicon (series 42000), are widely used in the synthesis of acetic acid (Othmer, Hoechst, Wacker, etc. processes) [8], for its transportation (in barrels or tank wagons) and storage, and in heaters of "glacial" acid.

Tanks in 3003 of 700-m^3 capacity have been in use for more than 30 years for the storage of acetic acid. Reaction vessels in aluminium are used for the manufacture of aspirin from acetic acid and salicylic acid [6].

■ **Propionic acid CH$_3$CH$_2$COOH**

The action of propionic acid on aluminium is somewhat more moderate than that of acetic acid. The decrease in thickness of aluminium is, for any concentration, less than

Table F.6.3. Dissolution rate in acetic acid (mm/year): influence of the concentration and the temperature [7]

Concentration (%)	Temperature		
	20 °C	50 °C	Boiling [1]
0.004	0.013		
0.03	0.045		
0.25	0.033		
1.0	0.013	0.11	16.9
3.0	0.011	0.11	9.1
5	0.011	0.11	8.7
10	0.009	0.11	7.8
20	0.009	0.11	7.3
30	0.008	0.11	6.8
40	0.008	0.11	6.4
50	0.008	0.11	6.0
60	0.008	0.10	5.7
70	0.008	0.09	5.1
80	0.008	0.08	4.1
85	0.005	0.06	3.4
90	0.004	0.04	2.5
95	0.003	0.01	1.4
98–99.8	0.001	0.007	0.17

[1] The boiling point of acetic acid is 118 °C, and that of its aqueous solutions between 100 and 118 °C.

50 μm per year at room temperature. As with acetic acid, the dissolution rate sharply increases with temperature (Figure F.6.3), and the limit for the use of aluminium is around 52 °C for solutions of propionic acid in concentrations not exceeding 50°. In concentrated acid, the decrease in thickness amounts to 0.5 mm per year at 100 °C.

■ Butyric acid $CH_3(CH_2)_2COOH$ and isobutyric acid $(CH_3)_2CHCOOH$

The action of butyric acid and of its isomer isobutyric acid on aluminium is weaker than that of formic acid, acetic acid and propionic acid. At room temperature, the dissolution rate is less than 0.50 μm per year (Figure F.6.4); it progressively increases up to the boiling point.

Aluminium is used for the transportation and storage of butyric acid, as well as for certain equipment for its production and distillation [9].

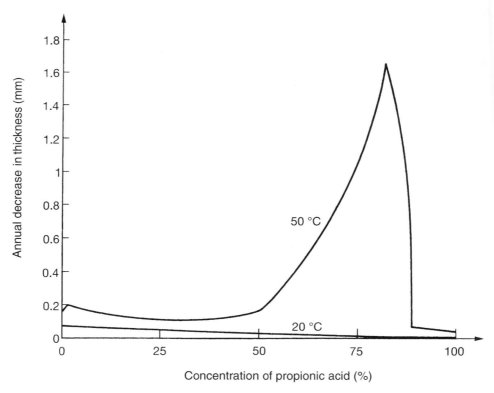

Figure F.6.3. Influence of the concentration of propionic acid and temperature on the dissolution rate of 1100 [6].

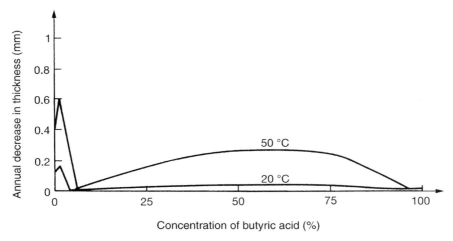

Figure F.6.4. Influence of the concentration of butyric acid and of the temperature on the dissolution rate of 1100 [6].

■ **Fatty acids**

□ **ADR number**

– Caproic acid $C_5H_{11}COOH$ [2829]

□ **General information**

These are saturated, acyclic monoacids higher than C_5. Many of them occur naturally as esters in plant essences and waxes, and in fats such as glycerides (Table F.6.4).

The most abundant ones are palmitic acid and stearic acid, the glycerides of which form the major part of fats, together with oleic acid.

Soaps and detergents are sodium and potassium salts of these acids.

These acids are virtually insoluble in water, and from C_7 are all solids at room temperature.

□ **Action on aluminium**

Whether cold or heated to their boiling point, the pure acids or mixtures thereof have no action on aluminium.

At the distillation temperature, fatty acids can give rise to severe pitting corrosion of aluminium. The decrease in temperature by using vacuum distillation, or the addition of traces of water contribute to reducing or cancelling this risk of attack.

Table F.6.4. Main fatty acids

Name	Formula	Natural occurrence
Caproic acid	$C_5H_{11}COOH$	Among butyric fermentation products of sugar, as an ester in essence of Palmarosa.
Enanthic acid	$C_6H_{13}COOH$	As an ester in essence of Calamus.
Caprylic acid	$C_7H_{15}COOH$	As a glyceride in butter of cow's milk and coconut oil, as an ester in wine.
Pelargonic acid	$C_8H_{17}COOH$	In essence of *Pelargonium roseum*, in sugar beet distillate and potato distillate.
Capric acid	$C_9H_{19}COOH$	As a glyceride in butter of cow's milk, coconut oil, Limbourg cheese, as an ester in wine
Lauric acid	$C_{11}H_{23}COOH$	As a glyceride in essence og bay tree berries, in coconut butter, Pichurim beans, spermaceti.
Myristic acid	$C_{13}H_{27}COOH$	As a glyceride in nutmeg butter, coconut butter, seeds of *Virola venezuelensis*.
Palmitic acid	$C_{15}H_{31}COOH$	As a glyceride in many animal and vegetal fats, as an ester in waxes
Stearic acid	$C_{17}H_{35}COOH$	As a glyceride in many animal and vegetal fats.
Arachidic acid	$C_{19}H_{39}COOH$	As a glyceride in peanut oil, colza oil, cacao oil, macassar oil.
Eicosan-l-carbocylic acid	$C_{20}H_{41}COOH$	In Japan wax, peanut oil, etc.
Behenic acid	$C_{21}H_{43}COOH$	As a glyceride in colza and peanut oil
Lignoceric acid	$C_{23}H_{47}COOH$	In peanut oil and wood tar
Cerotic acid	$C_{25}H_{51}COOH$	As an ester in beeswax and other waxes, in suint
Melissic acid	$C_{29}H_{59}COOH$	In Carnauba wax and beeswax
Psyllastearylic acid	$C_{32}H_{65}COOH$	In wax from the greenfly *Psylla alni*

During the production cycle of these acids, and with the exception of acidification or alkalinisation steps, equipment in alloy is widely used for condensation, filtration, crystallisation and storage [10–12].

Fatty acids do not stick to aluminium surfaces and will not be discoloured in contact with aluminium.

6.1.2.   Ethylenic monoacides

■ **ADR number**

— Methacrylic acid, stabilised C_3H_5COOH [2531]

■ Action on aluminium

The most common ones are

- acrylic acid $CH_2=CHCOOH$,
- methacrylic acid $CH_2=C(CH_3)COOH$,
- crotonic acid $CH_3CH=CHCOOH$.

They have no action on aluminium. They are stored or transported in aluminium containers.

The same applies to higher acids, C_{15} and higher, and to unsaturated fatty acids such as

- oleic acid $C_{17}H_{33}COOH$,
- ricinoleic acid $C_{17}H_{33}COOH$,
- linoleic acid $C_{24}H_{31}COOH$.

6.1.3. Monoacid alcohols

■ General information

Glycolic acid $CH_2OHCOOH$ is used for dyeing and tannery and as a bactericide or descaling agent.

Lactic acid $CH_3CHOHCOOH$ has a very large number of uses: tannery, bactericide, and cosmetics. It occurs naturally in dairy products.

■ Action on aluminium

Glycolic acid has no action on aluminium.

Lactic acid has a very moderate action at room temperature. The attack is uniform. The dissolution rate depends on the acid concentration, but remains largely acceptable at room temperature (Table F.6.5).

The increase in temperature accelerates the dissolution rate, which is 0.40 mm per year at 60 °C, and 1.50 mm per year at 80 °C, whatever the concentration of lactic acid.

Table F.6.5. Action of lactic acid at 20 °C

Concentration of lactic acid (%)	1	2	5	10	50	100
Decrease in thickness (mm/year)	0.01	0.01	0.02	0.04	0.16	<0.01

Much of the equipment for cheese making is in aluminium, such as draining boards and cheese strainers.

6.1.4. Saturated diacids

■ **General information**

The most commonly used saturated acids are listed in Table F.6.6.

All these acids are solids at room temperature. Their solubility in water decreases with increasing molecular mass.

■ **Action on aluminium**

At room temperature, a solution of 10% oxalic acid leads to uniform dissolution of 0.10 mm per year.

The other acids, as a solution or in concentrated form, only have an insignificant action on aluminium. The dissolution rate increases with temperature. For oxalic acid, it is multiplied by four between 20 and 100 °C.

These acids are transported and stored in aluminium vessels.

6.1.5. Unsaturated diacids

Maleic acid $COOH(CH)_2COOH$ and itaconic acid $C_3H_4(COOH)_2$ have practically no action on aluminium at room temperature. In solutions of maleic acid at 5, 15 and 30%, the decrease in thickness of 1100 is less than 10 μm per year, independent of the concentration. At higher temperatures, such as 100 °C or above, the dissolution rate significantly increases.

Maleic acid can be stored and transported in aluminium equipment, which is also used for the fabrication of itaconic acid.

6.1.6. Polyacids

–   Tartaric acid $COOH(CHOH)_2COOH$
–   Malic acid $COOHCHOHCH_2COOH$

Table F.6.6. Saturated diacids

Name	Formula	Use
Oxalic acid	$COOHCOOH$	Bleaching agents, descaling agents.
Malonic acid	$COOHCH_2COOH$	Intermediate for the synthesis of pharmaceuticals.
Succinic acid	$COO(CH_2)_2COOH$	Intermediate for the synthesis of dyestuffs and fragrances.
Adipic acid	$COOH(CH_2)_4COOH$	Basic compound for the synthesis of polyamide fibres (nylon).

– Citric acid $C_6H_{10}O_8$
– Aconitic acid $C_6H_8O_7$

These acids have a very moderate action on aluminium.

In solutions containing 5% (pH 1.7), 30% (pH 1.0), and 50% (pH 0.6) tartric or citric acid, the decrease of thickness is less than 10 μm per year at room temperature.

The dissolution rate increases with temperature. At 50 °C, the thickness decreases at a rate of 0.1 mm per year.

A great deal of equipment in aluminium is used for the production, storage and transportation of these acids. Aluminium modifies neither the aspect nor the taste of these acids or solutions thereof.

6.1.7. Aromatic acids

The most common aromatic acids are

– benzoic acid C_6H_5COOH,
– gallic acid $C_6H_2(OH)_3COOH$, H_2O,
– phenylacetic acid $C_6H_5CH_2COOH$,
– phthalic acid $C_6H_4(COOH)_2$,
– salicylic acid $C_6H_4OHCOOH$,
– tannic acid $C_{20}H_{16}O_{19}$,
– toluic acid $CH_3C_6H_4COOH$,
– vanillic acid $C_6H_3(OH)(OCH_3)COOH$,
– naphthoic acid $C_6H_{11}COOH$.

These are solids at room temperature, and they are practically insoluble in water. They are intermediates for the synthesis of dyestuffs, fragrances, sweeteners, pharmaceuticals, resins, etc.

At room temperature, they have no action on aluminium. As an example, the dissolution rate in naphthenic acid corresponds to a decrease in thickness of 2 μm per year. Like acyclic acids, at the boiling point and in the absence of humidity, they can lead to severe corrosion of aluminium.

Aluminium and aluminium alloys are used for production, storage and transportation equipment of these products

6.1.8. Various acids

■ ADR number

– Thioglycolic acid $HSCH_2COOH$ [1940]

■ **Action on aluminium**

Ascorbic acid (vitamin C) has no action on aluminium. It can be prepared and transported in aluminium equipment.

Acetylsalicylic acid (aspirin) has no action on aluminium and is not altered by contact with aluminium.

Thioglycolic acid, an intermediate in the preparation of cosmetics, severely attacks aluminium and its alloys. The dissolution rate is 2–4 mm per year, depending on the acid concentration. Moreover, it becomes turbid in contact with the metal, which renders the acid unfit for many uses.

Acids

–   Aminobenzenesulphonic acid $NH_2C_6H_4SO_3H$,
–   aminophtalenesulphonic acid $NH_2C_{10}H_6SO_3H$,
–   aminophenolsuphonic acid $NH_2C_6H_8(OH)SO_3H$,
–   aminotoluenesulphonic acid $CH_3(NH_2)C_6H_3SO_3H$,
–   aminobenzoic acid $C_6H_4NH_2COOH$,
–   aminosalicylic acid $NH_2C_6H_3(OH)COOH$

are important intermediates for the synthesis of dyestuffs, detergents and drugs. Most of them are insoluble in water. At room temperature, they have no action on aluminium.

6.2.   ACID ANHYDRIDES

■ **ADR numbers**

–   Acetic anhydride $(CH_3CO)_2O$ [1715]
–   Butyric anhydride $(C_3H_7CO)_2O$ [2739]
–   Isobutyric anhydride $(C_3H_7CO)_2O$ [2530]
–   Propionic anhydride $(C_2H_5CO)_2O$ [2496]
–   Maleic anhydride $(CHCO)_2O$ [2215]
–   Tetrahyrdrophthalic anhydride with more than 0.05% of maleic anhydride [2698]

■ **General information**

Acid anhydrides of general formula RCO–O–COR react with water, yielding the corresponding acid. While this reaction is rapid with acetic anhydride, it becomes very slow as the molecular mass of the anhydride increases. These are intermediates for organic synthesis. As an example, acetic anhydride is used for the fabrication of cellulose acetates.

■ Action on aluminium

□ Acetic anhydride (CH₃CO)₂O

At room temperature, the dissolution rate is 0.05 mm per year. At 50 °C, the decrease in thickness of aluminium is 0.50 mm per year, and 1.25 mm per year at the boiling point of pure anhydride (136 °C).

Mixtures of acetic acid and acetic anhydride at any ratio have only a very slight action on aluminium. At ambient temperature, the decrease in thickness is less than 0.05 mm per year [13]. At the boiling point of mixtures containing more than 10 wt% of acetic acid, the attack of aluminium becomes significant.

Acetic anhydride will be neither altered nor discoloured in contact with aluminium. Exchangers and reaction vessels in aluminium are commonly used in production units for this product. It is stored and transported in aluminium drums and tanks.

□ Propionic anhydride (C₃H₅O)₂O

Propionic anhydride (C₃H₅O)₂O has no action on aluminium up to 80 °C, even when it contains a high proportion of propionic acid: the decrease in thickness is less than 0.05 mm per year.

At the boiling point (168 °C), a very severe attack of aluminium is observed when the anhydride contains a small proportion of propionic acid (less than 2%).

It is stored and transported in aluminium tanks.

□ Butyric anhydride (C₄H₇O)₂O

Butyric anhydride (C₄H₇O)₂O only has a very moderate action on aluminium at room temperature: the decrease in thickness does not exceed 50 μm per year.

Mixtures of butyric anhydride and butyric acid do not have a stronger action on aluminium at room temperature than propionic anhydride. The same applies at the boiling point, whatever the acid concentration, under the condition that the mixture contains at least 0.25% water. Otherwise, in the absence of humidity, a very severe attack of aluminium is observed.

Butyric anhydride may be stored and transported in vessels and tanks made in aluminium.

□ Other anhydrides

Higher anhydrides such as valeric anhydride (C₄H₉CO)₂O and caproic anhydride (C₅H₁₁CO)₂O, which are used as fragrances and in the synthesis of certain plasticisers, have no action on aluminium.

Maleic anhydride $(CHCO)_2O$ has no action on aluminium and can be stored in liquid form (melting point at 56 °C) in aluminium vessels.

Succinic anhydride $(CH_2CO)_2O$ has no action on aluminium.

Benzoic anhydride $(C_6H_5CO)_2O$, used for the production of certain plastics, has no action on aluminium. The same applies to phthalic anhydride $(C_6H_4(CO)_2)O$, which is transported and stored, even in liquid form (melting point at 130 °C), in aluminium alloy vessels. These products are not discoloured in contact with aluminium.

Heptanoic acid $(CH_3(CH_2)_5CO)_2O$, used in brake fluids and lubricants, has no action on aluminium if the elastomers of the joints of the brake components are exempt of certain charges, in particular graphite.

6.3. SALTS OF ORGANIC ACIDS

■ ADR numbers

–   Mercurous acetate $(C_2H_8O_2)_2Hg_2$ [1629]
–   Mercuric acetate $(C_2H_8O_2)_2Hg$ [1629]
–   Lead acetate [1616]
–   Mercury benzoate $(C_6H_5CO_2)_2Hg,\cdot H_2O$ [1631]

■ General information

Salts of organic acids have the general formula

$$(ROO)_nM$$

where M is a metal, most commonly sodium.

They are used as intermediates for synthesis, as mordants in dyeing, as pharmaceuticals, as cleaning agents; some are insecticides.

Soluble organic salts are dissociated in water according to the reaction

$$(ROO)_nM \rightarrow n(ROO)^- + M^{n+}$$

Whether they are salts of a strong base (sodium or potassium hydroxide) or of a weak base (ammonium), as is frequently the case, their hydrolysis results in a slightly alkaline medium (pH 7–9), according to their nature and to the acetate concentration.

Stearates $(C_{18}H_{35}O_2X)$, oleates $(C_{18}H_{33}O_2X)$, and palmitates $(C_{15}H_{31}O_2X)$ are powders, mostly unctuous, and only slightly soluble in water. Stearates of aluminium, barium, calcium, sodium and zinc, used for impermeabilisation, as polymerisation agents and lubricants, have no action on aluminium. Zinc stearate is used as a lubricant for the drawing of aluminium wire.

Soaps are sodium or potassium salts of higher fatty acids based on animal fats (tallow, etc.) or palm or coconut oil. Their solutions are slightly alkaline.

■ Action on aluminium

□ Alkaline salts

Alkaline salts with sodium, potassium and ammonium in general have only a very moderate action on aluminium at room temperature, which leads to a decrease in thickness in the order of 50 μm per year.

□ Salts of heavy metals

Solutions of inorganic salts such as copper or tin are aggressive due to the presence of ions of heavy metals that are reduced in contact with aluminium; this leads to pitting corrosion of aluminium. This can be observed with the stearates of copper $Cu(C_{18}H_{35}O_2)_2$, mercury $Hg(C_{18}H_{35}O_2)_2$ and lead $Pb(C_{18}H_{35}O_2)_2$.

The products are very insoluble in water, which reduces the risk of pitting corrosion of aluminium.

□ Formates

Aluminium formate $Al(HCOO)_3$ has no action on aluminium. In solutions up to a concentration of 20% and up to 80 °C, it does not attack aluminium.

Formates with ammonium NH_4COOH, sodium $NaCOOH$ and potassium $KCOOH$ also have no action on aluminium at room temperature.

Formates with copper $Cu(COOH)_2 \cdot 4H_2O$, cobalt $Co(COOH)_2 \cdot 2H_2O$, mercury $Hg(COOH)_2$ and lead, which are used as antiseptics, attack aluminium because of the presence of heavy metals in the molecules.

□ Acetates

Diluted acetate solutions of sodium acetate CH_3CO_2Na, ammonium acetate $CH_3CO_2NH_4$ and potassium acetate CH_3CO_2K have no action on aluminium at room temperature. At temperatures between 60 and 70 °C, the metal blackens and will suffer superficial pickling: the degree of pickling depends on the acetate. For example, the dissolution rate in a solution of 10% potassium acetate amounts to 1.25 mm per year.

More concentrated solutions, with a concentration of 20% or more, lead to superficial attack, whatever the temperature. Sodium silicates inhibit this attack.

These acetates may be prepared in aluminium equipment [13].

Acetates

–   Calcium acetate $Ca(CH_3CO_2)_2$,
–   magnesium acetate $Mg(CH_3CO_2)_2$,
–   barium acetate $Ba(CH_3CO2)_2$,
–   basic aluminium acetates $(CH_3CO_2)Al_2O_3 \cdot 4H_2O$ and $(CH_3CO_2)_2AlOH$,
–   zinc acetate $Zn(CH_3CO_2)_2$

have no action on aluminium at room temperature.

When aluminium acetate $Al(CH_3CO2)_3$ is prepared from aluminium sulphate and lead acetate, it may lead to pitting corrosion of aluminium.

Acetates

–   Copper acetate, whether neutral $Cu(C_2H_3O_2)_2 \cdot H_2O$ or basic $2Cu(C_2H_3O_2)_2$,
–   cobalt acetate $Co(C_2H_3O_2)_2 \cdot 4H_2O$,
–   tin acetate $Sn(C_2H_3O_2)_2$,
–   nickel acetate $Ni(C_2H_3O_2)_2 \cdot 4H_2O$,
–   lead acetate $Pb_n(C_2H_3O_2)_m \cdot x\, H_2O$,
–   mercurous acetate $Hg_2(C_2H_3O_2)_2$,
–   mercuric acetate $Hg(C_2H_3O_2)_2$

are aggressive to aluminium, as is any salt of these metals.

■ Propionates

Propionates

–   sodium propionate $NaC_2H_5CO_2$,
–   calcium propionate $Ca(C_2H_5CO_2)_2$,
–   zinc propionate $Zn(C_2H_5CO_2)_2$

have no action on aluminium. Their solutions can be stored and transported in aluminium vessels.

■ Salts of fatty acids

Stearates

–   Aluminium stearate $Al(C_{18}H_{35}O_2)_3$,
–   barium stearate $Ba(C_{18}H_{35}O_2)_2$,

- calcium stearate $Ca(C_{18}H_{35}O_2)_2$,
- sodium stearate $NaC_{18}H_{35}O_2$,
- zinc stearate $Zn(C_{18}H_{35}O_2)_2$

have no action on aluminium.
Stearates

- Copper stearate $(C_{18}H_{35}O_2)_2Cu$,
- mercuric stearate $(C_{18}H_{35}O_2)_2Hg$,
- lead stearate $(C_{18}H_{35}O_2)_2Pb$

are very insoluble in water.

They can lead to pitting corrosion of aluminium because of the presence of heavy metals; however, because they are highly insoluble in water and the metal is bonded to a long organic chain, corrosion is not very aggressive.

Soaps lead to slightly alkaline solutions; they have no action on aluminium.

☐ Lactates and oxalates

Ammonium lactate $CH_3CHOH-COONH_4$,

- ammonium oxalate $(NH_4)_2C_2O_4 \cdot H_2O$,
- calcium oxalate $CaC_2O_4 \cdot H_2O$,
- potassium oxalate $K_2C_2O_4 \cdot H_2O$,
- sodium lactate $CH_3CHOHCOONa$,
- sodium oxalate $Na_2C_2O_4$,
- sodium bitartrate $C_4H_5O_6Na$

whether dry or in solution, have no action on aluminium at room temperature.

Boiling solutions can lead to more or less substantial pickling, depending on the nature and the concentration of the salt. This can be prevented by adding a corrosion inhibitor to the solution such as sodium silicate.

☐ Benzoates and naphthenates

Sodium benzoate $C_6H_5CO_2Na$, whether dry or in solution, has no action on aluminium.

Zinc naphthenate has no action on aluminium. Lead and copper naphtenates can lead to pitting corrosion of aluminium.

☐ **Sodium trichloroacetate**

Sodium trichloroacetate CCl_3CO_2Na, a weed-killer, is highly soluble in water. These solutions will severely attack aluminium at any concentration

☐ **Tetraethyl lead**

In the absence of humidity, tetraethyl lead $Pb(C_2H_5)_4$ has no action on aluminium. In the presence of humidity, the product becomes aggressive to aluminium and its alloys. Its addition to gasoline as an anti-knocking agent does not modify the resistance of aluminium to gasoline.

6.4. ESTERS

■ **ADR numbers**

– Methyl formate HCO_2CH_3 [1243]
– Ethyl formate $HCO_2C_2H_5$ [1190]
– Allyl formate $HCO_2C_3H_5$ [2336]
– Propyl formate $HCO_2C_3H_7$ [1281]
– Isobutyl formate $HCO_2C_4H_9$ [2393]
– *n*-butyl formate $HCO_2C_4H_9$ [1128]
– Amyl formate $HCO_2C_5H_{11}$ [1109]

– Ethylbutyl acetate $CH_3CO_2C_6H_{13}$ [1177]
– Methylamyl acetate $CH_3CO_2C_6H_{13}$ [1233]
– Methyl acetate $CH_3CO_2CH_3$ [1231]
– Ethyl acetate $CH_3CO_2C_2H_5$ [1173]
– *n*-propyl acetate $CH_3CO_2C_3H_7$ [1276]
– Isopropyl acetate $CH_3CO_2(i\text{-}C_3H_7)$ [1220]
– Vinyl acetate, stabilised $CH_3CO_2C_2H_3$ [1301]

– Methyl proprionate $C_2H_5CO_2CH_3$ [1248]
– Ethyl proprionate $C_2H_5CO_2C_2H_5$ [1195]
– Isopropyl proprionate $C_2H_5CO_2C_3H_7$ [2409]
– Butyl proprionate $C_2H_5CO_2C_4H_9$ [1914]
– Isobutyl proprionate $C_2H_5CO_2C_4H_9$ [2394]
– Methyl butyrate $C_3H_7CO_2CH_3$ [1237]
– Ethyl butyrate $C_3H_7CO_2C_2H_5$ [1180]
– Isopropyl butyrate $C_3H_7CO_2C_3H_7$ [2405]

- Amyl butyrate $C_3H_7CO_2C_5H_{11}$ [2620]
- Vinyl butyrate, stabilised $C_3H_7CO_2C_2H_3$ [2838]
- Methyl methacrylate, stabilised $C_3H_5O_2CH_3$ [1247]
- Ethyl methacrylate, stabilised $C_3H_5O_2C_2H_5$ [2277]
- Isobutyl methacrylate, stabilised $C_3H_5O_2C_4H_9$ [2283]

■ General information

Esters of the general formula R_1COOR_2 result from the reaction of an organic acid with an alcohol, or of an organic acid chloride with an alcohol.

Three main families of esters can be distinguished, which include many important industrial and natural products:

- Fruit esters are liquids with medium molecular weight that have a pleasant smell and taste. They make up many important natural or synthetic products used for fragrances or in drinks. For example, isoamyl acetate $CH_3COOC_2H_5$ is used in the formulation of pineapple, pear, raspberry, etc. fragrances;
- fats are glycerol esters of medium or higher carboxylic acids,
- waxes are inert products.

■ Action on aluminium

□ Formates

Methyl formate HCO_2CH_3 does not attack aluminium. This has been shown by immersion tests over 4 months with several alloys such as 1050A, 6062 and 5754. Only a slight blackening of the metal was observed. The presence of traces of water does not modify the resistance of aluminium. Methylformate may be stored and transported in aluminium vessels.

Ethylformate $HCO_2C_2H_5$ has a very similar action on aluminium.

The other esters of formic acid

- Propyl formate $HCO_2C_3H_7$,
- allyl formate $HCO_2C_3H_5$,
- isobutyl formate $HCO_2C_4H_9$,
- *n*-butyl formate $HCO_2C_4H_9$,
- amyl formate $HCO_2C_5H_{11}$,
- octyl formate $HCO_2C_5H_{17}$,
- cyclohexyl formate $HCO_2C_6H_{11}$,
- bornyl and linalyle formate $HCO_2C_{10}H_{17}$,

-     menthyle and citronellyl formate $HCO_2C_{10}H_{19}$,
-     vinyl formate $HCO_2C_2H_3$,
-     butylene glycol formate $HCO_2C_4H_8OH$,
-     ethylene glycol formate $HCO_2C_2H_4OH$

have no action on aluminium.

These products are stored and transported in aluminium vessels and do not suffer any discolouration in contact with aluminium [15].

□ Acetates

Methyl acetate $CH_3CO_2CH_3$ has no action on aluminium, even at the boiling point (57 °C). The same applies to ethyl acetate, which boils at 77 °C.

Tests over 3 months on 1100 and 5754 immersed in ethyl acetate containing 2% of thioglycolic acid have shown that this mixture has no action on aluminium. The decrease in thickness is below 1 μm per year.

In the reaction mixture used to produce ethyl acetate

-     ethyl alcohol: 30%,
-     acetic acid: 34%,
-     ethyl acetate: 34%,
-     water: 1%,
-     sulphuric acid: 0.5%,

the decrease in thickness of 3003 amounts to 0.2 mm per year in the temperature range from 100 to 150 °C.

Equipment made in aluminium alloy is used for the manufacture, storage, packaging and transportation of ethyl acetate.

Higher acetates

-     Propyl acetate $CH_3CO_2C_3H_7$,
-     butyl acetate $CH_3CO_2C_4H_9$,
-     amyl acetate $CH_3CO_2C_5H_{11}$,
-     hexyl acetate $CH_3CO_2C_6H_{13}$,
-     heptyle acetate $CH_3CO_2C_7H_{15}$,
-     octyl acetate $CH_3CO_2C_8H_{17}$,
-     nonyl acetate $CH_3CO_2C_9H_{19}$,
-     decyl acetate $CH_3CO_2C_{10}H_{21}$,
-     lauryl acetate $CH_3CO_2C_{12}H_{25}$,
-     cyclohexyl acetate $CH_3CO_2C_6H_{11}$,

- bornyl acetate $CH_3CO_2C_{10}H_{17}$,
- menthyl acetate or citronellyl acetate $CH_3CO_2C_{10}H_{19}$,
- vinyl acetate $CH_3CO_2C_2H_3$,
- ethylene glycol acetate $CH_3CO_2C_2H_4OC_2H_5$,
- butyl glycol acetate $CH_3CO_2C_2H_4OC_4H_9$

have no action on aluminium at room temperature and even at temperatures close to the boiling point of these products.

Aluminium equipment is used for the production, storage and transportation of these esters, which do not suffer any alteration in contact with aluminium.

☐ Propionates

The most common propionates, which are used in the formulation of solvents for resins, lacquers and fragrances

- methyl propionate $C_3H_7CO_2CH_3$,
- ethyl propionate $C_3H_7CO_2C_2H_5$,
- propyl propionate $C_3H_7CO_2C_3H_7$,
- butyl propionate $C_3H_7CO_2C_4H_9$,
- amyl propionate and isoamyl propionate $C_3H_7CO_2C_5H_{11}$,
- cyclohexyl propionate $C_3H_7CO_2C_6H_{11}$,
- heptyl propionate $C_3H_7CO_2C_7H_{15}$,
- octyl propionate $C_3H_7CO_2C_8H_{17}$,
- geranyl (or linalyl) propionate $C_3H_7CO_2C_{10}H_{16}$

have no action on aluminium.

Aluminium equipment is used: reaction vessels, heat exchangers and condensers for the fabrication, storage and packaging of these products, which do not suffer any alteration in contact with aluminium and aluminium alloys.

☐ Butyrates

The most common butyrates

- methyl butyrate $C_3H_7CO_2CH_3$,
- ethyl butyrate $C_3H_7CO_2C_2H_5$,
- propyl butyrate $C_3H_7CO_2C_3H_7$,
- butyl butyrate $C_3H_7CO_2C_4H_9$,
- amyl butyrate $C_3H_7CO_2C_5H_{11}$,

- cyclohexyl butyrate $C_3H_7CO_2C_6H_{11}$,
- octyl butyrate $C_3H_7CO_2C_8H_{17}$,
- geranyl butyrate (also called bornyl butyrate) $C_3H_7CO_2C_{10}H_{17}$,
- citronellyl butyrate $C_3H_7CO_2C_{10}H_{19}$,
- vinyl butyrate $C_3H_7CO_2C_2H_3$,
- glycol butyrate $C_3H_7CO_2C_2H_4OH$

have no action on aluminium.

Equipment in aluminium such as reaction vessels, heat exchangers and condensers is used for the fabrication, storage and packaging of these products, which do not suffer any alteration in contact with aluminium and aluminium alloys.

☐ Esters of fatty acids

The esters of higher fatty acids

- methyl caproates $C_5H_{11}CO_2CH_3$,
- ethyl caproates $C_5H_{11}CO_2C_2H_5$,
- butyl caproates $C_5H_{11}CO_2C_4H_7$,
- amyl caproates $C_5H_{11}CO_2C_5H_{11}$,

- methyl caprylates $C_7H_{15}CO_2CH_3$,
- ethyl caprylates $C_7H_{15}CO_2C_2H_5$,
- amyl caprylates $C_7H_{15}CO_2C_5H_{11}$,

- methyl laurates $C_{11}H_{23}CO_2CH_3$,
- ethyl laurates $C_{11}H_{23}CO_2C_2H_5$,
- butyl laurates $C_{11}H_{23}CO_2C_4H_9$,
- amyl laurates $C_{11}H_{23}CO_2C_5H_{11}$,

- methyl palmitates $C_{15}H_{31}CO_2CH_3$,
- ethyl palmitates $C_{15}H_{31}CO_2C_2H_5$,
- butyl palmitates $C_{15}H_{31}CO_2C_4H_9$,
- cetyl palmitates (also called spermaceti),

- methyl stearate $C_{17}H_{35}CO_2CH_3$,
- ethyl stearate $C_{17}H_{35}CO_2C_2H_5$,
- butyl stearate $C_{17}H_{35}CO_2C_4H_9$,
- amyl stearate $C_{17}H_{35}CO_2C_5H_{11}$

have no action on aluminium, neither at room temperature, nor at higher temperature.

These liquids have a nice smell and are essences of fruits and fragrances, solvents, plasticisers and intermediates for synthesis. They are produced, stored and packed in aluminium equipment, and do not suffer any alteration in contact with aluminium.

☐ **Other esters**

Esters of ethylenic acids such as

- methyl methacrylate $C_3H_5CO_2CH_3$,
- ethyl methacrylate $C_3H_5O_2CC_2H_5$,
- propyl methacrylate $C_3H_5CO_2C_3H_7$,
- butyl methacrylate $C_3H_5CO_2C_4H_9$

are used as plasticisers in the formulation of lacquers and resins. They have no action on aluminium. They can be stored and transported in aluminium vessels.

Esters of acid alcohols and diacids such as

- methyl lactates $C_3H_5CO_2CH_3$,
- ethyl lactates $C_3H_5CO_2C_2H_5$,
- butyl lactates $C_3H_5CO_2C_4H_9$,

- ethyl adipates $C_3H_5CO_2C_2H_5$,
- butyl adipates $C_3H_5CO_2C_4H_9$,
- amyl adipates $C_3H_5CO_2C_5H_{11}$,

- diethyl malonate $CH_2(C_2H_5CO_2)_2$,
- diethyl oxalate $(C_2H_5CO_2)_2$,
- methyl salicylate $HOC_6H_4CO_2CH_3$

have no action on aluminium, even at relatively high temperatures.

For example, butyl lactate leads only to slight pickling, corresponding to a decrease in thickness of 50 μm per year at 200 °C. Methyl oxalate has a similar effect at its boiling point (186 °C).

The three acetines, which are glycerine acetates

- monoacetin $CH_3COOC_3H_5(OH)_2$,
- diacetin $(CH_3COO)_2C_3H_5OH$,
- triacetin $(CH_3COO)_3C_3H_5$

have no action on aluminium, which is used for the preparation, storage and packaging of these esters.

☐ Cellulose esters

The most common ones are

–   cellulose acetate $[C_6H_7O_2(OCOCH_3)_x]_n$ with $1.5 < x < 3$,
–   cellulose triacetate $[C_6H_7O_2(OCOCH_3)_3]_n$

These esters have many applications: textile fibres, photographic films, fillers, coatings, paints, and lacquers. Whether as powders or in solution with appropriate organic solvents such as acetates, these cellulosic esters have no action on aluminium.

Aluminium alloys have an excellent resistance to the action of reagents used for the fabrication of cellulose acetate: acetic anhydride, concentrated acetic acid, butyric anhydride, and cotton. Many pieces of equipment in various aluminium alloys are used in production units for cellulose esters: 3003, 5052, A-S13 (44100), etc., for storage tanks, reaction vessels, dryers, evaporators, and valves.

☐ Aromatic esters

Aromatic esters such as

–   methyl benzoate $C_6H_5CO_2CH_3$,
–   ethyl benzoate $C_6H_5CO_2C_2H_5$,
–   propyl benzoate $C_6H_5CO_2C_3H_7$,
–   butyl benzoate $C_6H_5CO_2C_4H_9$,
–   amyl benzoate $C_6H_5CO_2C_5H_{11}$,
–   citronellyl benzoate $C_6H_5CO_2C_{10}H_{19}$

are used in perfumery and pharmaceutical products. They have no action on aluminium, which is used for the storage and bottling of these esters.

The same applies to esters such as

–   anisyl formate $HCO_2-C_6H_4OCH_3$,
–   benzyl formate $HCO_2-CH_2C_6H_5$,
–   phenyl formate $HCO_2-C_6H_5$,

–   anisyle acetate $CH_3CO_2-C_6H_4OCH_3$

– benzyl acetate $CH_3CO_2-CH_2C_6H_5$,
– phenyl acetate $CH_3CO_2-C_6H_5$,

– methyl naphthenates $C_{10}H_7OCH_3$,
– ethyl naphhtenates $C_{10}H_7OC_2H_5$,
– butyl naphthenates $C_{10}H_7OC_6H_5$.

☐ **Waxes**

Waxes mainly contain esters of higher monocarboxylic acids and higher monoalcohols. They always contain some free acid, and even higher hydrocarbons. As an example, the palmitic ester of myricyl alcohol is the main constituent of beeswax, which further contains 10–14% of ceritic acid and 12–17% of higher hydrocarbons.
These products are

– of plant origin:

• turpentines, obtained by spontaneous or induced exudation from certain conifers,
• pinene or terebene, a terpenic hydrocarbon used for the manufacture of many products such as synthetic camphor,
• carnauba wax
• copal resin,
• gum tragacanth, used in therapeutics and in tanning,
• gum arabic, which has many applications as a dressing of tissues and for the fabrication of glues,
• gamboge, used in therapeutics, and the fibres of which are used for the manufacture of ropes;

– of animal origin:

• beeswax.

Aluminium equipment is used for collecting gums, pitch and resins, as well as for the distillation, purification, storage and packaging of these products.

☐ **Fat**

Fats such as solid fat and oil are always glycerines, i.e. esters of glycerol trialcohol and medium and higher fatty acids.

These fats can be

- of *plant origin* such as colza oil, peanut oil, palm oil, cacao butter, or mixtures such as margarine,
- of *animal origin*, such as butter, tallow, lard.

So-called siccative oils such as linseed oil and hemp oil (generally not food-compatible), are transformed into resins under the influence of the oxygen contained in air. They are used for the preparation of varnish, paints, etc.

Aluminum is not attacked by these fats. Even after prolonged contact with aluminium, they are neither altered nor discoloured, and their taste does not change [16].

6.5. AMIDES

■ ADR numbers

- *N,N*-Dimethylformamide $HCON(CH_3)_2$ [2265]
- Acrylamide $CH_2CHCONH_2$ [2074]

■ General information

One distinguishes

- primary amides, of the general formula R_1CONH_2,
- secondary amides, of the general formula R_1CONHR_2,
- tertiary amides, of the general formula $R_1CONR_2R_3$.

Only formamide is a liquid at room temperature; the other amides are solids. The first amides (formamide, acetamide) are water soluble; their solubility rapidly decreases with increasing molecular weight. Amide solutions have a pH close to neutral.

■ Action on aluminium

The main amides

- formamide $HCONH_2$,
- dimethylformamide $HCON(CH_3)_2$,
- acetamide CH_3CONH_2,
- butylcetamide $CH_3CONHC_4H_9$,

–   acrylamide $CH_2CHCONH_2$,
–   benzamide $C_6H_5CONH_2$

have no action on aluminium, even at high temperatures.

Aluminium equipment is widely used for the production, purification, storage or packaging of these products such as

–   equipment for the production of formamide,
–   heat exchangers used in the distillation and transport of dimethylformamide (DMF),
–   equipment for the treatment of benzamide,
–   equipment for the production and purification of acetanilide.

6.6.   NITRILES AND NITROUS DERIVATIVES

■ ADR numbers

–   Acetonitrile CH_3CN [1648]
–   Acrylonitrile CH_2CHCN [1093]
–   Benzonitrile C_6H_5CN [2224]
–   Nitroethane $C_2H_5NO_2$ [2842]
–   Nitrobenzene $C_6H_5NO_2$ [1662]
–   Nitrotoluene (*o-*,*m-*,*p-*) $NO_2C_6H_4CH_3$ [1664]
–   Nitrophenol (*o-*,*m-*,*p-*) (picric acid) $(NO_2)_3C_6H_2OH$ [1664]
–   Nitrocresols $(CH_3O)(HO)C_6H_3NO_2$ [2446]
–   Nitrocellulose solution, flammable $C_{12}H_{16}N_4O_{18}$ [2059]
–   Nitroxylenes (*o-*,*m-*,*p-*) $(CH_3)_2C_6H_3NO_2$ [1665]

■ General information

Acyclic nitriles correspond to the general formula $R-C\equiv N$, and aromatic nitriles to the formula $Ar-C\equiv N$. These neutral substances are liquids at room temperature.

Acyclic nitriles are intermediates in the synthesis of dye stuffs and certain synthetic fibres (acrylonitrile).

■ Action on aluminium

Acyclic nitriles do not attack aluminium, even at high temperatures. They are prepared in aluminium vessels. The most common ones are

- acetonitrile CH_3CN,
- acrylonitrile CH_2CHCN,
- hydracrylonitrile HOC_2H_4CN.

Most aromatic nitriles also do not attack aluminium, for example:

- benzonitrile, C_6H_5CN,
- phtalonitrile $C_6H_4(CN)_2$

The following nitrous derivatives

- nitromethane CH_3NO_2,
- nitroethane $C_2H_5NO_2$,
- nitropropane $C_3H_7NO_2$,
- pentyl nitrate $C_5H_{11}ONO_2$,
- pentyl nitrite $C_5H_{11}NO_2$,
- nitrobenzene $C_6H_5NO_2$,
- nitrotoluene $NO_2C_6H_4CH_3$,
- nitrophenol (picric acid) $(NO_2)_3C_6H_2OH$,
- nitrocresol $NO_2CH_3-C_6H_3O_{11}$,
- nitroglycerin $C_3H_5(ONO_2)_3$,
- nitrobenzoic acid $NO_2C_6H_4COOH$,
- nitrocellulose $C_{12}H_{16}N_4O_{18}$

have no action on aluminium, even at high temperatures.

Aluminium resists nitric acid very well and does not trigger a spark on impact. It is widely used for the manufacture of reaction vessels used for the production of these explosive products, some of which are particularly unstable, such as nitroglycerin and nitrocellulose.

They are stored, transported and packed in aluminium equipment.

REFERENCES

[1] Leyerzapf H., *Formic acid. Corrosion Handbook*, vol. 4, Dechema, 1989.
[2] Zaritskii V.I.D., Corrosion and strenght behavior of aluminium in the formic acid–water system, *Fisiko Khimicheskaya Mekhanika Materialov*, no 2, 1990, p. 110–113.
[3] Elsner G., *Acetic acid. Corrosion Handbook*, vol. 6, Dechema, 1990.
[4] Swandby R.K., Corrosion charts. Guides to materials selection, *Chemical and Engineering*, vol. 69, 1962, p. 186.

[5] Skvortsov I.I., An investigation of the corrosion resistance of welded joints in aluminium, *Corrosion*, vol. 16, 1960, p. 159a.

[6] McKee A.B., Binger W.W., Using aluminum alloys with short chain aliphatic acids and anhydrides, *Corrosion*, vol. 13, 1957, p. 786t–792t.

[7] Ritter F., *Korrosionstabellen metallischer Werkstoffe*, 4th edition, Springer, Wien, 1957, p. 88.

[8] *Ullmanns Encyklopädie der technischen Chemie*, 4th edition, vol. 11, 1976, 57.

[9] Rabald E., *Essigsäure*, Dechema Werkstofftabelle, 1955.

[10] Friend W.Z., Mason J.F., Corrosion tests in the processing of soap and fatty acids, *Corrosion*, vol. 5, 1949, p. 355–368.

[11] Hasenberg L., *Alkanecarboxylic acids. Corrosion Handbook*, vol. 4, Dechema, 1989.

[12] L'Hoir G., Les grands réservoirs de stockage en aluminium. L'aluminium et les acides gras, *Revue de l'Aluminium*, no 95, 1937, p. 923–926.

[13] Ritter F., *Korrosionstabellen metallischer Werkstoffe*, 4th edition, Springer, Wien, 1957, p. 88.

[14] Hasenberg L., *Acetates. Corrosion Handbook*, vol. 1, Dechema, 1987.

[15] Hasenberg L., *Carboxylic acids esters. Corrosion Handbook*, vol. 10, Dechema, 1991.

[16] Loury M., Defromont C., Étude sur la corrosion des métaux par l'oléine, *Revue française des corps gras*, 1956, p. 436–444.

Chapter F.7
Other Organic Products

Chapter F.7
Other Organic Products

7.1. ALKALOIDS AND HETEROCYCLIC COMPOUNDS

■ ADR numbers

- Acridine $(C_6H_4)_2CHN$ [2713]
- Furan C_4H_4O [2389]
- Picolines C_6H_7N [2313]
- Pyridine C_5H_5N [1282]
- Quinoline C_9H_7N [2656]

■ General information

Alkaloids are nitrogen-containing organic compounds that are slightly alkaline. They are natural products and occur in plants.

Like heterocyclic compounds, they have a complex structure. They are used in pharmaceuticals and perfumery, and for the fabrication of certain dyestuffs, etc.

■ Action on aluminium

Acridine $C_6H_4CH-N-C_6H_4$ (which is the starting compound for the synthesis of certain dyestuffs),

- atropine $C_{17}H_{23}O_3$,
- bergaptol $C_{11}H_6O_4$,
- borneol $C_{10}H_{18}O$,
- caffeine $C_8H_{10}N_4O_2$,
- carone $C_{10}H_{16}O$,
- carotine $C_{40}H_5O$,
- carvone $C_{10}H_{14}O$,
- cocaine $C_{17}H_{21}NO_4$,
- coumarin (or 1,2-benzopyrone),
- fenchone $C_{10}H_{16}O$,
- furan C_4H_4O and its derivatives such as furfural $C_5H_4O_2$,
- indole C_8H_7N (or benzopyrrole),
- nicotine $C_{10}H_{14}N_2$ and its derivatives such as nicotine acid or nicotine sulphate,

–   morphine $C_{17}H_{19}NO_3$ and its derivatives such as codeine,
–   papaverine $C_{20}H_{21}O_4N$,
–   picolines C_6H_7N and their numerous derivatives used in perfumery,
–   pyridine C_5H_5N,
–   quinine $C_{20}H_{22}O_2N_2 \cdot 3H_2O$,
–   quinoline C_9H_7N,
–   thiophene C_4H_4S and its derivatives,

whether concentrated or in solution (aqueous or alcoholic), have no action on aluminium, even at high temperatures (100–150 °C).

Aluminium equipment is widely used for the extraction, distillation, packaging or storage of these products, which are not altered in contact with aluminium.

This list has been limited to cases cited in more than one publication. It is likely that a very large number of other alkaloids and heterocyclic compounds are prepared or stored in aluminium equipment.

Pure pyridine does not attack aluminium; its solutions are alkaline and lead to a very superficial pickling of aluminium. Equipment in aluminium is used for the extraction of pyridine. The azeotropic mixture water–pyridine (60:40) has only a very weak action on aluminium: the dissolution rate is in the order of 25 μm per year with alloys 3003, 5052 and 6061 [1].

Uric acid $C_5H_4N_4O_3$, although very insoluble in water, leads to slight pitting on aluminium.

Aluminium does not modify the properties of the numerous variants of penicillins, and does not destroy these substances.

7.2.   GLUCIDES

Glucides are very important in the food industry, but also as raw materials for the manufacture of numerous products such as glues, textile fibres, and wood pulp.

All "-oses" (glucoses and other sugars) have no action on aluminium, whatever their concentration in water, and whatever the temperature. A great deal of aluminium equipment is used in confectionery.

In addition, holoside sugars such as maltose, lactose, and saccharose have no action on aluminium.

Cellulose, starch and its derivatives, including methyl-cellulose, as well as cellulosic glues have no action on aluminium.

REFERENCE

[1] Somekh G.S., Water–pyridine a azeotrope is an excellent Rankine cycle fluid, *IECEC* 75, p. 1443–1447.

Part G

The Effect of Other Environments

Chapter G.1

Corrosion in Soil

Chapter G.1
Corrosion in Soil

The corrosion behaviour of aluminium in soil is not only a complex issue, but also an important one because of many relevant applications: cables for electricity and telecommunications, water and gas distribution grids, embeddings of street signs, street lamps and various supporting structures, etc.

Predicting the corrosion resistance of a metal and assessing the aggressiveness of a given soil are very difficult.

1.1. TYPES OF SOIL

The concept of soil is different for a geologist, an agronomist, a civil engineer and a corrosion expert. Indeed, soil can be very different, even at two locations only a few hundred meters apart, including natural soil in rural areas. The nature of a soil varies as a function of depth, while the nature of successive layers depends on the local geology. The uppermost layer is generally constituted by humus. Differences can be even more apparent in urban and industrial areas, where urbanisation often has deeply transformed the soil. Soil can partially consist in backfill.

For soil, as for waters, the corrosion expert needs to know its physicochemical characteristics in order to be able to establish a relationship between these characteristics and the level of aggressiveness towards a metal or an alloy.

1.1.1. Constituents of soil

Natural soil results from the crumbling of rocks over geological time. There are several types of constituents with specific physicochemical properties: clay, marl, limestone, sand, gravel, etc.

Artificial soils are constituted from backfill, industrial slag, mining residues, etc. The composition and structure of artificial soils often have nothing in common with that of natural soils, and they can vary substantially from one point to another, depending on their origin. Besides inorganic constituents, all organic constituents of plant and animal origin, which are very important in arable soil, need to be taken into account. Bacterial decomposition of organic matter forms humus.

1.1.2. *Physical chemistry of soil*

Soil is a very heterogeneous, more or less humid medium. The humidity level depends on the soil's nature and on the volume of precipitation, and thus on the local climate. Water is retained mainly by capillary action.

The main parameters of a soil are

– *Physical parameters*: the shape and size of inorganic constituents and their plasticity, on which both water drainage and aeration depend. Clays or silty soils have a fine texture that retains water; they are poorly drained and aerated. On the other hand, sandy soils are more aerated, and water is easily evacuated.
– *Chemical parameters*: mainly the composition of the water that soaks the soil, which depends on the climatic conditions and on the layers crossed by the water. The main inorganic constituents in water are:
  - cations: Na^+, K^+, Ca^{2+}, and Mg^{2+}
  - anions: Cl^-, NO_3^-, SO_4^{2-}.

The mineral content of water in soil varies between 0.5 and 1.5 $g \cdot l^{-1}$.

On industrial sites, even on former industrial sites, other inorganic elements can be found, the nature of which is related to the industrial activity.

Soil also contains gases: oxygen, nitrogen, and carbon dioxide, originating mainly from the decomposition of organic matter, etc.

– *Organic and bacteriological parameters*: natural soils can exhibit intense biological activity, mainly in the uppermost layer constituted from arable earth and humus, which is acidic. This layer contains organic acids, the nature of which is more or less well known [1].

A soil is characterised by a pH value and an electrical resistivity that is closely related to the nature of salts dissolved in the humidity. The pH also depends on the quantity of inorganic and organic acids, and on the carbon dioxide (CO_2) level, as well as on possible contamination by industrial or household wastewater. In general, soils have an acidic pH, between 3.5 and 4.5. The pH increases with depth (Figure G.1.1).

1.2. THE INFLUENCE OF THE NATURE OF SOIL ON THE CORROSION BEHAVIOUR OF ALUMINIUM

The corrosion resistance of a metal depends on several more or less related parameters:

– the water content,
– the structure of the soil,

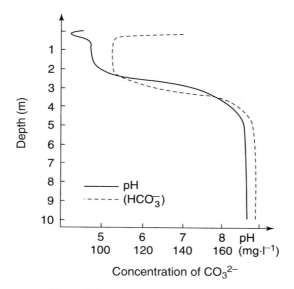

Figure G.1.1. pH as a function of depth [2].

- the concentration of dissolved oxygen, which depends on the depth and the structure of the soil: a sandy and permeable soil is more oxygenated by rain water that can infiltrate easily than a dense, clayey soil,
- the concentration of inorganic salts,
- the concentration of organic acids and organic products from the decomposition of animal and plant organisms,
- the pH, and
- the resistivity of the soil, which itself depends on the water content and the concentration of inorganic salts.

Experience has taught that the aggressiveness of soils is related to their resistivity (Figure G.1.2 and Table G.1.1). Experiments in Canada have led to the embedding of unprotected aluminium limited to 1500 $\Omega \cdot$cm [3].

Resistivity is not the only criterion for the aggressiveness of a soil. Its structure, its acidity and the organic matter that it contains must also be taken into account [4].

In spite of corrosion tests in many different soils over almost a century, based on a great number of samples of different metals, it has not been possible to come up with a relationship between the typology of soils and the corrosion resistance of metals [5]. Consequently, the corrosion resistance of aluminium in soils is also very difficult to predict, because of the great diversity in the composition of soils [6, 7].

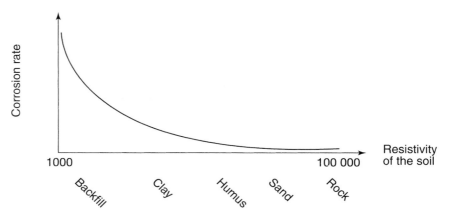

Figure G.1.2. Relationship between resistivity and aggressiveness of soils [8].

1.3.   FORM OF ALUMINIUM CORROSION IN SOILS

The corrosion of metals in soils is electrochemical in nature, as a result of the presence of water.

   In soil, unprotected aluminium can exhibit the following forms of corrosion:

–   pitting corrosion;
–   galvanic corrosion, if in contact with other metals: steel, copper, lead, etc. Often, severe galvanic corrosion is observed if a totally or partially embedded structure is earthed with a copper strap;
–   corrosion by stray currents (see Chapter G.2).

1.4.   ALUMINIUM CORROSION RESISTANCE IN SOILS

The use of unprotected aluminium is limited, mainly to North America where between 1948 and 1963, 626 km of pipeline made of aluminium alloys (of which 462 km were

Table G.1.1. Resistivity and aggressiveness of soils [9]

Resistivity (Ω·cm)	Aggressiveness to metals
<500	Very corrosive
500–1000	Corrosive
1000–2000	Moderately corrosive
2000–10 000	Slightly corrosive
>10 000	Less and less corrosive

totally unprotected) were buried: 6063, 3003, 3003–7072. They were used for the transportation of crude oil and water [10].

Aluminium is used unprotected for underground passages (under motorways or railways) as corrugated sheet covered with backfill. In spite of a few limited experiences, some successful, aluminium should not be buried without protection, especially cables for the transportation of electrical energy and telecommunications, pipes for water adduction, heat exchange circuits for heat pumps, etc.

1.5. PROTECTION AGAINST CORROSION IN SOIL

Several modes of protection can be envisioned:

–   *continuous extrusion sheathing with polymers*: PVC, polyethylene, etc. This mode of protection is used for electrical and telecommunication cables, pipes, and other products manufactured in long lengths. The sheath should not be damaged by stones or sharp objects when buried in a trench. In order to prevent these incidents, it is recommended to cover the sheath completely with a layer of sand before filling. A localised rupture of the protection becomes an anodic zone where the density of the corrosion current (or of a stray current) can be quite high, leading to severe corrosion that can severely damage the proper working condition of the installation;
–   *paint* is used, especially on components such as the foot of lamps, road signs, etc;
–   *cathodic protection* is normally used in addition to classic protections [11, 12].

Aluminium and its alloys, when placed on the ground, will undergo a slight surface attack, whose intensity depends on the nature of the soil (and the products that have been poured out there: fertilisers, plant-care products, etc.). Experience with aluminium tubes in irrigation installations shows that this corrosion in contact with soil is generally very superficial.

REFERENCES

[1] Elsner G., Jänsch-Kaiser G., Sharp D.H., *Soil underground corrosion. Corrosion Handbook*, vol. 1, Dechema, 1987.
[2] Matthess G., Pekdeger A., Chemical and biochemical reactions during ground water regeneration, *Wasser/Abwasser*, vol. 121, 1980, p. 214.
[3] Whiting J.F., Wright T.E., Tests to five indicate aluminum alloys pipe gives an economical service, *Materials Protection*, vol. 1, 1962, p. 36–46.
[4] Thomas R.F., Corrosion of metals in New Zealand soils, *Corrosion Australasian*, vol. 15, 1990, p. 8–11.

[5] Romanoff M., *Underground corrosion*, Rapport NBS Circulaire 579 du 01/04/57.

[6] Sprowls D.O., Carlisle M.E., Resistance of aluminum alloys to underground corrosion, *Corrosion*, vol. 17, 1961, p. 125t–132t.

[7] Latin A., Metallurgical considerations in the use of aluminium for cable shielding. *Metallurgia*, vol. 44, p. 167–173, 231–238.

[8] Haynes G.S., Baboian R., A comparative study of the corrosion resistance of cable-shielding materials, *Materials Performance*, vol. 18, 1979, p. 45–56.

[9] Haynes G.S., Hessler G., Gerdes R., Bow K., Baboian R., A method for corrosion testing of cable-shielding materials in soils, *ASTM STP*, 1989, p. 144–155.

[10] Wright T.E., New trends in buried aluminum pipelines, *Materials Performance*, Sept. 1976, p. 26–28.

[11] Snodgrass J.S., Soil corrosion of aluminum in underground electric plasma, *Corrosion NACE 75*, paper No. 130.

[12] Wright T.E., The corrosion behavior of aluminium pipe, *Materials Performance*, vol. 22, 1983, p. 9–12.

Chapter G.2

The Effect of Stray Currents and Alternating Current

Chapter G.2
The Effect of Stray Currents
and Alternating Current

2.1. STRAY CURRENTS

Stray currents are electrical currents, direct or alternating, that pass through the soil. They stem either from electrical lines from transportation systems such as railways or tramways or from fixed or mobile industrial installations such as welding posts. Stray currents can also arise from lightning [1]. They propagate preferentially through metallic casings of embedded electrical cables and telecommunication cables, pipelines and even metallic skeletons of buildings, which are all paths of low resistance.

A metallic structure embedded in the soil can suffer corrosion due to stray currents; the humidity of the soil acts as the electrolyte. The higher the mineralisation of the humidity, the more severe the corrosion [2].

Stray currents lead to localised corrosion in the form of craters at the zone where they leave the metallic structure in the soil: the smaller the surface area of the exit zone, the more severe the localised corrosion is. On aluminium, corrosion can occur whatever the polarity of the current that flows from the metal into the soil [3].

When a very long, unprotected or poorly protected embedded structure such as a pipeline crosses different types of soil, the dissolution potentials of the metal with respect to the soil are not consistently the same. This leads to the circulation of currents, which results in localised corrosion at the exit zones into the soil. This is also observed with immersed or semi-immersed structures such as ship hulls. For this reason, the return current should not flow through the hull, as one would be inclined to do on a small craft with battery-powered electric equipment. One conductor for each polarity is required if the system is distributing direct current, and one conductor per phase (plus one for the neutral, if required) for alternating current.

2.2. ALTERNATING CURRENTS

The corrosive effects of industrial alternating current of 50 or 60 Hz on embedded lead and steel have been known since 1907 [4]. Embedded or immersed aluminium can undergo pitting corrosion under the influence of alternating current [5], according to a mechanism proposed by French [6].

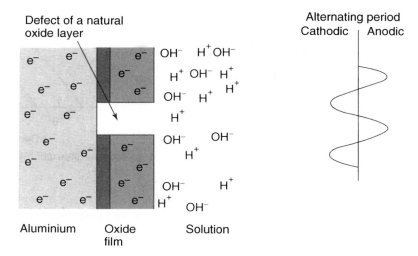

Figure G.2.1. Surface of aluminium in a humid medium under the influence of an alternating current [6].

Figure G.2.1 represents an aluminium surface in a humid environment subjected to an alternating current. During the anodic period, i.e. the one that leads to the displacement of Al^{3+} ions from the metal into the solution, Al^{3+} ions will hydrolyse according to the reaction:

$$2Al^{3+} + 3H_2O + 6e^- \rightarrow Al_2O_3 + 3H_2$$

and form a gelatinous alumina film. Simultaneously, OH^- ions are pushed towards the metal's surface, which becomes alkaline (Figure G.2.2).

During the cathodic period (Figure G.2.3), H^+ protons are pushed towards the surface of the metal, while hydroxyl ions will take the inverse path. Since the natural oxide film is

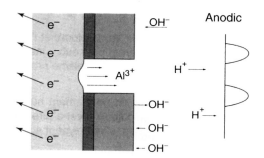

Figure G.2.2. Action of alternating current, anodic period.

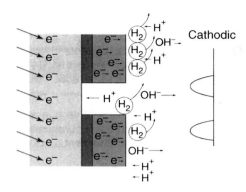

Figure G.2.3. Action of alternating current cathodic period.

an n-type semiconductor, it will supply the electrons necessary for reduction protons. This leads to the depletion of H^+ in the solution in contact with aluminium:

$$2H^+ \rightarrow H_2$$

This leads to an excess of hydroxyl OH^- ions, and thus to local alkalinisation. As a result, after two full periods, the metal surface will be alkaline, dissolving the natural oxide film

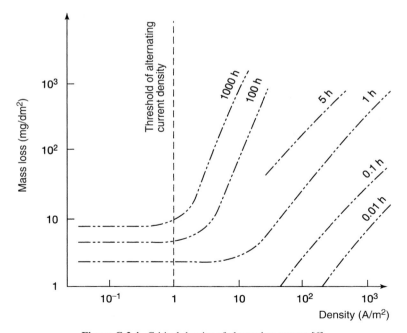

Figure G.2.4. Critical density of alternating current [6].

according to the equation:

$$Al_2O_3 + 2OH^- \rightarrow 2AlO_2^- + H_2O$$

The corrosion of aluminium is not directly proportional to the intensity of the alternating current. There is a threshold current density at which corrosion is initiated (Figure G.2.4). This threshold depends on the exposure time: it is approximately 1 $A \cdot m^{-2}$ for a duration of at least 100 h, and on the order of 100 $A \cdot m^{-2}$ for a duration of a few minutes; these periods of time are cumulative. Other tests have confirmed the existence of a threshold of the current density that is 10 times higher: 10 $A \cdot m^{-2}$ [7]. These results show that it is possible to delay corrosion resulting from industrial alternating currents of 50 or 60 Hz by increasing the noninsulated surface.

REFERENCES

[1] Smith C.A., Mech F.I., Soil in the corrosion process, *Anti-Corrosion*, Feb. 1981, p. 5–8.
[2] Al Abbassi A.M., Basu S., Aluminium lamp posts: corrosion problems and solutions, *Fifth International Aluminium Extrusion Technology Seminar*, Chicago, May 1992, p. 503–510.
[3] Sprowls D.O., Carlisle M.E., Resistance of aluminum alloys to underground corrosion, *Corrosion*, vol. 17, 1961, p. 125t–132t.
[4] Hayden J.L.R., *AIEE Transactions*, vol. 26, part I, 1907.
[5] French W.H., Lightner R.B., Aluminium underground electrical system, *International Conference on Aluminium Industrial Products*, Pittsburgh, October 1973.
[6] French W.H., Alternating current corrosion of aluminum, *IEEE Transactions Power Apparatus and Systems*, vol. 92, 1973, p. 2053–2062.
[7] Serra E.T., de-Araujo M.M., Mannheimer W.A., On the influence of alternating current on the corrosion of aluminium and copper in contact with the soil, *Corrosion 79*, paper No. 55.

Chapter G.3

Fertilisers and Herbicides

Chapter G.3
Fertilisers and Herbicides

3.1. FERTILISERS

Fertilisers are mixtures of inorganic salts, the most common ones of which are

- ammonium nitrate $(NH_4)NO_3$,
- potassium nitrate KNO_3,
- ammonium phosphates $NH_4H_2PO_4$ and $(NH_4)_2HPO_4$,
- ammonium chloride NH_4Cl,
- calcium dihydrogen phosphate $Ca(H_2PO_4)_2 \cdot H_2O$
- calcium hydrogen phosphate $CaHPO_4 \cdot 2H_2O$.
- potassium chloride KCl, and
- potassium sulphate K_2SO_4.

Their action on aluminium is discussed in Chapter E.6.
There are several types of fertilisers:

- *complex* or *ternary* fertilisers, which comprise a variable proportion of salts of three base elements:
  - N (nitrogen),
  - P (phosphoros),
  - K (potassium).
  They are designated by three figures expressing, in that order, the percentage of each of the elements N, P and K. For example, a fertiliser 18-22-18 comprises 18% nitrogen, 22% phosphoros, and 12% potassium. An absent element is designated by a zero, so the fertiliser 24-24-0 contains no phosphoros.
- *superphosphates*, with a phosphate content in excess of 20%, such as 24-24-0, 18-22-12, etc.
- *superpotassic fertilisers*, rich in potassium chloride,
- nitrate fertilisers based on ammonium nitrate (also called ammononitrate), ammonium sulphate and urea.

Fertilisers can be:

- *solids*: they are present as granules with a diameter of about 2–5 mm.
- *liquids*: these are solutions, as concentrated as possible, and of high density (1.3–1.4).

The most commonly used liquids are solutions of nitrate fertilisers (containing between 36 and 40 kg nitrogen per hectolitre), but solutions of superphosphate 14-48-0 (14 kg nitrogen and 48 kg phosphoros per hectolitre) are also used. These solutions are generally slightly acidic (pH between 6 and 7).

The resistance of aluminium and aluminium alloys depends on the form of the fertiliser and on its composition.

3.1.1. Solid fertilisers

Solid fertilisers are more or less hygroscopic and deliquescent, depending on their composition. In the presence of humidity, the prolonged contact with the granules or fertiliser dust can lead to a pitting attack on aluminium. This attack is slight with nitrate fertilisers, and the more severe the attack is, the richer the fertiliser is in phosphoros. Superphosphates are very aggressive.

The prolonged storage of aluminium equipment such as irrigation tubes, should be made in a location that is free from fertiliser dust. Otherwise, severe attack of the equipment can be observed during storage.

3.1.2. Liquid fertilisers

Solutions of complete fertilisers (NKP) increase in aggressiveness when their phosphate (P) content is higher. Aluminium will suffer uniform pickling and possibly pitting corrosion.

Solutions of nitrate fertilisers, even highly concentrated, have no action on aluminium, as long as their pH is maintained between 6 and 7 [1]. The annual decrease in thickness of 3003 in contact with a solution containing 400 $g \cdot l^{-1}$ of nitrogen is below 10 μm, even at slightly elevated temperatures of 40–50 °C (Figure G.3.1).

Urea solutions, even concentrated, have only a very weak action on aluminium. For example, with a solution of 51% urea, the annual decrease in thickness amounts to a few micrometers.

3.1.3. Aluminium applications

The good resistance of aluminium to nitrate and ammonia environments makes it useful in production plants, namely for buildings, as internal cladding of granulation towers, etc. [2].

Granulated fertilisers are transported in bulk in aluminium tippers. Solutions of nitrate fertilisers can be transported and stored in aluminium tanks.

When the contact is intermittent, as in irrigation equipment made of aluminium, the shorter the contact time of the fertiliser solution, the lesser is the risk of pitting corrosion. Nevertheless, these installations should be rinsed carefully after being used with fertiliser solutions.

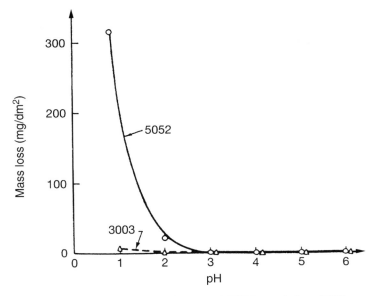

Figure G.3.1. Influence of the pH on the dissolution rate of 3003 in a solution of 83% ammonium nitrate at 90 °C [1].

3.2. PLANT-CARE PRODUCTS

The scope of this book does not allow listing all plant-care products that have been tested in contact with aluminium alloys [3, 4].

Products containing copper salts or other heavy metals (cobalt, mercury, etc.) can lead to severe pitting corrosion of aluminium.

REFERENCES

[1] Wyma B.H., Wagner R.H., Aluminium use in the fertilizer industry, *Chemical Engineering Progress*, vol. 60, 1964, p. 55–62.
[2] The Aluminum Association, *Aluminium for storage of nitrogen fertilizer solutions*, 1970.
[3] Marshall T., Neubauer L.G., Corrosion of aircraft structural materials by agricultural chemicals I. Laboratory tests with fertilizers compounds, *Corrosion*, vol. 11, 1955, p. 84t–92t.
[4] Schreiber C.F., Corrosion of aircraft structural materials by agricultural chemicals II. Effect of insecticides, herbicides, fungicides and fertilizers, *Corrosion*, vol. 11, 1955, p. 119t–130t.

Chapter G.4

Construction Materials

Chapter G.4
Construction Materials

The use of aluminium in the building industry puts it in contact with most materials used for constructions: concrete, plaster, wood, polymers, etc. Aluminium is also used for the transportation of concrete, as coffering material [1] and for levelling rulers for screed and walls.

4.1. THE EFFECT OF CONCRETE

Aluminium resists contact with concrete and mortar very well, in spite of their highly alkaline character, with a pH close to 12. When concrete starts setting, there will always be a slight pickling less than 30 μm thick. This attack, however, ceases after a few days of contact (Figure G.4.1). It leads to a very localised decrease in pH (to 8) and to the formation of a protective film of calcium aluminate $3CaOAl_2O_3 \cdot 8H_2O$ on the aluminium surface [2].

The attack leaves more or less grey traces on the surface of the metal, which is why aluminium needs to be protected against splashing of concrete and roughcast if its appearance is to be preserved.

Prolonged contact with concrete, even in the presence of humidity, leads only to a superficial attack, shown by testing results with concrete blocks with embedded unprotected aluminium (Table G.4.1) that were exposed to various media, including marine atmosphere [3]. The zone of emergence in the air is not prone to corrosion (see Section B.2.7).

Anodising layers do not resist alkaline media: they are dissolved during the setting of concrete and therefore do not improve the resistance of aluminium in contact with concrete [4].

4.1.1. Influence of chlorides
It is common to add up to 3% calcium chloride to concrete, in order to accelerate setting and thus avoid freezing in the cold of winter. This leads to a substantial decrease in the resistivity of concrete: adding 3% calcium chloride leads to a drop in resistivity by a factor of 1000. Since calcium chloride is hygroscopic, it may keep the concrete from drying. In the presence of humidity, the conductivity of the medium increases, and electrical corrosion reactions are facilitated. In the absence of humidity, stray currents and contact with steel, adding chlorides does not significantly modify the resistance of aluminium in concrete [7].

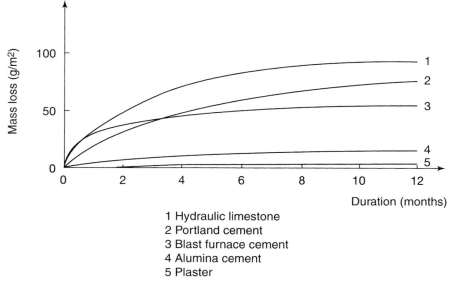

1 Hydraulic limestone
2 Portland cement
3 Blast furnace cement
4 Alumina cement
5 Plaster

Figure G.4.1. Mass loss in contact with concrete [6].

Experience has shown that adding up to 0.5% calcium chloride does not modify the resistance of aluminium.

4.1.2. Contact with steel
Since humid concrete is a conductive medium, experience shows that the more humid the concrete and the higher its chloride content, the more will be the galvanic corrosion of aluminium in contact with steel [8]. Tests performed in corrosion testing stations at the seacoast on concrete blocks with embedded aluminium profiles and steel reinforcement rods (Figure G.4.2) have shown that severe galvanic corrosion occurs when aluminium is

Table G.4.1. Decrease in thickness (μm) after 10 years in contact with concrete [5]

Medium	Profile 6082	Profile 6060	Sheet 3003	Tube 3003
Dry concrete	50	65	100	100
Atmosphere (bad weather)	100	140	140	125
Permanent humidity	140	125	50	75
Humid soil	90	230	185	180
Water	0	40	0	25

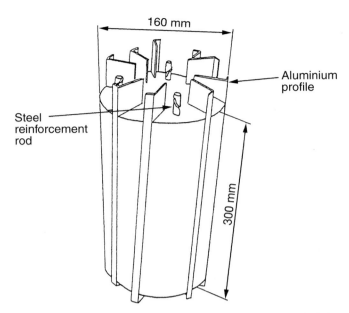

Figure G.4.2. Concrete testing block with embedded aluminium profiles and steel reinforcement rods [9].

in contact with steel (or electrically connected to it) [9]. The corrosion products of aluminium may lead to cracks in the concrete [10].

The intensity of galvanic corrosion of aluminium depends to a slight degree on the ratio of the surface areas of the two metals in contact or on the ratio of the volume of concrete and the aluminium surface area.

These tests as well as experience shows that any contact between aluminium and steel (such as reinforcements) should be avoided in concrete. Neither anodising nor painting prevents the risk of galvanic corrosion.

4.1.3. The action of stray currents

As in the soil, alternating or direct stray currents can lead to corrosion of aluminium (and other metals) that are embedded in concrete. The higher the level of humidity and chlorides, the stronger this effect is. For alternating currents of industrial frequency (50 or 60 Hz), a threshold exists below which no corrosion of aluminium is observed (Figure G.4.3).

4.2. THE EFFECT OF PLASTER

The resistance of aluminium in contact with plaster is excellent [11]. As in concrete, the contact with plaster leads to a very superficial and short attack of aluminium, even

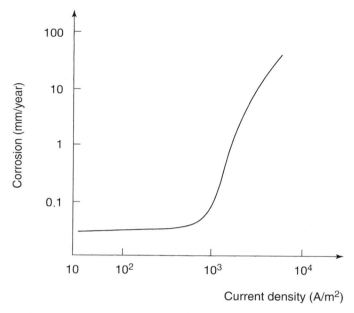

Figure G.4.3. Influence of the intensity of alternating stray currents in concrete [23].

anodised. This leads to a permanent alteration of the surface appearance of aluminium, which becomes white and mat.

Any alteration of the surface during the setting of plaster can be avoided by protecting them with appropriate resins or lacquers.

The adherence of aluminium to plaster is very good, twice as high as that of steel [12]. The linear expansion coefficient of plaster $(25 \times 10^{-6}\,K^{-1})$ is very close to that of aluminium $(24 \times 10^{-6}\,K^{-1})$.

4.3.   THE EFFECT OF WOOD

In buildings and assemblies, contacts between aluminium and wood are very frequent (windows, etc.).

Wood is generally acidic: its pH varies between 3 and 6 [13], and it releases minute quantities of formic acid and acetic acid. Tests have shown that aluminium will pickle only very superficially, about 10 μm, in water vapour containing 1% acetic acid [14].

Whether in seawater or in rural or urban atmospheres, long-term tests over 5–10 years have shown that aluminium screws have a very good resistance to corrosion in all common wood species: pinewood, fir, beech wood, and mahogany, unless these screws are made of copper-containing alloys of the 2000 or 7000 series [15–17].

Aluminium will be corroded in contact with wood only in the presence of humidity. This corrosion is very superficial. It mainly results from the structure of wood and its capacity to absorb moisture rather than its chemical composition.

Impregnation products for wood can have an influence on the corrosion resistance of aluminium if they contain copper salts (sulphate, naphthenate, etc.). They tend to increase the density, but not the depth of pitting [18].

Wood is dried in furnaces including aluminium equipment (partitions, inserts, etc.).

4.4. THE EFFECT OF POLYMERS

Upon heating and ageing, polymers can release minute amounts of more or less volatile products, the nature of which depends on the polymer (Table G.4.2).

Tests over 21 days at 35 °C in a saturated atmosphere at 100% humidity have shown that phenolic resins are the only ones capable of creating an atmosphere that is slightly aggressive to aluminium or that leads to a slight pickling of aluminium in contact with these resins [19]. Polyurethane foam has only a very moderate action on aluminium [20, 21]. Moisture trapped inside the foam or at the interface between the two materials is likely to be the main cause of this superficial corrosion, which is seen after delamination or separation of aluminium from the foam.

In buildings, rigid insulating panels consisting of aluminium bonded to polystyrene or other polymers are used. Flexible packaging materials are made of aluminium bonded to PVC films.

Aluminium is widely used in plastics and resin industries [22]. Since products of possible corrosion are white, aluminium neither stains nor modifies the surface appearance of plastics during their storage or transformation. Storage silos, tank trucks and tank wagons for polymer granules are made of aluminium alloys 5754, 5083, 5086, etc.

Table G.4.2. Emanations of polymers

Polymer	Released products
Phenolic resins	Ammonia, formaldehyde
Polyvinyl chloride	Hydrochloric acids, plasticisers
Polyvinyl acetate	Acetic acid
Cellulose acetate	Acetic acid
Epoxy resins	Hydrochloric acid, ammonia, catalysts
Polyester	Maleic acid
Teflon	Hydrochloric acid, hydrofluoric acid
Formol-urea resins	Formic aldehyde, ammonia
Rubber	Sulphur compounds

576

REFERENCES

[1] Coffrage en aluminium pour la construction de la centrale hydro-électrique de Revin, *Revue suisse de l'aluminium*, vol. 24, 1974, p. 52–54.

[2] Walton C.J., McGeary F.L., Englehart E.T., Compatibility of aluminium with alkaline building products, *Corrosion*, vol. 13, 1957, p. 807t–816t.

[3] Guilhaudis A., Les alliages légers exposés à l'atmosphère marine, *Corrosion et Anticorrosion*, vol. 10, 1962, p. 80–85.

[4] Tronstad L., Veimo R., Korrosion von Aluminium in verschiedenen Mörtelmaterialen, *Aluminium*, vol. 12, 1939, p. 839–842.

[5] Jenks J.H., Wright T.E., Godard H.P., Die Beständigkeit von Aluminium gegen Korrosion durch Beton, Verputz und Mörtel, *Baugewerbe*, 1962, p. 18.

[6] Bukowiecki A., Über das Korrosionsverhalten von Eisen und Nichteisenmetallen gegenüber verschieden Zementen und Mörteln, *Scheizerarchiv*, 1965, p. 9.

[7] McGeary F.L., Performance of aluminium in concrete containing chlorides, *Journal of the American Concrete Institute*, vol. 62, 1966, p. 247–265.

[8] Copenhagen W., Castello J.A., Corrosion of aluminium alloys balusters in reinforced concrete bridge, *Materials Protection and Performance*, vol. 9, 1970, p. 31–34.

[9] Bacaert J.P., Lyonnet C., Comportement de l'aluminium et de ses alliages au contact des bétons, *Revue de l'Aluminium*, 1976, p. 175–190.

[10] Porter F.C., Aluminium embedded in building mortars and plasters. Ten years tests, *Metallurgia*, vol. 65, 1962, p. 65-71.

[11] Foucault M., La corrosion des métaux par le plâtre, *Colloque RILEM*, 1977, p. 489–507.

[12] Le comportement des métaux au contact du plâtre, *Revue de l'Aluminium*, 1977, p. 73–76.

[13] Gray V.R., The acidity of the wood, *Journal Institute Wood Science*, 1958, p. 58–64.

[14] Clarke S.G., Longhurst E.E., The Corrosion of metals by acid vapors from wood, *Journal Applied Chemistry*, vol. 11, 1961, p. 435–443.

[15] Wright T.E., Godard H.P., Jenks I.H., The performance of Alcan 65S-T6 aluminium alloys embedded in certain woods under marine conditions, *Corrosion*, vol. 12, 1957, p. 481t–87t.

[16] Panseri C., Luft G., *Influenza della natura del legno sulla corrosione in aqua di mare delle viti lega leggera*, ISML, rapport no. 233, série X, 1960.

[17] Simm D.W., Button H.E., The corrosion behavior of certains metals in Cca-treated timber. Environmental tests at 100 relative humidity, *Corrosion prevention & control*, vol. 32, 1985, p. 25–35.

[18] Farmer R.H., Porter F.C., The corrosion of aluminium in contact with wood, *Metallurgia*, 1963, p. 161–167.

[19] Knotkova-Cermakova D., Vilcava J., Corrosion effects of plastics, rubber and wood on metals in confined spaces, *British Corrosion Journal*, vol. 6, 1971, p. 17–22.

[20] Van Delinder D.S., Hilado C.J., Corrosion protection for metals in contact with rigid polyurethane foam, February, *Journal of Cellular Plastics*, 1975, p. 25–35.

[21] Elliott R.W., Corrosivity of polyurethane foam toward metals, *Journal of Thermal Insulation*, vol. 10, 1986, p. 142–146.

[22] The Aluminium Association, *Aluminium alloys in the plastics and resins industry*, 1970.

[23] McGeary, *Journal of American Concrete Institute*, Fev. 1966, p. 247.

Chapter G.5

Food Industry

Chapter G.5
Food Industry

The first applications of aluminium in the food industry date back to the early 20th century, in breweries and as kitchen utensils for household and industrial use, where it has progressively replaced tin-coated copper.

Several properties of aluminium explain its outstanding position on this market: its good thermal conductivity, 7–12 times higher than that of steel, its good formability and capability to deep drawing. It is inert in contact with food, does not modify its organoleptic properties, and is nontoxic. Since products of possible corrosion are white, they do not discolour liquid or solid food.

5.1. ALUMINIUM AND HEALTH

Ever since kitchen utensils made of aluminium have been used, the question of the toxicity of aluminium has been posed, and the issue has never actually ceased being discussed, sometimes in a rather polemical manner.

5.1.1. Uptake of aluminium

Aluminium is one of the most abundant elements in the earth's crust, amounting to about 8%; it is the third most abundant element. Aluminium is ubiquitous: in soil, clay, rocks, and even in water. It occurs in the form of chemical compounds (in the oxidation state Al (III+)), but never as an unoxidised metal (oxidation state 0).

Because of its high concentration in the earth's crust, aluminium is necessarily found in the air and water. It thus penetrates into the food chain and is found in plants and all animals, as well as in humans (30–50 mg for a body weight of 70 kg). Exposure to aluminium through natural foodstuffs is thus daily and unavoidable. The ingested quantity is estimated at 10 mg per day, and its origin can be broken down as follows [1]:

- foodstuffs: 9.6 mg per day, for 96% of total intake
- kitchen utensils and packaging: 0.1–0.4 mg, i.e. 1–4% of total intake,
- ambient air: 5 µg.

The aluminium content of many foodstuffs has been evaluated by the US Food and Drug Agency (FDA) [2]. In general, vegetables contain more aluminium than meat. For example, small peas contain 0.6 mg of aluminium per 100 g of foodstuff, while chicken contains only 0.5 g of aluminium per 100 g of meat. Tea is particularly high in aluminium: its leafs contain $400-600$ mg·kg^{-1}, but because of the very low solubility of alumina, only $0.7-1.7$ mg·l^{-1} are found in brewed tea.

The daily aluminium uptake varies from one country to another, because it depends on both eating habits and food additives (generally aluminium salts) used as preservatives, fermenting agents and emulsifiers in the food industry. For example, in the United States, the daily aluminium uptake is estimated at $2-35$ mg, compared to $2.5-6.8$ mg in Italy.

Aluminium sulphate is widely used for the treatment of drinking water, in order to flocculate organic matter in suspension. The aluminium concentration of treated water is generally below 200 μg·l^{-1}.

Several drugs contain aluminium hydroxide, especially antacids used to treat gastric acidity. The treatment of certain gastric disorders may lead to the ingestion of more than 1000 mg aluminium hydroxide per day. A buffered aspirin tablet contains between 10 and 20 mg of aluminium hydroxide.

Aluminium is attacked by certain highly acidic foods such as tomatoes, rhubarb or cabbage, and will pass into the food: this is called migration [3]. The quantity of aluminium ingested in this way is on the order of 0.1 mg per 100 g of foodstuff [4, 5].

The FAO (the Food and Agricultural Organisation in the United Nations) has set a limit for the ingestion of aluminium to 7 mg per week and per kilogram of body weight, corresponding to 420 mg per week for an adult weighing 60 kg [6]. For drinking water, WHO states:

> No health-based guideline value is derived. However, a concentration of aluminium of 0.2 mg·l^{-1} in drinking-water provides a compromise between the practical use of aluminium salts in water treatment and discoloration of distributed water.

5.1.2. The harmlessness of aluminium

The toxicity of aluminium, alumina and other inorganic aluminium salts has never been proven. At the beginning of the 1950s, the FDA declared aluminium as inoffensive and rated it as GRAS (generally recognized as safe). Aluminium foil received the same rating in 1996.

During the 1970s, several publications in newspapers alleged that aluminium could lead to early senility and could be one of the factors responsible for Alzheimer's disease. In fact, the only reliable data show that aluminium has led to certain medical disorders in patients suffering from renal insufficiency and treated by renal dialysis [7]. These disorders have totally disappeared since dialysis techniques were modified.

Large epidemiological studies were undertaken in Europe and in the United States [8]. They showed no significant difference in the aluminium content of the brains of deceased patients suffering from Alzheimer's disease and those of deceased patients who did not suffer from this disease [9, 10].

All these studies have been the basis for the following statement from the World Health Organization:

There is no evidence to support a primary causative role for aluminium in Alzheimer's disease and aluminium does not induce Alzheimer's disease pathology in vivo in any species, including humans [12].

While it is necessary to substantiate the argument concerning the harmlessness of aluminium by further medical and toxicological studies, a common sense argument that had been formulated already in 1915 by M.A. Trillat should be recalled: "Nobody can have a doubt about the harmlessness of aluminium salts, if one remembers that this metal is used daily and everywhere in the household. Saucepans made of aluminium are more and more commonly used and tend to replace those made in copper. This use has become very widespread in Germany, Austria and in the United States, among all the classes of population" [13].

One could argue that for this author, in 1915, i.e. hardly two decades after the introduction of aluminium as a food-contact material, it was still too early for such a peremptory statement. Today, at the end of the 20th century, the situation is quite different: while the use of aluminium is much more widespread in the food industry than in 1915, no indication of any health-related disorder has been reported.

It is well known nowadays that the noxious effect of certain types of diet (or lifestyle) can have fatal consequences for public health. Rich Romans paid for this, using beautiful pottery decorated with lead-containing enamels that were attacked by foodstuffs that were allowed to macerate in this pottery.

This led to the unconscious, daily ingestion of lead, which has severe, noxious effects on the nervous system. Saturnism (i.e. lead poisoning) thus became a widespread disease afflicting members of the Roman high society, to such an extent that for a long time, epilepsy was called "le haut mal" (the upper-class disease) because its symptoms are similar to those of acute saturnism.

5.2. FOOD COMPATIBILITY OF ALUMINIUM: EUROPEAN STANDARDISATION

Two European standards

- EN 601: Aluminium and aluminium alloys—Castings—Chemical composition of castings for use in contact with foodstuffs
- EN 602: Aluminium and aluminium alloys—Wrought products—Chemical composition of semi-finished products used for the fabrication of articles for use in contact with foodstuffs

define the food compatibility of aluminium and aluminium alloys.

5.2.1. Unalloyed aluminium

For wrought aluminium, the mass percentage of other elements present shall not exceed the following limits:

–   iron + silicon: ≤ 1%;
–   chromium, manganese, nickel, zinc, titanium, tin: ≤0.10% each;
–   copper: ≤0.10%. Copper is permitted in a proportion greater than 0.10% but not more than 0.20% and provided that neither the chromium nor manganese content exceeds 0.05%;
–   other elements: ≤0.05% each.

5.2.2. Alloyed aluminium

The maximum content of added elements or impurities is given in Table G.5.1 These limits are such that wrought alloys of the 2000 and 7000 series as well as castings in copper-containing alloys of the 40000 series and alloys of the 20000 and 70000 series are not deemed food-compatible.

According to a decree of the French Ministry of Economy and Finances from August 27, 1987, anodised wrought and cast products in aluminium alloys may be used in contact with foodstuffs, if anodising has been carried out in a diluted bath containing one or more of the following acids:

Table G.5.1. Food compatibility of wrought aluminium alloys (table 2 of EN 602)

Element	Maximum content (wt%)
Silicon	13.5
Iron	2.0
Copper	0.6
Manganese	4.0
Magnesium [1]	11.0
Chromium	0.35
Nickel	3.0
Zinc	0.25
Antimony	0.2
Tin	0.10
Strontium	0.3
Zirconium	0.3
Titanium	0.3
Other elements [2]	0.05 each, 0.15 total

[1] Alloys containing more than 5% magnesium shall not be used for the production of pressure-resisting parts in pressure cooking applications.

[2] For some alloying elements (e.g. Ag) as mentioned under "Other elements" the maximum content is limited to 0.05% because of insufficient knowledge about behaviour in contact with food. Higher limits may be introduced when more information is available.

– sulphuric acid,
– sulfomaleic acid,
– sulfosalicylic acid,
– oxalic acid, and
– phosphoric acid.

Anodic coatings may be coloured by using dyestuffs and pigments that are approved as food-compatible by applicable regulations.

The anodic oxidation coating must be sealed unless the coating has been produced in a bath containing only diluted phosphoric acid, or unless the product on which the anodic oxidation coating has been produced will receive another food-compatible organic or inorganic coating prior to its use in contact with foodstuffs. Sealing shall be done in distilled or demineralized water, which may contain one or both of the following additives: nickel acetate in a concentration not exceeding $8 \; g \cdot l^{-1}$, and cobalt acetate in a concentration not exceeding $1 \; g \cdot l^{-1}$. All the technical parameters of the sealing process, especially the temperature of the bath and the duration of the treatment, must be chosen so as to minimise the absorptive power of the anodic oxidation coating, and to maximise its chemical resistance.

5.3. APPLICATIONS OF ALUMINIUM IN CONTACT WITH FOODSTUFFS

Aluminium is used in the food industry either in food processing equipment: vats, trays, kitchen utensils, cutting equipment, mixers, etc. or in packaging.

In packaging, aluminium can be

– unprotected, or
– protected by a lacquer or a plastic film.

The lacquer thickness varies between 5 and 15 μm depending on the nature of the foodstuff. It aims at protecting aluminium against prolonged action of aggressive foodstuffs such as tomato ketchup. In practice, most cans for fruit juice, beer, sodas (beverage cans), etc. are lacquered.

Anodising is used only for aluminium jewellery.

Thin foil in 1050A, 1100, 1200, 8011, etc. is used for short-term food packaging (household foil). Flexible packaging products are laminated complexes using thin aluminium foil and polymer films, in which aluminium ensures impermeability.

A great variety of packaging products in aluminium are known: food cans, beverage cans, aerosols, flexible tubes, and capsules. Given its very good thermal conductivity and its good formability, aluminium is widely used in the dairy industry [13] and cheese making, as well as in fish rooms on fishing boats [14].

Aluminium has no action on brewers' yeast and does not alter the fermentation or the taste of beer; therefore, it is widely used in breweries [15, 16].

Since 1950, in certain food industries such as dairies, aluminium has been dropped in favour of stainless steel for two reasons: blackening and cleaning (see Chapter G.6).

5.4.   THE RESISTANCE OF ALUMINIUM IN CONTACT WITH FOODSTUFFS

At room temperature, aluminium has good resistance to diluted or highly diluted organic acids, which produces more or less substantial pickling, depending on the nature of the acid (see Table G.5.2 [17]).

It is not possible to draft an exhaustive list for the resistance of aluminium in contact with foodstuffs. This has been done in great detail in several specialised publications [18–21].

In general, neither the appearance nor the taste of natural or processed foods are altered in contact with aluminium, with the exception of wine, liquors or alcohols [22, 23]. Aluminium may suffer pitting corrosion in these media.

Table G.5.2. Action of diluted organic acids (decrease in thickness in $\mu m \cdot h^{-1}$)

Acid	At 25 °C			At the boiling point 0.1N
	1N	0.1N	0.01N	
Acetic	1.7	1.3	0.7	31.0
Aconitic	1.0	0.4	0.1	3.0
Adipic		0.2	0.1	0.0
Butyric		1.4	0.3	14.0
Citric	0.6	0.5	0.4	6.0
Formic	6.3	2.5	0.8	31.4
Glutamic		0.4	0.1	4.4
Glutaric		0.1	0.2	0.1
Glycolic		2.0	0.8	16.4
Lactic	1.9	1.8	1.4	25.1
Maleic	1.0	0.8	0.3	3.3
Malic	2.7	0.8	0.3	6.6
Malonic	3.3	1.6	0.3	4.4
Mandelic		0.5	0.8	3.6
Oxalic	16.8	2.0	0.1	5.8
Propionic	0.7	0.7	0.3	7.6
Succinic	0.7	0.7	0.3	0.0
Tannic			0.3	0.6
Tartric	1.0	1.0	0.2	6.7

REFERENCES

[1] Buclez B., L'aluminium au contact des aliments et la santé, *Annales des falsifications de l'expertise chimique et toxicologique*, no. 940, 1997, p. 207–216.

[2] Pennington J.A.T., Schoen S.A., Estimates of dietary exposure to aluminium, *Food Additives and Contaminants*, vol. 12, 1995, p. 119–128.

[3] Schmidt M.P., *Migration from aluminium*, Pechiney Rhenalu, avril 1997.

[4] Greger J.L., Goetz W., Sullivan D., Aluminium levels in foods cooked and stored in aluminium pans, trays and foils, *Journal of the Food Protection*, vol. 48, 1985, p. 772–777.

[5] Muller J.P., Steinnegger A., Schlatter C., Contribution of aluminium from packaging materials and cooking ustensils to the daily aluminium intake, *Z Lebensm Unters Forsch*, vol. 197, 1993, p. 332–341.

[6] Provisional tolerable weekly intake (PTWI) for aluminium, in *Toxicological evaluation of current food additives and contaminants*, WHO-FAO Expert Committee on Food Additives, Genève, 1989, p. 113–154..

[7] Galle P., La toxicité de l'aluminium, *La Recherche*, no. 178, 1986, p. 766–776.

[8] Buclez B., Y a-t-il une neurotoxicité de l'aluminium? *Connaissances actuelles*, Le Concours Médical, 30 mai 1998, p. 1482–1485.

[9] Lovell M.A., Ehmann D., Markesbery W.R., Laser microprobe analysis of brain aluminium in Alzheimer disease, *Annals of neurology*, vol. 33, 1993, p. 36–42.

[10] Bjertness E., Candy J.M., Torvik A. et al., Content of brain aluminium is not elevated in Alzheimer disease, *Alzheimer Disease and associated disorders*, vol. 10, 1996, p. 171–174.

[11] World Heath Organization (WHO), International Programme on Chemical Safety. Environmental Health Criteria for Aluminium (Collective views of the task group IPCS-EHC-95-96).

[12] Trillat M.A., Emploi de l'aluminium dans les industries alimentaires, *Bulletin de la Société d'encouragement de l'industrie nationale*, no. 122, 1915, p. 555–574.

[13] Prévot P., L'emploi de l'aluminium dans le matériel d'industrie laitière, *Revue de l'Aluminium*, 1947, p. 121–128.

[14] L'aménagement des cales à poisson des chalutiers modernes, *Revue de l'Aluminium*, no. 252, 1958, p. 317–324.

[15] Bailly J., Le matériel de brasserie. Cuves de fermentation, tank de garde, *Revue de l'Aluminium*, no. 48, 1932, p. 1741–1752.

[16] Les cuves de fermentation et de stockage de la brasserie Arthur Guiness à Saint James Gate. 170 000 de réservoirs en aluminium, *Revue de l'Aluminium*, no. 303, 1962, p. 1237.

[17] Poe C.F., Warnock R.M., Wyss A.P., Action of dilute acids on aluminium, *Industrial and Engineering Chemistry*, vol. 27, 1935, p. 1505–1507.

[18] Bryan J.M., *Aluminium and aluminium alloys in the food industry with special reference to corrosion and its prevention*, Cambridge, Department of Science and industry, Food investigation, special report no. 50, 1948.

[19] Baudart A., L'aluminium dans les industries alimentaires, *Revue de l'Aluminium*, 1967, p. 521–594.

[20] Kunze E., Corrosive properties of various types of foodstuff and the like, *Aluminium*, vol. 52, 1976, p. 296–301.

[21] EWAA, *Resistance of aluminium to attack by natural foodstuffs and some organic acids*, report by Food Legislation Study Group, 1974.

[22] Ash C.S., Metals in wineries, *Industry Engineering and Chemistry*, vol. 27, 1935, p. 1243–1244.

[23] Walter E., The action of metals on alcoholic liquids, *Alkohol industry*, vol. 64, 1951, p. 235–237.

Chapter G.6

Cleaning of Aluminium

Chapter G.6
Cleaning of Aluminium

Aluminium is cleaned for several reasons:

- *hygiene*, for domestic and industrial kitchen utensils and in the food industry: dairies, breweries, canning, storage tanks, transport tanks, etc.
- *maintenance*, for buildings and transport equipment (commercial vehicles, aircraft, railways, etc.) Experience has shown that maintenance is necessary to preserve a good appearance and avoid corrosion. It is necessary to regularly remove accumulated inorganic dust that may contain chlorides. In the presence of moisture, these deposits can form an acidic medium, with a pH below 4, that attacks aluminium.
- *quality*, for the transportation of certain industrial or food products. Residues of preceding loads need to be removed, because even in small quantities, they can be problematic for the transformation of plastics or for the organoleptic properties of foodstuffs.

In the following, cleaning prior to surface treatments is not discussed, because it refers to industrial processes that are beyond the scope of this book.

6.1. CLEANING OF METALLIC SURFACES

Cleaning aims at removing inorganic or organic residues that have accumulated on the surface of the metal. This is a short operation, not exceeding a few minutes, that aims at obtaining an appearance that is in conformity with the applicable criteria.

Four conditions need to be controlled during cleaning:

- the *chemical effect*: one or more components of the cleaning agent dissolve dirt and inorganic salts without attacking the substrate. Cleaning products are very often based on acids or hydroxides;
- the *physical effect*: dirt needs to be unstuck from the surface of the metal by modifying its surface tension, the reason why cleaning products contain surfactants;
- the *mechanical effect*: brushing or high-pressure water jets contribute to eliminating dirt that is insoluble in the cleaning products and increasing the effect of surfactants; and

–   the *temperature*: it is well known that an increase in temperature increases the rate of chemical reactions and thus the effectiveness of cleaning. (For some reactions, the reaction rate doubles for every 10 °C increase in termperature). However, the stability thresholds of certain cleaning product constituents should not be exceeded.

6.2.   THE CHOICE OF CLEANING PRODUCTS

There is no such thing as a universal cleaning agent capable of cleaning all metallic surfaces such as ordinary steel, stainless steel, aluminium or copper. In fact, each metal (or alloy) reacts differently to acidic or alkaline media that form most cleaning products. For example, steel resists nitric acid poorly, but resists sodium hydroxide or sodium carbonate solutions well, whereas the situation is the inverse for aluminium.

A cleaning product is a complex and well-balanced mixture of several agents; it may include up to 20 constituents in different concentrations, from 1% to 1 ppm. They have several different functions, acting as pickling agents, detergents, degreasing agents, and inhibitors. Cleaning products must be

–   compatible with aluminium and aluminium alloys;
–   nontoxic for users;[1]
–   nonpolluting (otherwise, they must be treated in accordance with applicable regulations before their release into the environment).

Only the first of these aspects will be discussed here.

6.3.   COMPATIBILITY WITH ALUMINIUM

Cleaning aluminium has the same goal as cleaning other metallic surfaces:

–   eliminate inorganic and organic contamination,
–   no attack of the metal surface.

Aluminium and alumina are amphoteric, i.e. can be attacked both by acids, mainly hydracids such as HF and HCl, and by bases such as sodium hydroxide (NaOH), potassium hydroxide (KOH) and sodium carbonate (Na_2CO_3). For this reason, the dissolution rate

---

[1] Producers of industrial cleaning products must supply material safety data sheets according to a format specified by the European Directive 91-555. These data sheets specify the nature of the hazardous product or impurity and give advice about storage and handling, as well as toxicological information.

of aluminium and alumina is higher in higly acidic or alkaline pH, as can be seen from Figure B.1.18.

Cleaning products based on alkaline products such as sodium hydroxide, potassium hydroxide, sodium carbonate, even if containing corrosion inhibitors such as sodium silicate, should be used with precaution, because

– aluminium and aluminium alloys are very sensitive to alkaline media, in particular sodium or potassium hydroxide;
– the oxide film formed after attack in highly alkaline medium is highly hydrated, grey and difficult to modify (except by rinsing with a solution containing 50% nitric acid).

The surface of aluminium alloy products should be cleaned preferably with products based on phosphoric acid, which form with aluminium a stable, highly insoluble and transparent salt, aluminium phosphate.

Rinsing is a very important operation that aims at eliminating residual traces of cleaning agents. In particular, zones that are difficult to access should be rinsed carefully so that acid is not retained, which may lead to localised corrosion. Rinsing is most often carried out with tap water. It should contain as few chlorides as possible. Deionised water has the advantage of preventing the formation of deposits after drying.

6.4. CLEANING MATERIALS USED WITH FOOD

Contamination is generally fatty and must, therefore, be saponified with cleaning agents containing a strong base with a pH above 10 such as sodium hydroxide or sodium carbonate. Cleaning products must also contain bactericides.

Aluminium does not resist well in such a medium: the thickness of the attack depends on concentration, contact time, and temperature. In addition, aluminium can also blacken (see Section D.1.5).

Blackening of aluminium is only an alteration of the surface appearance. While in certain cases it may degrade the esthetic aspect of an equipment, it does not imply a lack of maintenance and cleaning of the equipment. It does not favour bacterial growth.

The inhibiting effect of sodium silicate has been known since 1922 [1] and has been the subject of many studies [2, 3].

Bactericides and disinfectants have no action on aluminium in the usual concentrations.

Specific cleaning products for food equipment are available that are compatible with aluminium and avoid corrosion and blackening.

6.5. CLEANING ANODISED SURFACES

Cleaning should not attack the anodic layer that has dissolved in acidic and alkaline media, for example, the natural oxide layer. Normalised procedures can assess products for cleaning anodised surfaces [4, 5].

6.6. SAND BLASTING

Sand blasting eliminates surface deposits (such as paint or various contaminations) by using an abrasive jet of calibrated particles. Compared to liquid cleaning products, it provides the advantage of being a dry method, without water, acids or bases. It can be used for large surfaces of shaped structures such as tanks, hulls of ships, etc.

In the case of aluminium, non-recycled corundum must be used, in order to prevent any risk related to the use of corundum enriched with iron that has been used previously for steel.

Blasting with steel powder must not be used. The incrustation of iron particles on the aluminium surface is a source of galvanic corrosion that will lead, in the presence of moisture, to superficial micropitting. Experience shows that the pitting depth reaches a few hundredths of a millimetre.

6.7. CLEANING BY PROJECTION OF ICE OR CARBONIC ICE

Ice or carbonic ice projection is a recent cleaning technology that blasts calibrated ice particles at high speed and cryogenic temperatures. These ice particles can be prepared from demineralised water. The surface cleaning produces three effects:

- *thermal*: the rapid and strong cooling solidifies certain kinds of dirt, and the differential expansion between the substrate and the dirt can lead to unsticking;
- *mechanical*: the high-speed impact of ice particles removes dirt particles;
- *dynamic*: the dirt particles are removed by the water resulting from melting of the ice particles.

These methods can be expected to be used more widely in the future, because they have several advantages compared to traditional sand blasting: they are easy to use, they do not lead to dust formation, and they do not damage the metal surface.

REFERENCES

[1] Seligman R., Williams P., Cleaning of aluminium utensils, *Journal of Institute of Metals*, 1922.

[2] Dauphin G., Kerherve L., Labbe J.-P., Pagetti J., Corrosion par les produits de net-toyage et de désinfection dans les industries alimentaires, *Matériaux et Techniques*, nov., 1978, p. 379–382.

[3] Kerherve L., Dauphin G., Michel F., Corrosion des alliages d'aluminium et des aciers inoxydables par les produits de nettoyage et de désinfection, *Industries Alimentaires et Agricoles*, vol. 101, 1984, p. 779–787.

[4] Norme NF A 91-451: Traitements de surface. Aluminium et alliages d'aluminium. Qualification des produits d'entretien.

[5] Hinüber H., Cleaning of anodized and coated aluminium façades, *Aluminium*, vol. 66, 1990, p. 1148–1152.

Chapter G.7

Behaviour in Fire

Chapter G.7
Behaviour in Fire

This section attempts to answer two questions related to the behaviour of aluminium castings and wrought aluminium semi-products in fire:

–   What is their fire resistance, i.e. the changes in their physical properties as the temperature increases?
–   What is their reaction to fire, i.e. the behaviour of solid and liquid metal during a fire? Is aluminium flammable, can it thus maintain fire?

This section does not deal with finely divided aluminium in the form of powder or granules. It is well known that aluminium, like many other metals, is flammable in the finely divided state. Parenthetically, aluminium powder has pyrotechnic applications and is used in solid rocket boosters (Ariane, etc.).

Regulations on the use of metallic materials as a building material, in ship building, transport, etc. are also beyond the scope of this book.

7.1. FIRE RESISTANCE

The melting point of aluminium is 660 °C. Only two other common metals have a lower melting point: magnesium (650 °C) and zinc (419 °C) (Table G.7.1). In practice, as the temperature exceeds 660 °C, aluminium structures will melt but not burn.

7.1.1. Linear expansion coefficient, thermal conductivity and mass thermal capacity
The linear expansion coefficient of aluminium is twice as high as that of steel. It increases with temperature (Figure G.7.1). The same applies to its thermal conductivity (Figure G.7.2) and mass thermal capacity (Figure G.7.3).

7.1.2. Young's modulus and yield strength
As shown in Figures G.7.4 and G.7.5, the Young's modulus and the yield strength decrease with increasing temperature. For example, 50% of the elastic strength is lost when the following temperature is reached:

Table G.7.1. Physical properties

Property	Alloy 5083 H11	Non-alloyed steel	Stainless steel Z7CN18-09, annealed
Melting point (liquid) (°C)	660	1450	1450
Linear expansion coefficient (10^{-6} °C^{-1})	23.1	11.7	17.5
Mass thermal capacity (J·kg^{-1}·°C^{-1})	900	420	5002
Thermal conductivity (W·m^{-1}·°C^{-1})	237	46	15

– 250 °C for aluminium,
– 500 °C for structural steel (250 °C for stainless steel).

The decrease in the mechanical strength of aluminium leads to a reduction in its load-bearing capacity if the structure that is exposed to heat is under load. As in the case of steel, it is necessary to provide adequate thermal insulation in compliance with applicable regulations, in order to guarantee the long-term mechanical integrity of loaded structures in case of fire.

This applies especially to fireguard walls made of aluminium, which need to be protected, as in the case of steel or any other metal, in order to resist a fire for a given period of time.

After prolonged heating of age-hardenable aluminium alloys (series 2000, 6000 and 7000), their mechanical characteristics should be checked to see that they have not been altered as a consequence of heating.

Since aluminium has a good thermal conductivity, heat is dissipated more rapidly than with other metals. This avoids hot spots. Heat diffusion is accompanied by deformation, because of aluminium's high linear expansion coefficient.

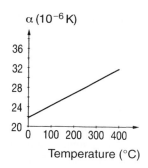

Figure G.7.1. Linear expansion coefficient of alloys 5083 and 6061.

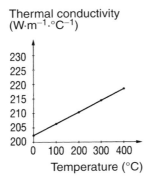

Figure G.7.2. Thermal conductivity of aluminium.

7.2. REACTION TO FIRE

There is no proof for the self-ignition of solid or liquid aluminium in fire [1]. Experience and many research results have shown that under normal fire conditions, aluminium will not ignite.

In pure oxygen and with an atmospheric pressure of 1013 mbar, the inflammation point of aluminium is above 1000 °C and thus higher than that of the other common metals: 930 °C for steel and 900 °C for zinc. The order of the inflammation points does not depend on the order of the melting point [2]. For some metals, the ignition point is below the melting point and vice versa (Table G.7.2).

It is difficult to ignite aluminium because of its natural oxide film that blocks the reaction of the metal with air or oxygen, since it insulates the liquid metal in a more or less

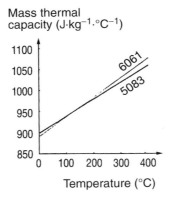

Figure G.7.3. Mass thermal capacity of aluminium.

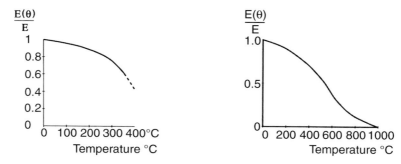

Figure G.7.4. Young's modulus of aluminium and steel.

airtight envelope. In other words, oxidation and combustion of aluminium are two competing processes.

In oxygen–argon mixtures, aluminium ignites only if the temperature is higher than the melting point of alumina, and self-sustained combustion of aluminium occurs only as the temperature reaches the boiling point of aluminium, i.e. 3073 °C [3]. If the conditions are unfavourable for the formation of oxide film, the ignition point can be decreased to about 1400 °C [4].

Among alloying elements and additives, only magnesium significantly changes the fire resistance of aluminium. Alloys containing more than 10% magnesium can ignite at 550 °C [5, 6] because magnesium can burn at that temperature.

Water can be sprayed on molten aluminium. Only a small quantity of water will be decomposed since the metal's reactivity is decreased by the formation of a natural oxide film. Very little hydrogen is released and thus there is no explosion hazard.

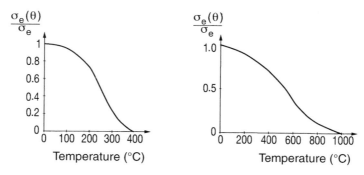

Figure G.7.5. Yield strength of alloy 5083 and of steel.

Table G.7.2. Inflammation points and melting points

Metal	Inflammation point (°C)	Melting point (°C)
Magnesium	623	650
Iron	930	1535
Molybdenum	750	2620
Aluminium	1000	660
Zinc	900	419
Lead	870	327

7.3. THE CLASSIFICATION OF ALUMINIUM ALLOYS

A decree of the French Home Secretary from June 30, 1983 on the classification of construction materials according to their fire resistance and on the definition of test methods has rated common metals such as aluminium alloys and steel "M0", which means that under the conditions of a fire, these materials are considered to be incombustible.

Aluminium is approved as a material for the construction of ferries under the International Convention for the Safety of Life at Sea of the International Maritime Organisation [7].

Frictional sparking cannot occur on aluminium, which is why aluminium equipment has long been used in coal mines. Fire fighting equipment can be made of aluminium: valves, ladders, tool boxes, water tanks, etc.

REFERENCES

[1] Hill V.J., *Résistance à l'incendie des constructions en aluminium*, rapport du Comité international de développement de l'aluminium 7132, 1971.

[2] Grosse A.V., Conway J.B., Combustion of metals in oxygen, *Industrial and Engineering Chemistry*, vol. 50, 1958, p. 663–672.

[3] Bourrianes R., *Température de l'aluminium pendant sa combustion dans les mélanges oxygène/ argon, dans l'azote et dans l'air*, compte-rendu de l'Académie des sciences, Paris, vol. 275, 1972, p. 717–720.

[4] Kuehl D.K., Ignition and combustion of aluminium and beryllium, *AIAA Journal*, vol. 12, 1966, p. 2239–2247.

[5] Drouzy M., Mascre C., The oxidation of liquid non-ferrous metals in air or oxygen, *Metallurgical Review*, 1969, p. 25–46.

[6] Thiele W., L'oxydation de l'aluminium en fusion, *Aluminium*, 1962, p. 707.

[7] West E.G., The fire risk in aluminium alloys ships structures, *The Metallurgist and Materials Technologist*, 1982, p. 395–398.

Products that may be Dangerous in Contact with Aluminium

The contact of aluminium with certain inorganic or organic products may lead to dangerous reactions, including explosion, splashes, etc. Several lists are available that include substances capable of producing such reactions with aluminium. In fact, a close analysis of such lists shows that there is very often a confusion between finely divided aluminium such as powder or granules and metal in the form of cast products or wrought semi-products: sheet, plate, profiles, bars, tubes, etc. This distinction is very important. Here are a few well-known facts:

– The reactivity of massive metal such as that used commonly in the form of plates, sheets or profiles is in no way comparable with what can be deduced from certain thermodynamic data. The reason is simple: when aluminium is put in contact with the oxygen contained in air, it will be covered immediately by a continuous oxide film consisting of alumina, the thickness of which, between 5 and 10 nm, is sufficient to slow down and even annihilate the reactivity of aluminium with many products (see Section B.1.8), including air and oxygen, even at high temperatures.

– On the other hand, finely divided aluminium in the form of powder or granules has a very high reactivity with air, oxygen and numerous inorganic or organic products. During its manufacture, aluminium powder receives a passivation treatment that increases its stability at room temperature. Nevertheless, great precaution must be taken for the transportation, storage and use of finely divided aluminium. This issue is beyond the scope of the present volume.

– Turnings, chips, etc. resulting from machining of aluminium and aluminium alloys can, under certain circumstances (temperature, particle size), react violently with certain products such as organic solvents containing chlorine. For this reason, stabilised organic solvents have to be used in degreasing units, and any accumulation of fine turnings or machining chips at the bottom of the degreasing tank must be avoided.

Eventually, in contact with massive aluminium and aluminium alloys, only rather few inorganic or organic products are capable of provoking violent reactions that could present a hazard to workers or equipment.

603

Without aiming at (or being capable of) being exhaustive, the following products must be kept in mind:

- concentrated inorganic hydracids, which even at room temperature react very actively with aluminium:
  - Hydrochloride acid HCl,
  - Sulphuric acid H_2SO_4,
  - Hydrofluoric acid HF,
- concentrated inorganic bases
  - Sodium hydroxide NaOH,
  - Potassium hydroxide KOH,
- Halogens (in the presence of humidity)
  - Chlorine Cl_2,
  - Fluorine F_2,
  - Bromine Br_2,
  - Iodine I,
- the following organic acids:
  - Formic acid HCOOH,
  - Chloroacetic acids (mono-, di-, and trichloroacetic acid) $CH_2ClCOOH$, $CHCl_2COOH$, CCl_3COOH, and their derivatives such as sodium chloroacetates,
- the following halogenated derivatives:
  - Methylchloride CH_3Cl,
  - Methylbromide CH_3Br,
  - Methyliodide CH_3I,
  - Chloroform $CHCl_3$,
  - Carbon tetrachloride CCl_4,
  - Propylene chloride $CH_3CHCl-CH_2Cl$,
  - Ethyl bromide CH_3CH_2Br.

All these products can react violently with aluminium and aluminium alloys, even at room temperature.

The following chlorinated organic derivatives can react with massive aluminium and aluminium alloys around 50 °C, and this reaction can accelerate fiercely as the temperature approaches the boiling point of the product under consideration:

- Ethyl chloride CH_3CH_2Cl,
- Ethylene chloride $CHClCH_2Cl$,
- Ethylidene chloride CH_3CHCl_2,
- Vinyl trichloride $CHCl_2CH_2Cl$,
- Tetrachloroethane $CHCl_2CHCl_2$,

- Pentachloroethane Cl_3CHCl_2,
- Hexachloroethane CCl_3CCl_3,
- Propylene chloride $CH_3CHClCH_2Cl$,
- Dichloroethylene $CHClCHO$,
- Trichloroethylene CCl_2CHCl,
- Tetrachloroethylene CCl_2CCl_2,
- Benzyl chloride $C_6H_5CH_2Cl$,
- Benzylidene chloride $C_6H_5CHCl_2$.

This information is given for information only, the reader is, therefore, invited to refer to official documents published by official organisations.

General References

JOURNALS

Chemicals Abstracts, 1907–1997.
Revue de l'Aluminium, 1924–1983.
World Aluminium Abstracts, 1971–1996.

BOOKS IN ENGLISH

Trethwey K.R., Chamberlain, J., *Corrosion for Science and Engineering*, 2nd edition, Longman, Harlow, UK, 1996.
Nitsche J., Lang G., *Glossary of Technical Terms. Metallurgy of Non-ferrous Metals*, Aluminium Verlag, Düsseldorf, 1994.
Schreir L.L., Jarman R.A., Burnstein G.T., *Corrosion*, 3rd edition, Butterworth-Heinemann, Oxford, 1993.
Fontana M.G., Greens N.D., *Corrosion Engineering*, McGraw-Hill, New York, 1978.
Scully J.C., *The Fundamentals of Corrosion*, 2nd edition, Pergamon International Library, Oxford, 1975.
Godard H.P., Bothwell M., *The Corrosion of Light Metals*, John Wiley and Sons, New York, 1967.
Combating Chemical Corrosion with Alcoa Aluminium, Aluminium Company of America, Pittsburgh, 1933.
Wernick S., Pinner R., Sheasby P.G., *The Surface Treatment and Finishing Aluminium and Its Alloys*, 5th edition, ASM International, Ohio/Finishing Publications Ltd, Teddington, UK.
Kammer C., *Aluminium Handbook, Vol. 1: Fundamentals and Materials*, Aluminium-Verlag, Düsseldorf, 1999.
Kammer C., *Aluminium Handbook, Vol. 2: Forming, Casting, Surface Treatment, Recycling and Ecology*, Aluminium-Verlag, Düsseldorf, 2003.

BOOKS IN GERMAN

Bakalov A.S., Beneke H., Wulff I., *Wörterbuch für den Korrosionsschutz*, Vulkan-Verlag, Essen, 1994.
Aluminium Taschenbuch, Aluminium Verlag, Düsseldorf, 1994.

BOOKS IN FRENCH

Philibert J., Vignes A., Bréchet Y., Combrade P., *Métallurgie : du minerai au matériau*, Masson, Paris, 1998.

Arnaud P., *Cours de chimie organique*, 16ᵉ édition, Dunod, Paris, 1996.

Talbot J., *Les éléments chimiques et les hommes*, SIRPE, Paris, 1995.

Perrin R., Scharff J.-P., *Chimie industrielle*, Masson, Paris, 1993.

Atkins, P., *Chimie générale*, InterÉditions, Paris, 1992.

Mémento technique de l'eau, Degrémont, Paris, 9ᵉ édition, 1989.

Vargel C., *Le comportement de l'aluminium et de ses alliages*, Dunod, Paris, 1979.

Altenpohl, D., *Un regard á l'intérieur de l'aluminium. Une introduction à la métallurgie structurale et à la transformation de l'aluminium*, Aluminium Verlag, Düsseldorf, 1976.

Junière P., Sigwalt M., *Les applications de l'aluminium dans les industries chimiques et alimentaires*, Eyrolles, Paris, 1962.

Karrer P., *Traité de chimie organique*, Dunod, Paris, 1948.

PUBLICATIONS VARIOUS

Le transport des matières dangereuses par route ADR TMDR 1997, Le Règlement, Éditions ADIIS, Lyon.

Demi-produits aluminium, Pechiney Rhenalu, 1997.

Guide de la chimie, Union des industries chimiques, 1994-1995.

Annuaire produits chimiques, Union des industries chimiques, 1993.

Société des annuaires internationaux, *Lexique technique des produits chimiques*, Éditions Rousset, Vincennes, 1978.

Aluminium with Food and Chemical Compatibility Data, The Aluminium Association, New York, 1969.

Renseignements pratiques sur l'aluminium et ses alliages, L'aluminium Français, 5ᵉ édition, 1939.

Glossary

English	French	German
Adhesion	Adhérence	Haftvermögen
Age hardening	Durcissement structural	Aushärtung
Air quenching	Trempe à l'air	Luftabschrecken
Alloy	Alliage	Legierung
Alloying element	Elément d'addition	Legierungselement
Alternating immersion test	Essai d'immersion émersion alternées	Weschseltauchversuch
Alternating current	Courant alternatif	Wechselstrom
Aluminium	Aluminium	Aluminium
Aluminium alloy	Alliage d'aluminium	Aluminiumlegierung
Aluminium oxide	Oxyde d'aluminium	Alumniumoxid
Aluminium refined	Aluminium raffiné	Reinstaluminium
Anion	Anion	Anion
Annealing	Recuit	Glühung
Anode	Anode	Anode
Anodic	Anodique	Anodisch
Anodic current	Courant anodique	Anodenstrom
Anodic dissolution	Dissolution anodique	Anodsische Auflösung
Anodizing	Anodisation	Anodisation
Aqueous corrosion	Corrosion dans l'eau	Wasserkorosion
Artificial ageing	Revenu	Warmauslagerung
Artificially aged	Etat revenu	Warmausgehärtet
As-quenched condition	Brut de trempe	Abgeschreckt
As-quenched condition	Trempe fraîche	Frische Abschreckärtung
As-quenched temper	Etat trempé	Abgeschreckter Zustand
Atmospheric corrosion	Corrosion atmosphérique	Atmosphärische Korrosion
Atmospheric exposure station	Station de corrosion	Bewitterungsstation
Auxiliary electrode	Contre électrode	Gegenelektrode
Blackening, Blackening	Noircissement	Schwärzung, Trübung
Bonding	Collage	Kleben
Brushing	Brossage	Bürsten

609

Buffing	Polissage au disque	Schwabbeln
Calomel electrode	Electrode au calomel	Kalomelelektrode
Casting	Moulage	Guß
Casting alloy	Alliage de moulage	Gußlegierung
Cathode	Cathode	Kathode
Cathodic current	Courant cathodique	Kathodenstrom
Cathodic protection	Protection cathodique	Korrosionsschutz
Cation	Cation	Kation
Cavitation	Cavitation	Kavitation
Cavitation corrosion	Corrosion par cavitation	Kavitationskorrosion
Chalking	Farinage	Kreiden
Chemical brightening	Brillantage chimique	Chemisches Glänzen
Chemical conversion	Conversion chimique	Chemische Konversion
Chemical polishing	Polissage chimique	Chemisches Polieren
Clad alloy	Alliage plaqué	Plattierte Legierung
Cladding	Placage	Plattierung
Coating	Revêtement	Beschichten
Coil coating	Revêtement en bande	Bandbeschichtung
Cold working	Ecrouissage	Kaltverfestigung
Compound intermetallic	Composé intermétallique	Intermetallische Verbindung
Condensation	Condensation	Kondensation
Corrosion	Corrosion	Korrosion
Corrosion behavior	Comportement à la corrosion	Korrosionsverhalten
Corrosion product	Produit de corrosion	Korrosionsprodukt
Corrosion resistant	Résistant à la corrosion	Korrosionsbeständing
Corrosion susceptibility	Susceptibilité à la corrosion	Korrosionsempfindlichkeit
Corrosion testing	Essai de corrosion	Korrosionsprüfung
Corrosivness	Corrosivité	Korrosivität
Crevice corrosion	Corrosion caverneuse	Spaltkorrosion
Crevice corrosion	Corrosion sous dépôt	Belagkorrosion
Critical quenching rate	Vitesse critique de trempe	Kritische Abschreckgeschwindigkeit
Critical strain	Ecrouissage critique	Kritischer Verformungsgrad
Current density	Densité de courant	Stromdichte
Degreasing	Dégraissage	Entfettung
Degree of moisture	Humidité relative	feuchtegrad

Degree of oxydation	Degré d'oxydation	Oxidationsgrad
Desensitization	Désensibilisation	Desensibilisierungsglühung
Diffusion	Diffusion	Diffusion
Direct current	Courant continu	Gleischstrom
Dissolution potential	Potentiel de dissolution	Lösungspotential
Duplex ageing	Double revenu	Stufenaushärtung
Electrochemical brightening	Brillantage électrochimique	Elekrochemisches Glänzen
Electrochemical cell	Pile	Element galvanisches
Electrochemistry	Electrochimie	Elektrochemie
Electrode	Electrode	Elektrode
Electrode half-cell potential	Potentiel d'électrode	Elektrodenpotential
Electrolysis	Electrolyse	Elektrolyse
Electron	Electron	Elektron
Electronegativ	Electronegatif	Elektronegativ
Electropositive	Electropositif	Electronpositiv
Elongation	Allongement	Dehnung
Erosion	Erosion	Erosion
Erosion corrosion	Corrosion érosion	Erosionskorrosion
Etching	Attaque chimique	Ätzen
Etching	Gravure (chimique)	Chemische Gravierung
Exfoliation corrosion	Corrosion exfoliante Corrosion feuilletante	Schichtkorrosion
Filiform corrosion	Corrosion filiforme	Filigrankorrosion
Finish	Finition de surface	Oberflächenfinish
Fretting corrosion	Corrosion par frottement	Reibkorrosion
Galvanic corrosion	Corrosion galvanique	Galvanische Korrosion
Galvanic series	Echelle des potentiels	Elektrochemische Spannungsreihe
Grain	Grain	Korn
Guinier-Preston zones	Zones de Guinier-Preston	Guinier-Preston zonen
Half hard temper	Etat demi dur	Halbhart-Zustand
Hard temper	Etat quatre quart dur	Hart-Zustand
Hardness	Dureté	Härte
Hart anodizing	Anodisation dure	Hartanodisation
Heat treatment	Traitement thermique	Wärmebehandlung
Heat-affected zone	Zone affectée thermiquement	WärmeeinflußZone

Heat-treatable alloy	Alliage à traitement thermique	Aushärtbare Legierung
Heat-treatable alloy	Alliage trempant	Aushärtbare Legierung
Homogenization	Homogoénéisation	Homogenisierung
Hot rolled temper (F)	Etat brut de laminage à chaud (F)	Warmwaldzzustand (F)
Hydrogen electrode	Electrode à hydrogène	Wasserstoffelektrode
Hydrogen embrittlement	Fragilisation hydrogène	Wasserstoffversprödung
Immersion test	Essai d'immersion	Tauschversuch
Impurities	Impuretés	Verunreinigungen
Industrial atmosphere	Atmosphère industrielle	Industrieatmosphäre
Inhibitor	Inhibiteur	Korrosionsinhibitor
Intercrystalline corrosion	Corrosion intercristalline	Interkristalline Korrosion
Intergranular corrosion	Corrosion intergranulaire	Korngrenzenkorrosion
Ion	Ion	Ion
Ion conductivity	Conductibilité ionique	Ionenleitfähigkeit
Laquering	Vernissage	Transparentlackieren
Long transverse direction	Sens travers long	Längs-Querrichtung
Longitudinal direction	Sens long	Längsrichtung
Loss of electrons	Perte d'électrons	Electronenverlust
Marine atmosphere	Atmosphère marine	Seeatmosphäre
Mass weight loss	Perte de masse	Massenverlust
Mechanical polishing	Polissage mécanique	Mechanisches Polieren
Mechanical properties	Caractéristique mécaniques	Mechanische Eigenschaften
Metal	Métal	Metall
Metallization	Métalisation	Metallisierung
Microbiological corrosion	Corrosion bactériologique	Mikrobiologische korrosion
Modulus of elasticity	Module d'élasticité	Elastizitätsmodul
Natural ageing	Maturation	Kaltauslagerung
Natural oxide film	Film d'oxyde naturel	Natürliche Oxidhaut
Natural oxyde film	Oxyde naturel	Natürliche Oxidschicht
Negative pole	Pole négatif	Negativer Pol
Non heat treatable alloy	Alliage non trempant	Nichtaushärtbare Legierung
Over-ageing	Sur revenu	Überalterung
Overheating	Brûlure	Überhitzen
Oxidation	Oxydation	Oxidation

Painting	Peinture	Decklackieren
Partial annealing	Restauration	Anlassen auf Zustand, Erholung
Partially annealed	Etat restauré	Rückgeglühter Zustand
Passivation	Passivation	Passivation
Passivity	Passivité	Passivität
Pickling	Décapage	Beizen
Pit depth	Profondeur de piqûre	Lochtiefe
Piting	Piqûre (de corrosion)	Lochfraβstelle
Pitting corrosion	Corrosion par piqûre	Lochfraβkorrosion
Pitting potential	Potential de piqûre	Lochfraβpotential
Polarization	Polarisation	Polarisation
Polarization curve	Courbe de polarisation	Polarizationskurve
Polarization resistance	Résistance de polarisation	Polarisationswiderstand
Polishing	Polissage	Polierung
Positive pole	Pole positif	Positiver Pol
Potential measurement	Mesure de potentiel	Potentialmessung
Pre-ageing	Pré revenu	Vorauslagerungsbehandlung
Press quenching	Trempe sur presse	Abschrecken aus der Preßhitze
Primer	Primaire	Primer
Proof strength (Rp)	Limite conventionnelle d'élasticité (Rp)	Dehngrenze bei nichtproportionaler Verlängerung (Rp)
Protective anodizing	Anodisation de protection	Schutzanodisation
Quarter hard temper	Etat quart dur	Viertelhart-Zustand
Quenching	Trempe	Abschrecken
Rate of dissolution	Vitesse de dissolution	Lösegeschwindigkeit
Recovery annealing	Recuit de restauration	Erholungsglühen
Recrystallisation annealing	Recuit de recristallisation	Rekristallisationsglühung
Recrystallization	Recristallisation	Rekristallisation
Reduction	Réduction	Reduktion
Reduction potential	Potentiel de réduction	Reduktionspotential
Resistance to stress corrosion	Résistance à la corrosion sous contrainte	Spannungskorrosion-beständigkeit
Rural atmosphere	Atmosphère rurale	landatmosphäre
Sacrificial anode	Anode sacrificielle	Galvanische Anode
Salt spray	Brouillard salin	Salzsprühnebel
Sample	Echantillon	Muster

Sand blasting	Sablage	Sandstrahlen
Saturated-calomel electrode	Electrode au calomel saturé	Gesättigte Kalomelelektrode
Scratch-brushed finish	Fini brossé	Gebürstet
Sealing	Colmatage	Verdichten
Semifinished product	Demi-produit	Halbzeug
Short transverse direction	Sens travers court	Kurz-Querrichtung
Shot blasting, Blast cleaning	Grenaillage	Strahlen
Soft annealing	Recuit d'adoucissement	Weichglühung
Soft temper (O)	Etat recuit (O)	Geglüht-Zustand (O)
Solubility	Solubilité	Löslichkeit
Solution	Solution	Lösung
Solution treatment	Mise en solution	Lösungsglühen
Stabilized temper	Etat stabilisé	Stabilisierter Zustand
Standard	Norme	Norm
Standard electrode potential	Potentiel standard	Standardelektroden-potential
Storage	Stockage	Legerung
Strain hardened temper	Etat écroui	Kaltverfestigter Zustand
Strain hardening	Durcissement par écrouissage	Verfestigung
Strain hardening	Ecrouissage	Kaltverfestigung
Strain-hardening alloy	Alliage à durcissement par écrouissage	Kaltverfestigende Leguirung
Stray current	Courant vagabond	Streustrom
Stress corrosion	Corrosion sous contrainte	Spannungsrißkorrosion
Surface condition	Etat de surface	Oberflächenbeschaffenheit
Surface milling	Surfaçage	Fräsung
Surface treatment	Traitement de surface	Oberflächenbehandlung
Surface treatment	Traitement de surface	Oberflächenspannung
Temper	Etat (métallurgique)	Werkstoffzustand
Test	Essai	Prüfung
Test conditions	Conditions d'essai	Versuchsbedingungen
Test piece	Eprouvette	Probe
Three quarter hard temper	Etat trois quart dur	Dreiviertelhart-Zustand
Traffic marks	Fretting corrosion	Transportscheuerstellen
Transverse direction	Sens travers	Querrichtung

Tropical atmosphere	Atmosphère tropicale	Tropenatmosphäre
Ultimate tensile strength	Charge de rupture	Zugfestigkeit
Under-ageing	Sous revenu	Unteralterung
Urban atmosphere	Atmosphère urbaine	Stadtatmosphäre
Valence	Valence	Wertigkeit
Water line corrosion	Corrosion à la ligne d'eau	Wasserlinienkorrosion
Water soluble	Soluble dans l'eau	Wassrerlöslich
Water stain	Corrosion lors du stockage	Wasserflecken
Water staining	Ternissement	Trübung
Weather proof	Résistant aux intempéries	Wretterbeständig
Weathering	Comportement aux intempéries	Bewitterungsverhalten
Weathering	Corrosion atmosphérique	Atmosphärische Korrosion
Weight-loss measurement	Mesure de perte de poids	Gewichtsverlustemessung
Welding	Soudage	Schweißen
Work hardening	Corroyage	Verschmiedungsgrad
Wrought alloy	Alliage de corroyage	Knetlegierung
Wrought product	Produit corroyé	Kneterzeugnis

Index